3rd International Conference on Chemical and Biological Sciences 2016

MATEC Web of Conferences Volume 60 (2016)

Amsterdam, Netherlands
23 - 25 March 2016

Editors:

Ioana Demetrescu
Keimei Oh
Nagendra Kumar Kaushik
Alina Butu
Faizah Othman

ISBN: 978-1-5108-2532-1

Printed from e-media with permission by:

Curran Associates, Inc.
57 Morehouse Lane
Red Hook, NY 12571

Some format issues inherent in the e-media version may also appear in this print version.

Printed by Curran Associates, Inc. (2016)

For additional information, please contact EDP Sciences – Web of Conferences at the address below.

EDP Sciences – Web of Conferences
17, Avenue du Hoggar
Parc d'Activité de Courtabœuf
BP 112
F-91944 Les Ulis Cedex A
France

Phone: +33 (0) 1 69 18 75 75
Fax: +33 (0) 1 69 28 84 91

contact-edps@webofconferences.org

Additional copies of this publication are available from:

Curran Associates, Inc.
57 Morehouse Lane
Red Hook, NY 12571 USA
Phone: 845-758-0400
Fax: 845-758-2633
Email: curran@proceedings.com
Web: www.proceedings.com

TABLE OF CONTENTS

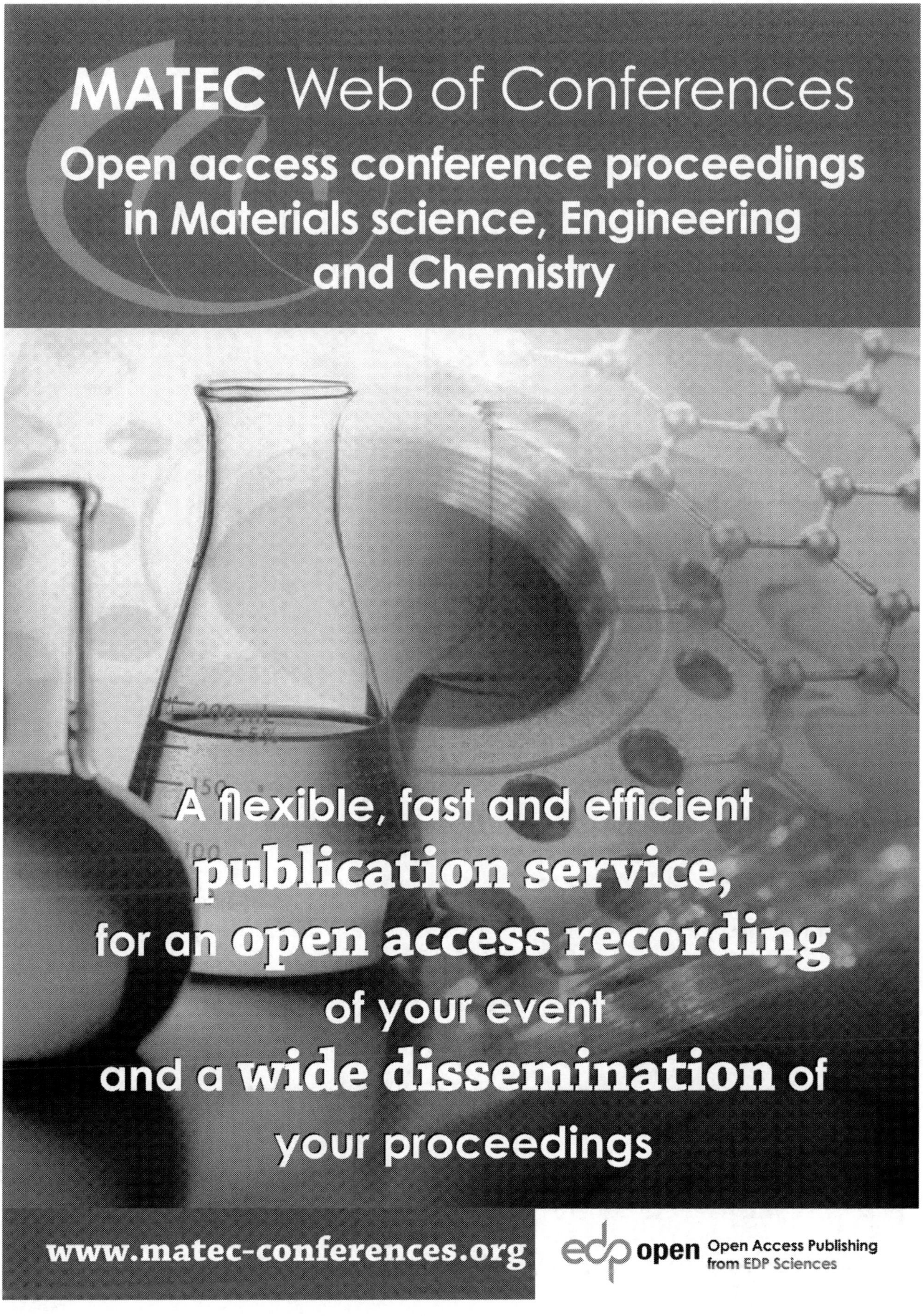
MATEC Web of Conferences
Open access conference proceedings in Materials science, Engineering and Chemistry
A flexible, fast and efficient
publication service,
for an open access recording
of your event
and a wide dissemination of
your proceedings
www.matec-conferences.org
edp open
Open Access Publishing from EDP Sciences

ISSN (electronic edition): 2261-236X

Conference Chair

Prof. Ioana Demetrescu,
University Politehnica Bucharest, Romania

Program Chair

Prof. Keimei OH,
Aki Prefectural University, Japan

Contact Chair

Ms. Flora Feng,
Asia-Pacific Chemical, Biological & Environmental Engineering Society, Hong Kong

Technical Committee

Prof. Alina BUTU
National Institute of Research, Bucharest, Romania

Prof. Nagendra Kumar Kaushik
Kwangwoon University, Seoul

Prof. Keimei OH
Aki Prefectural University, Japan

Associate Prof. Faizah Othman
Universiti Kebangsaan, Malaysia

Associate Prof. Norzana Abd Ghafar
Universiti Kebangsaan Malaysia

Dr. Waseem Shehzad Abbasi
International Academy of New Zealand

Dr. Lupa Lavinia
Afrodita Politehnica University of Timisoara, Romania

Dr. Xianfeng Fan
School of Engineering, The University of Edinburgh, UK

Dr. Rodrigo J. Alvarez-Mendez
Universidad de Xalapa, Mexico

Dr. Bassam B. J. ALKINDY
University of Franche-Comte, France

Prof. K.S.K. Rao Patnaik
University College of Technology, Osmania University, India

Assoc. Prof. Ananda Kumar C.S
Visvesvaraya Technological University, Karnataka, India

Dr. Abhilash Thakur
National Institute of Technical Teachers Training and Research, India

Dr. Suriani Mat Jusoh
School of Ocean Engineering, Universiti Malaysia

Dr. Lakhveer Singh
Universiti Malaysia Pahang, Malaysia

PREFACE

2016 3rd International Conference on Chemical and Biological Sciences has been successfully held in Amsterdam, Netherlands during March 23-25, 2016. ICCBS 2016 was organized by CBEES, and the objective of the conference is to provide a forum for the discussion of new developments, recent progress, and innovations in the chemical and biological sciences. It addresses all aspects of the related fields. The emphasis is on current and future challenges in research and development in both academia and industry. It emphasizes long-term, fundamental research aimed at discovering novel phenomena, processes, and tools.

The proceedings present a selection of 33 papers submitted to the conference from universities, research institutes and industries. All of the papers were subjected to peer-review by conference committee members and international reviewers. The papers selected for publishing in the proceedings depended on their quality and their relevancy to the conference. The proceedings tend to present to the readers the recent advances in the field of chemical and biological sciences and various related areas, including materials science and application, environment and ecology, applied chemistry, renewable energy and energy engineering

We are grateful to all the authors for submissions to this conference and for sharing their research fruits and valuable experiences they've acquired. And we would like to express our special thanks to the conference chair, Prof. Ioana Demetrescu and the program chair, Prof. Keimei OH.

Finally, we are very delighted to welcome you to ICCBS next year.

Organizing Committees
April 2016

edp sciences

MATEC Web of Conferences

Statement of Peer review

In submitting conference proceedings to MATEC Web of Conferences, I certify to the Publisher that I adhere to the **Policy on Publishing Integrity** of the journal in order to safeguard good scientific practice in publishing.

1. All articles have been subjected to peer review administered by the proceedings editors.
2. Reviews have been conducted by expert referees, who have been requested to provide unbiased and constructive comments aimed, whenever possible, at improving the work.
3. Proceedings editors have taken all reasonable steps to ensure the quality of the materials they publish and their decision to accept or reject a paper for publication has been based only on the merits of the work and the relevance to the journal.

Title, date and place of the conference

2016 3rd International Conference on Chemical and Biological Sciences

March 23-25, 2016 Amsterdam, Netherlands

Proceedings editor(s):

Prof. Ioana Demetrescu, University Politehnica Bucharest, Romania

Prof. Keimei OH, Aki Prefectural University, Japan

Prof. Nagendra Kumar Kaushik,Kwangwoon University, Korea

Prof. Alina BUTU, National Institute of Research and Development for Biological Sciences,Romania

Associate Prof. Faizah Othman, Universiti Kebangsaan Malaysia

Date and editor's signature

21/04/2016

Effect of Deposition Time on the Photoelectrochemical Properties of Cupric Oxide Thin Films Synthesized via Electrodeposition Method

Chong Siang Yaw[1], Ai Kah Soh[2], Meng Nan Chong[1,3]

[1]School of Engineering, Chemical *Engineering* Discipline, Monash University Malaysia, Jalan Lagoon Selatan, Bandar Sunway, Selangor DE 47500 Malaysia

[2]School of Engineering, Mechanical *Engineering* Discipline, Monash University Malaysia, Jalan Lagoon Selatan, Bandar Sunway, Selangor DE 47500 Malaysia

[3]Sustainable Water Alliance, *Advanced* Engineering Platform, Monash University Malaysia, Jalan Lagoon Selatan, Bandar Sunway, Selangor DE 47500 Malaysia

Abstract. The main aim of this study was to investigate the effect of deposition time on the physicochemical and photoelectrochemical properties of cupric oxide (CuO) thin films synthesized via electrodeposition method. Firstly, the electrodeposition of amorphous CuO films on fluorine-doped tin oxide (FTO) working electrodes with varying deposition time between 5 and 30 min was carried out, followed by annealing treatment at 500 °C. Resultant nanocrystalline CuO thin films were characterised using field emission-scanning electron microscopy (FE-SEM), photocurrent density, and photoluminescence measurements. Through FE-SEM analysis, it was observed that the surface of thin films was composed of irregular-sized CuO nanocrystals. A smaller CuO nanocrystals size will lead to a higher photoactivity due to the increase in overall catalytic surface area. In addition, the smaller CuO nanocrystals size will prolongs the electron-hole recombination rate due to the increase in copious amount of surface defects. From this study, it was revealed that the relationship between deposition time and CuO film thickness was non-linear. This could be due to the detachment of CuO thin films from the FTO surface at an increasing amount of CuO mass being deposited. It was observed that the amount of light absorbed by CuO thin films increased with film thickness until a certain extent whereby, any further increase in the film thickness will result in a reduction of light photon penetration. Therefore, the CuO nanocrystals size and film thickness have to be compromised in order to yield a higher catalytic surface area and a lower rate of surface charge recombination. Finally, it was found that the deposition time of 15 min resulted in an average CuO nanocrystals size of 73.7 nm, optimum film thickness of 0.73 μm, and corresponding photocurrent density of 0.23 mA/cm^2 at the potential bias of - 0.3 V (versus Ag/AgCl). The PL spectra for the deposition time of 15 min has the lowest rate of recombination of photogenerated electron-hole pairs by referring to its lowest PL intensity.

1 Introduction

Photoelectrochemical (PEC) water splitting is a photo-driven conversion process of water molecules into hydrogen and oxygen molecules through the use of semiconductor metal oxides photocatalysts [1]. In a PEC cell, minority charge carriers will be generated when the amount of light photons being absorbed by the semiconductor photocatalyst used is greater than its corresponding bandgap energy level [2]. Therefore, the type of semiconductor photocatalysts used in the PEC water splitting process will affect the efficiency of water-to-hydrogen conversion process due to their inherent bandgap characteristics.

To date, the synthesis of nanostructured semiconductor metal oxides photocatalysts have attracted much attention due to their promising properties in different fields of application ranging from sensors, optoelectronics, and electronics to catalysis. Binary copper semiconductors have attracted much research interests owing to their high photoactivity, inexpensive and that copper is a relatively abundant element [2]. Cupric oxide (CuO) is one of the *p*-type binary copper semiconductors with a narrow bandgap of 1.3 to 1.9 eV, and a corresponding theoretical photocurrent density of 35 mA/cm^2 [3, 4].

The physical and chemical properties of CuO thin films are strongly dependent on the method of preparation. There are various synthesis methods used to synthesize CuO thin films, such as pulsed laser deposition [5], electrodeposition [6], chemical vapor deposition [7], and sol-gel method [8]. Among all, electrodeposition is a simple and straightforward method that can be operated at ambient operating pressure and temperature, as well as being an inexpensive method. The electrodeposition method is also enabling the direct control of film thickness by systematically manipulating the synthesis parameters [9]. It was reported that controlling the CuO film thickness could have a profound effect on the photocurrent density, and subsequently leading to a higher efficiency in the water-to-hydrogen conversion process [10].

Previously, Patake et al. [11] have prepared porous and amorphous CuO thin films via cathodic electrodeposition method, and subsequently studied for their potential application as supercapacitor and their characteristics. Lim et al. [4] studied the deposition of Cu_2O and CuO thin films by a sol-gel spin-coating process, and enhanced the PEC performance by depositing a thin layer of NiO_x co-catalyst. Wang et al. [6] investigated the effect of deposition voltage on the morphology of CuO thin films. Although these previous studies have discussed on the growth, characterisation and application of CuO thin films, there is less information available that detailed on the relationship between deposition time, nanocrystals size, film thickness, and photocurrent density.

Thus, the main aim of this study was to systematically investigate the effect of deposition time on the physicochemical and PEC properties of CuO thin films synthesized via electrodeposition method. Initially, the *p*-type CuO thin films of different film thicknesses were synthesized on fluorine-doped tin oxide (FTO) working electrodes by varying the deposition time between 5 and 30 min, followed by annealing treatment at 500 °C. The resultant *p*-type CuO thin films were further characterised using field emission-scanning electron microscopy (FE-SEM), photocurrent density, and photoluminescence (PL) measurements.

2 Materials and methods

Materials. Copper (II) sulphate ($CuSO_4$), di-potassium hydrogen phosphate (K_2HPO_4), potassium hydroxide (KOH), lactic acid ($C_3H_6O_3$), and sodium sulfate (Na_2SO_4) were purchased from Merck Malaysia. All the reagents are of analytical grade and used without further purification.

Preparation of CuO thin films. Electrodeposition of CuO thin films was carried out in a three-electrode PEC cell that comprised of FTO, a platinum (Pt) rod, and a saturated silver/silver chloride (Ag/AgCl) electrode used as the working, counter, and reference electrodes, respectively. Before the electrodeposition synthesis process, FTO glass slides (ChemSoln, USA; resistivity of 14 Ω/sq) were cleaned ultrasonically using acetone, ethanol, and deionised water for 20 min, before being dried in the oven. The electrolyte used for the electrodeposition synthesis process was made up of 0.2M $CuSO_4$ and 3 M lactic acid solution in deionised water with 0.5M K_2HPO_4 buffer. Following this, the electrodeposition process was carried out at pH 9 by adjusting the pH of the solution using 2M KOH. The electrodeposition synthesis process was conducted potentiostatically with an applied potential of -0.4 V versus Ag/AgCl. The deposition time was varied between 5 and 30 min. After the electrodeposition process, the FTO glass slides were washed with deionised water to remove the presence of any ion followed by annealing treatment at 500 °C for 30 min.

Characterisation of CuO thin films. Different thicknesses of CuO thin films were examined using FE-SEM (FEI Nova NanoSEM; accelerating voltage 1kV). Photocurrent density measurements were carried out in a standard three-electrode assembly using an applied potentiostat/galvanostat (Metrohm PGSTAT 204 model, Netherlands). Similarly, the saturated Ag/AgCl electrode and Pt rod were used as the reference and counter electrodes, respectively. The photocurrent density measurements were conducted inside a dark box at room temperature. A 100 W halogen lamp restricted at a frequency of 0.05 Hz was used as the light source during the photocurrent density measurements. The intensity of incident light flux was measured using a digital light meter to be 8700 mW/m^2 at a light source-to-sample distance of 10 cm. The FTO working electrode was estimated to be approximately 1.32 cm^2. All the photocurrent experiments were conducted in a 0.5M aqueous Na_2SO_4 solution. PL measurements were carried out using a fluorescence spectrophotometer (LS 55, Perkin Elmer) at room temperature. The excitation wavelength was 325 nm with a scanning speed of 500 nm/min, while the emission spectra were scanned from 600 to 700 nm.

3 Results and Discussion

FE-SEM.

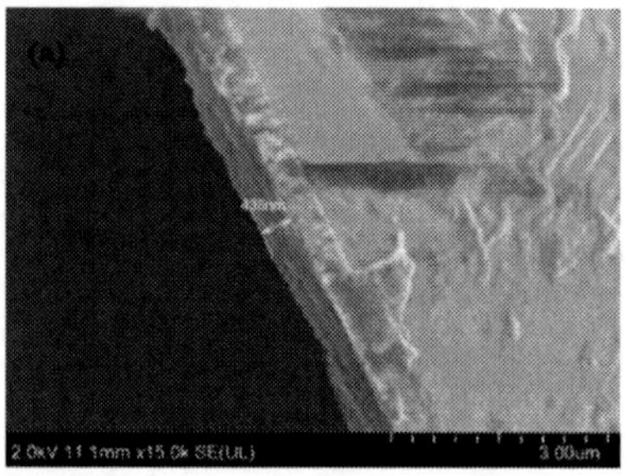

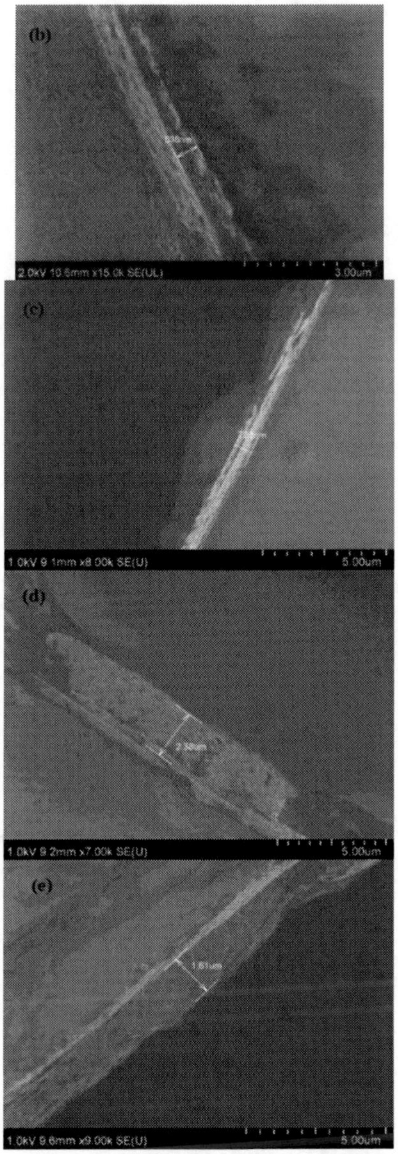

Figure 1. Cross-sectional FE-SEM images of CuO thin films at different deposition time of: (a) 5 min; (b) 10 min; (c) 15 min; (d) 20 min and ; (e) 30 min.

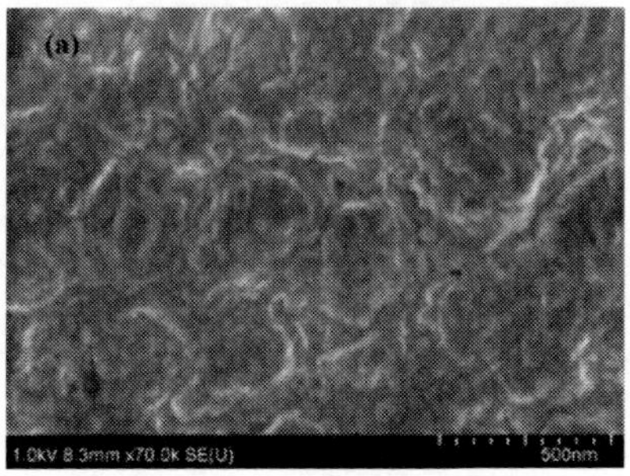

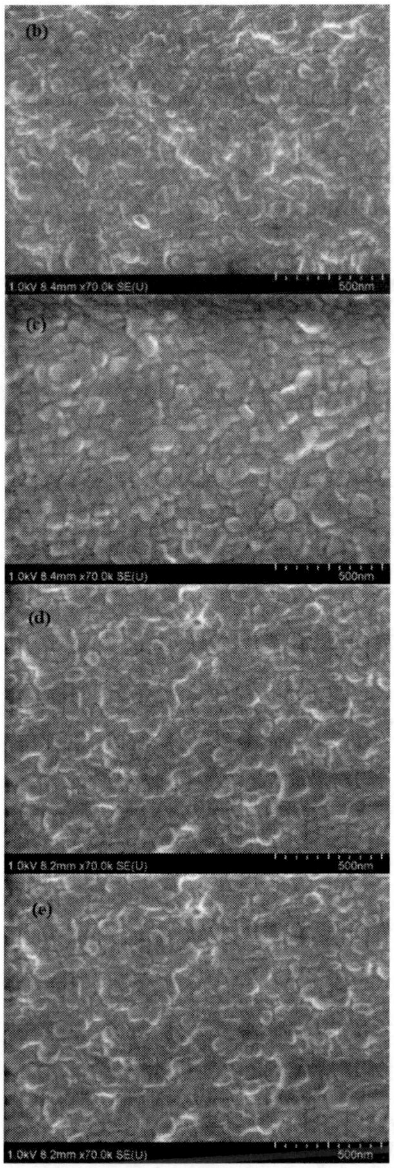

Figure 2. Top view FE-SEM images of CuO thin films at different deposition time of: (a) 5 min; (b) 10 min; (c) 15 min; (d) 20 min and; (e) 30 min.

Table 1. Relationship between deposition time, CuO film thickness and nanocrystals size.

Deposition time (min)	Film thickness (μm)	Nanocrystals size (nm)
5	0.43	53.8
10	0.56	74.6
15	0.73	73.7
20	2.57	74.0
30	1.61	83.6

Figure 1 and 2 show the FE-SEM images used to determining the CuO film thickness and nanocrystals size at varying deposition time. From the FE-SEM images, it was observed that the surface of thin films is composed of irregular-sized CuO nanocrystals. Moreover, it was observed that the longer the deposition time, the rougher the CuO film surface. This is expected because the larger the CuO grains size will results in a rougher surface. Table 1 shows how the variation in deposition time affects the CuO film thickness and nanocrystals size. From Table 1, it was observed that CuO films with different thicknesses were synthesized by effectively manipulating the deposition time. Generally, the CuO film thickness was found to increase with increasing deposition time due to that more Cu^{2+} ions being diffused to the FTO surface, underwent reduction reaction, and subsequently deposited on it as shown in Eq. (1).

$$2Cu^{2+} + H_2O + 2e^- \rightarrow Cu_2O + 2H^+ \qquad (1)$$

However, the relationship between deposition time and CuO film thickness was non-linear. It was observed that the CuO film thickness decreased when the deposition time was increased to more than 20 min. This could be due to the porous structure formation of CuO films, and caused the falling-off of porous deposits [1]. On the other hand, the CuO nanocrystals size was found to increase with increasing deposition time. This trend is expected due to the coalescence of growing CuO grains after they have attained a given size.

Photocurrent Density Measurements.

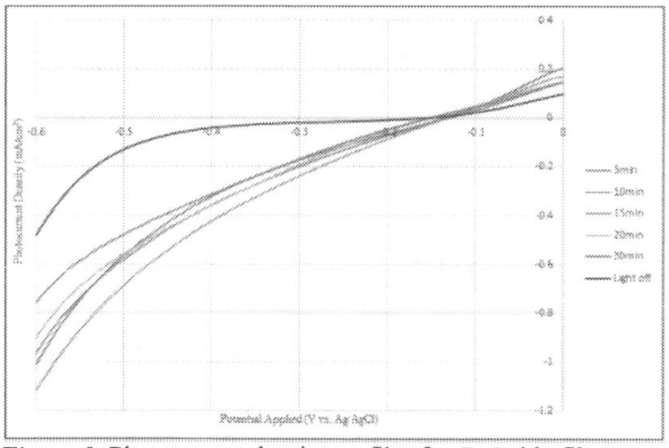

Figure 3. Photocurrent density profiles for CuO thin films under dark and irradiated conditions.

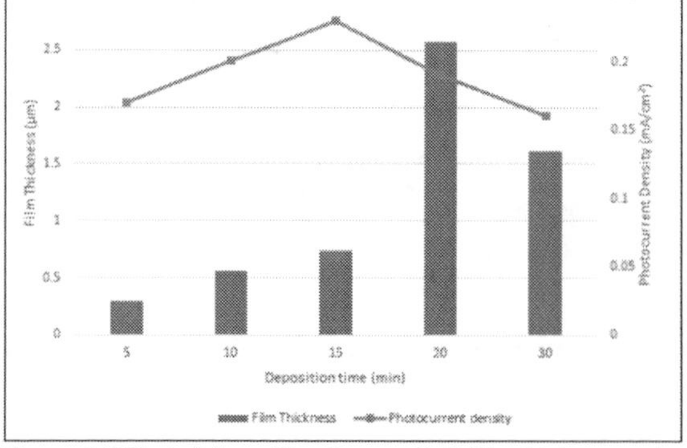

Figure 4. Effect of deposition time on CuO film thickness and photocurrent density.

Figure 3 shows the photocurrent density profiles for CuO thin films under dark and irradiated conditions. As the potential swept towards the negative direction from 0 V to −0.6 V (vs. Ag/AgCl), cathodic photocurrents were observed. Generally, the photocurrent density of CuO thin films is increasing with the increase in applied potential. This is due to the higher band-bending at the working electrode/electrolyte interface, which leads to a higher mobility of the photogenerated excitons and expansion of the depletion width with increasing applied potential [2]. In this study, the potential bias of -0.3 V (vs. Ag/AgCl) on the CuO thin films was used for the comparison of photocurrent density for different CuO thin films formed under varying deposition time. From Figure 4, it can be seen that the photocurrent density of CuO thin films increased with increasing film thickness. This could be due to the formation of steep band bending that enhances the charge separation of electrode and reduced carrier recombination in the bulk [3]. However, the photocurrent density started to decrease when the deposition time used was more than 15 min. As the CuO film thickness increases, more light can be absorbed until a certain thickness, whereby the CuO films will reduce the light photon penetration. This will increase the electron-hole pair recombination rate and results in a lower water-to-hydrogen conversion efficiency [4].

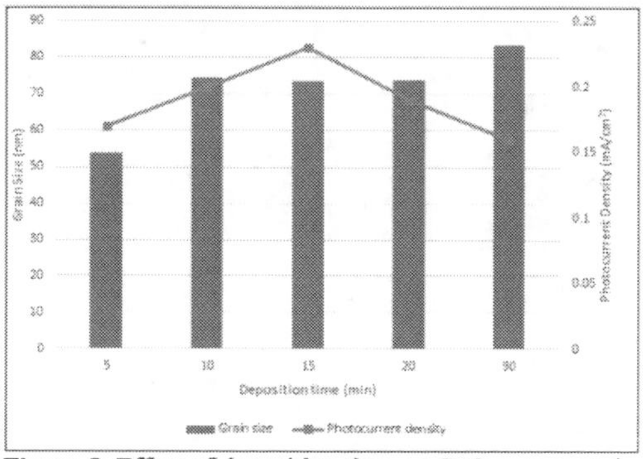

Figure 5. Effect of deposition time on CuO nanocrystals size and photocurrent density.

From Figure 5, it was found that the photocurrent density increased with increasing CuO nanocrystals size. It was observed that the CuO nanocrystals size remained almost constant for the depositon time of 10 - 20 min, while the corresponding photocurrent densities varied. Therefore, it can be concluded that the change in the photocurrent densities was not solely due to the change in CuO nanocrystals size. The photocurrent density started to decrease when the CuO nanocrystals size was larger than 73.7 nm. This observation could be explained by the larger CuO nanocrystals size and correspondingly lower catalytic surface area. This will result in a longer average minority carrier transport distance to the electrolyte interface and therefore, lower the collection efficiency.

Photoluminescence Measurements.

PL quenching effect is an efficient method to understand the emission mechanism and the charge transfer of the material. The PL intensity is representing the rate of recombination of photogenerated electron-hole pairs. Therefore, a lower PL intensity indicates that more photoinduced electrons are trapped and stably transferred through the interface. Otherwise, the more intense PL spectrum dictates the fast recombination rate of electron-hole pairs. Figure 6 shows the PL spectra of CuO thin films synthesized at varying deposition time of 5 - 30 min. It was found that the PL spectra are composed of a sharp emission peak in the red region (662 nm). The emission peak at 662 nm arises due to the band-to-band transition of CuO and other defects arising in the nanostructured mixture. The PL spectra for the deposition time of 15 min has the lowest rate of recombination of photogenerated electron-hole pairs by referring to its lowest intensity shown in Figure 6. A further comparison study of PL data and the photocurrent density measurements showed that the obtained results are consistent with each other.

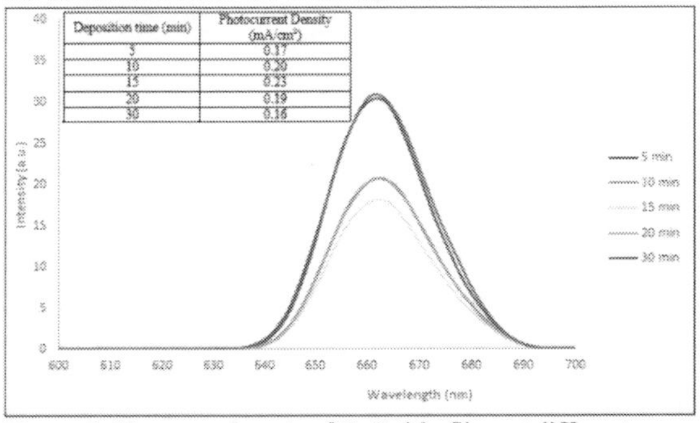

Figure 6. PL measurements of CuO thin films at different deposition time.

4 Conclusion

In conclusion, the effect of deposition time on the physicochemical and PEC properties of CuO thin films synthesized via electrodeposition method was systematically investigated. A smaller CuO nanocrystals size will lead to a higher photoactivity due to the increase in overall catalytic surface area. In addition, the smaller CuO nanocrystals size will prolongs the electron-hole recombination rate due to the increase in copious amount of surface defects. From this study, it was revealed that the relationship between deposition time and CuO film thickness was non-linear. This could be due to the detachment of CuO thin films from the FTO surface at an increasing amount of CuO mass being deposited. It was observed that the amount of light absorbed by CuO thin films increased with film thickness until a certain extent whereby, any further increase in the film thickness will result in a reduction of light photon penetration. Therefore, the CuO nanocrystals size and film thickness have to be compromised in order to yield a higher catalytic surface area and a lower rate of surface charge recombination. Finally, it was found that the deposition time of 15 min resulted in an average CuO nanocrystals size of 73.7 nm, optimum film thickness of 0.73 µm, and corresponding photocurrent density of 0.23 mA/cm2 at the potential bias of - 0.3 V (versus Ag/AgCl). The PL spectra for the deposition time of 15 min has the lowest rate of recombination of photogenerated electron-hole pairs by referring to its lowest PL intensity.

Acknowledgement

The authors are grateful to the financial support provided by the Fundamental Research Grant Scheme (FRGS) (Project Reference Code : FRGS/1/2014/SG06/MUSM/02/1) from Ministry of Education (MOE), Malaysia. Similar gratitude also goes to the Advanced Engineering Platform, Monash University Malaysia.

References

1. Zhu. T, M.N. Chong, and E.S. Chan, Nanostructured tungsten trioxide thin films synthesized for photoelectrocatalytic water oxidation: a review. ChemSusChem, 7, 11 (2014): p. 2974-2997.
2. A. Paracchino, V. Laporte, K. Sivula, M. Gratzel, E. Thimsen, Highly active oxide photocathode for photoelectrochemical water reduction. Nature materials, **10**, 6 (2011): p. 456-461.
3. M. Izaki, M. Nagai,K. Maeda, F. Mohamad, K. Motomura, J. Sasano, S. Watase , Electrodeposition of 1.4-eV-bandgap p-copper (II) oxide film with excellent photoactivity. Journal of The Electrochemical Society, **158**, 9 (2011): p. D578-D584.
4. Y.F. Lim, C.S. Chua, C.J.J. Lee, D. Chi, Sol–gel deposited Cu_2O and CuO thin films for photocatalytic water splitting. Physical Chemistry Chemical Physics, 2014. **16**, 47 (2014): p. 25928-25934.
5. A. Chen, H. Long, X. Li, Y. Li, G. Yang, P. Lu, Controlled growth and characteristics of single-phase Cu_2O and CuO films by pulsed laser deposition. Vacuum, **83**, 6 (2009): p. 927-930.
6. Y. Wang, T. Jiang, D. Meng, J. Yang, Y. Lim Q. Ma, J. Han, Fabrication of nanostructured CuO films by electrodeposition and their photocatalytic properties. Applied Surface Science, **317** (2014): p. 414-421.
7. M.R. Johan, M.S.M. Suan, N.L. Hawari, H.A. Ching, Annealing effects on the properties of copper oxide thin films prepared by chemical deposition. Int. J. Electrochem. Sci, **6** (2011): p. 6094-6104.
8. Akhavan, O. and E. Ghaderi, Cu and CuO nanoparticles immobilized by silica thin films as antibacterial materials and photocatalysts. Surface and Coatings Technology, **205**, 1 (2010): p. 219-223.
9. S. Wei, Y. Ma, Y. Chen, L. Liu, Z. Shao, Fabrication of WO_3/Cu_2O composite films and their photocatalytic activity. Journal of hazardous materials, **194** (2011): p. 243-249.
10. S. Ito, S.M. Zakeeruddin, R. Humphry-Baker, P. Liska, R. Charvet, P. Comte, M.K. Nazeeruddin, P. Pechy, M. Takata, H. Miura,High-Efficiency Organic-Dye-Sensitized Solar Cells Controlled by Nanocrystalline-TiO_2 Electrode Thickness. Advanced Materials, **18**,9 (2006): p. 1202-1205.
11. V. Patake, S. Joshim C. Lokhande, O.S. Joo, Electrodeposited porous and amorphous copper oxide film for application in supercapacitor. Materials Chemistry and Physics, 2009. **114**, 1 (2009): p. 6-9.
12. Kwong, W., N. Savvides, and C. Sorrell, Electrodeposited nanostructured WO_3 thin films for

photoelectrochemical applications. Electrochimica Acta, **75** (2012): p. 371-380.
13. M. Takahashi, K. Tsukigi, T. Uchino, T. Yoko, Enhanced photocurrent in thin film TiO_2 electrodes prepared by sol–gel method. Thin Solid Films, 2001. **388, 1** (2001): p. 231-236.
14. K.R. Lee, Y.P. Hsu, J.L. Chang, S.W. Lee, C.J. Tseng, J.S.C. Jang, Effects of Spin Speed on the Photoelectrochemical Properties of Fe_2O_3 Thin Films. Int. J. Electrochem. Sci, **9** (2014): p. 7680-7692.

Water Absorbing Plantation Clay for Vertical Greenery System

Lih-Jiun Yu[1], Aegil De Vera Santos[1], Khang-Wei Tan[1] and Mou'ad A Tawawneh[2]

[1]Faculty of Engineering, Technology and Built Environment, UCSI University, Kula Lumpur Campus (North Wing,) Lot 12734, Jalan Choo Lip Kung, Taman Tayton View, 56000 Cheras, Kuala Lumpur, Malaysia.
[2]Department of Physics, College of Science, Al-Hussein Bin Talal University, Ma'an, P.O Box 20, Jordan.

Abstract. With the arises of environmental conscious, the usage of vertical garden system has become more popular in urban cities. Citizens can enjoys the benefits of energy and cost saving besides ornamental effect. More investigations have been conducted on green facades led to the cities ecological enhancement.However, limited plants species can be planted for green facades systems as this system does not provide sufficient soil and nutrients for common plants. Alternative plantation methods such as planted box and felt system required additional maintenance attention. The idea of using clay composite which consists of nutritious soil, water absorbing polymer and flexible cement clay potentially become alternative vertical greenery systems that offers economic and sustainable plantation platform for more variety of plants.The fabricating of clay composite involved three processes, they are: mixing, moulding and drying. Physical properties characterisation (density, pH, compression test, aging test and water immersion test) were tested on the dried fabricated clay composite to ensure their sustainability in tropical climate. The results showed that clay composite with 1.5 wt% of cement and 0.3 wt% superabsorbent polymer shows optimum water absorbing properties. This system are expected to enable more agriculture activities in urban living.

1 Introduction

Using plants and greenery is an ecological solution to reduce ambient temperatures, mitigation of urban heat island effect and building energy saving [1]. Plants on vertical landscaping absorb dust and clean air, and in this way they work as natural air filtration [2,3]. Moreover, the use of plants offers natural advantages as plants are clean source and it has the ability to control noise and use as barriers for noise abeyance [4]. The development in vertical greenery system is emerging, as a results of limited green area in developing cities and growing of environmental consious. In some area, installment of vertical greeney system are receiving encouragement from authorities. For example, the German Jugendstil movement (Art Nouveau) from the early 20th century encouraged the integration of the house with the garden and emerged some incentive programs for the installation of green facades [5].

However, there are potential problems associated with linking the building with living organism. A supportive structures together with well designed dranaige system are essential to enable the sustainability of plant growth which variable along the time and space as the plants grows. Moreover, vegetation is usually linked to drawbacks such as higher initial investment, maintenance costs, and possible damages to the building [6]. Furthermore, the challenge when using growing media for vertical greenery systems is larger as the vertical orientation of the growing media increases the risks of uneven distribution of water and air, making the gravitational important [7]. The conventional vertical garden system by using green facaedes are limited to application of climbing plants species along the wall. In contrast, the living wall system, especially modular living wall system, is getting more popular in urban cities application. This system consists of growing media where it is supported by complementary structure, such as: planted box or felt system, allows uniform growth of light weight plants along the vertical wall. However, these systems required varies of investment cost and additional maintanance attention, subjected to the types of system employed [6].

A good growing media is one of the key aspects in constructing vertical greenery systems. They must be able to maintain moisture, firm, allow good aeration, free from pests, low salinity, enable high cation exchange capacity for nutrients, consistent in quality and economic available. The postivie effect of superabsorbent polymer (or known as hydrogels) application as soil additives for the retention of water in soil has been reported widely [8-10]. The optimum superabosorbent polymer usage for plant growth is 0.3% [9].

Developing a successful system for vertical greenery systems requires solid knowledge of the properties of the growing media. Hence, selecting a proper growing media is one of the key factors in building vertical greenery systems. The idea of using clay composite which consists of nutritious soil, water absorbing polymer and flexible cement clay potentially become alternative vertical greenery systems that offers economic and sustainable plantation platform for more variety of plants. The combination of superabsorbent polymer with plantation clay composites are expected to beneficial from low mainatenance needs in the aspects of reducing irrigation frequency and pipes replacement frequency. Cement is a potential material that readily available, low cost yet effective to enhance viscidity is being considered as growing media. However, the control of cement content is crucial as the plant growth and flower colors are affected by the alkalinity of the soil resulting from alkaline agent application [11,12]. Thus, the balancing of cost efficiency with the sustainability of plant growth becomes challenging.

In this paper, we study the effect of cement content in clay composite on water absorbing properties. The sustainability of plants growth in the clay composite were investigated.

2 Methodology

2.1 Materials

Malaysia plantation grade clayed soil, black soil and superabsorbent polymer (sodium polyacrylate) were supplied by SJH Nursery & Landscaping Sdn. Bhd. Ordinary portland cement were supplied by Lafarge, Malaysia.

2.2 Sample preparation

The clay composites of different composition were mixed according to the ratio as in Table 1. The soil mixtures were maintained at 1:4 of black soil: clayey soil respectively. The clay mixtures were added with 50 ml of water and moulded into the mould with dimension 13 cm x 13 cm x 12 cm (length x width x height), leaving a hollow space of 2 cm diameter in the center for plantation. Then, the mixtures were compressed, allowed for cure and dry naturally under sunlight for three days.

Table 1. Mixing composition of clay composite.

Sample	Composition (wt%)		
	Soil mixture (Black soil & Clayey soil)	Superabsorbent polymer (Sodium polyacrylate)	Cement
A	100	0	0
B	98.3	0.30	1.5
C	96.7	0.30	3.0
D	95.2	0.30	4.5

2.3 Characterization

Physcical tests such as: pH test, water absorbing test, density, compression test, water immersion test were conducted to ensure the clay composite materials are able to sustain in tropical climate before starts planting the plants. Water absorbing test was performed by using tea-bag method [13]. The chosen plant species for plantation was *Cuphea hyssopifolia* (Mexican Heather). Then, plant sustainability was evaluated based on plant's growth length with controlled irrigation (100 ml per day) and ultraviolet rays (1350 W/m^2, 10 hours).

3 Results and discussion

3.1 Density

The densities of raw material for clay composite are listed in Table 2. The theoretical density of clay composites with various cement content was calculated based on weight percent of composition as listed in Table 1. The actual density of all samples was measured by Archimedes principles were compared with theoretical density in Figure 1. From Figure 1, it is clearly shown that the actual densities of clay composite are lower than the theoretical densities. Lower densities of clay composite could be attributed to the larger volume appear in all samples due to high porosity and the air trapped in samples during sample preparation. It was found that the sample A (without cement) appear to have lower density due to absence of cement to bind the materials in to close pack arrangement. Some loose soil particles were separated easily from the bulk sample.

Table 2. Density of dry materials.

Sample	Density (g/cm^3)
Clayey	1.32
Black soil	1.41
Sodium polyacylate	1.23
Cement	3.15

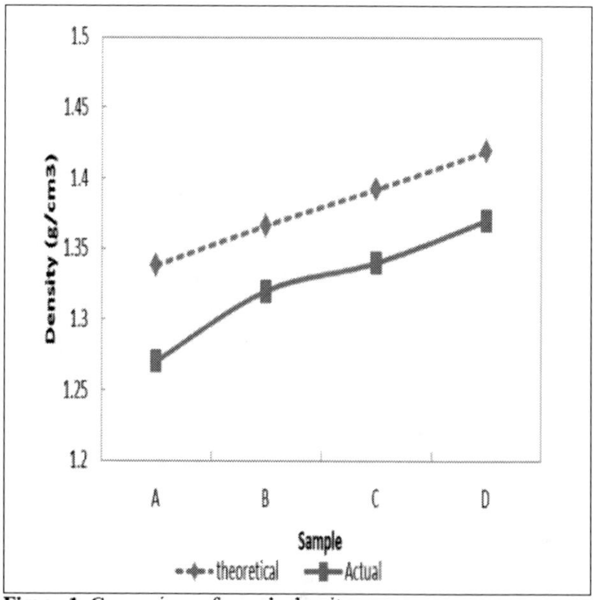

Figure 1. Comparison of sample density.

3.2 pH

Table 3 showed pH values of all samples. It was observed that an increment of cement content in the samples contributing to the increment of pH values. The pH of damp soil mixture without the presence of cement is almost similar to the pH of tap water in the range of 7.5-7.8. The presence of cement content from 1.5-4.5% was increasing pH upto 11.5. This results are in line with [14] in which showed that pH in original soil was increased from 9.5 to 11.0 when the cement content is between 1.0 to 2.0%. The optimum cement content for the synthetic soils in rock slope application was reported at 1.5% cement [9]. Based on the results in Table 3, the pH values of clay composites degrade as time extends. The degradation of pH values are attributed to carbonation process. Calcium hydroxide, which contains in the cement, reacts with carbon dioxide to form calcium carbonate and water. This process is naturally diluting the alkalinity of the clay composite from pH 11 to pH 8 in three weeks. These changes are favorable for plants growth as the clay composite condition became close to neutral.

Table 3. pH values of samples.

	pH			
Sample	**Week 1**	**Week 2**	**Week 3**	**Week 4**
Water	7.5	-	-	-
A	7.8	7.7	7.6	7.6
B	11.3	10.8	7.9	7.9
C	11.3	11.1	8.1	8.0
D	11.5	11.1	8.0	7.9

3.3 Water absorbing properties and compress stress test

Figure 2 showed water absorbed by clay composite and maximum compression stress experienced by clay composite before failure occurs. Line chart indicates water absorbed per gram of dry samples. Water absorbed by soil mixtures (sample A) is 0.46 ml per mass of dry soil. The amount of water absorbed in sample B after introducing cement and superabsorbent polymer almost 1.8 folds of samples A. The hydrophilic functional chains in superabsorbent polymer can hold a large amount of water and swell. The cross-linked chains in superabsorbent polymers tends to maintain their physical dimensional structure [15]. The postivie effect of superabsorbent polymer application as soil additives for the retention of water in soil are consistent with [8-10, 16] However, further increment of cement content after 1.5 wt% led to decreasing water absorbing properties. Water permeability of cement is reduced after cured, thus improving water resistance [17]. The presence of higher cement content in the clay composite led to a barrier effect and restricted water absorbs into the soil and water absorbent polymers. Hence, reducing the water absorbing capacity of the clay composite.

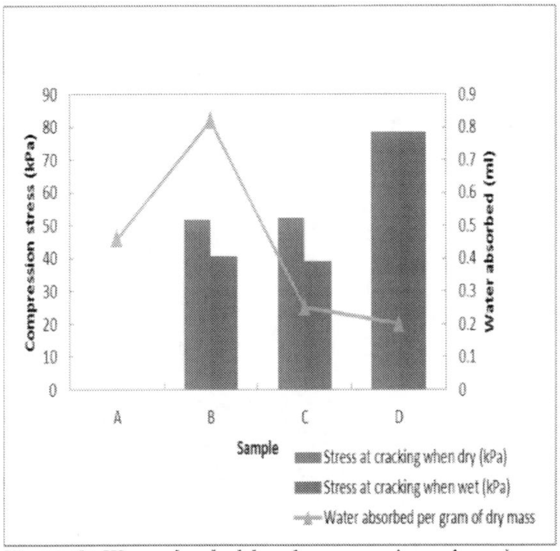

Figure 2. Water absorbed by clay composite and maximum compression stress experienced by clay composite before failed.

The benefits of cement enhancing soil through increasing soil strength, durability, rigidity and decreasing the soil permeability and deformability [11, 18, 19]. Bar chart in Figure 2 shows maximum compression stress applied on the clay composite before crack exists. It was found that the data is invalid for sample A, because compacted soil mixtures are prone to compression test either in dry or wet condition. With the addition of cement, the clay composites showed improvement in resisting the material failure upon apply compression stress. Cracks are visible on the applied clay composite surface around 52 kPa (under dry condition) and 40 kPa (under wet condition) for both sample B and C. While sample C with highest cement content (3wt%) showed highest compression stress resistance at 78.4 kPa for both wet and dry condition. The presence of surface imperfection on the clay composite surface serves as water duct. They feed the crack surface and damage the interface with water for further migration into the material by capillary suction [20]. Hence brittle fracture was accelerating under damp condition.

3.4 Plant growth

Figure 3 shows *Cuphea hyssopifolia* (Mexican Heather) plant's height planted in different clay composites. All plants are able to survive and continue to grow with fixed amount irrigation per day. Sample A shows highest average grow rate with 0.82 cm/week. Meanwhile, the average growth rates of plants are around 0.56 to 0.66 cm/week for all samples with of cement. The plants growth rates are dependent on the plant species. For some plants which are favor to grow in acidic environment, the present of heavy metals in cement may disturb metabolic process of plants [21]. Slower grow rate of *Cuphea hyssopifolia* plant could be attributed to the alkalinity environment resulting from cement. This revealed that cement has negative impact towards *Cuphea hyssopifolia* plant growth regardless to the water absorbing properties

DOI: 10.1051/matecconf/20166001002

of the growing media. It was observed that the *Cuphea hyssopifolia* plant average growth rate for all samples were increased (0.53 to 0.96 cm/week) after three weeks. Growth rate of plant was inclined, when the pH of clay composites reduced to pH of range 7.6 to 8.1. In conclusion, the plants growth rate is retarded by high alkalinity environment. On the contrary, the plants grow more rapidly at neutral environment.

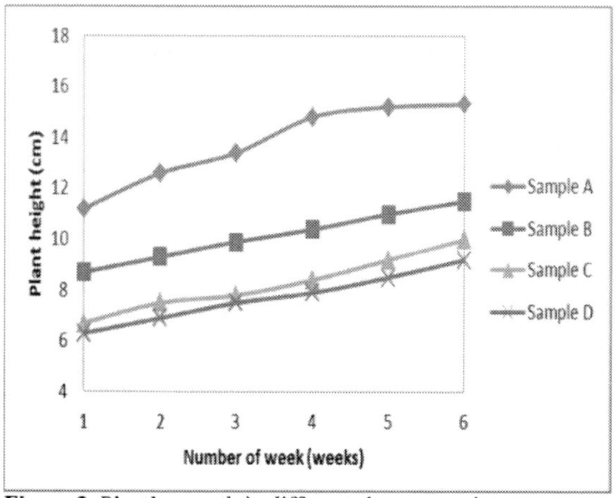

Figure 3. Plant's growth in different clay composites.

4 Conclusion

Clay composite samples were prepared by mixing of soil mixture and superabsorbent polymer at different cement concentration. Incremental of higher cement percentage in clay composites improved compression stress for both dry and damp condition. The plant's growth rate is increased when natural degradation of pH occurred after three weeks. The results showed that the system consists of nutritious soil, 0.3 wt% water absorbing polymer and 1.5 wt% of flexible cement clay is potentially offers economic and sustainable plantation platform. Further works are needed to investigate the clay composite's water retention properties, impact of weather on the clay composites and the sustainability of different plants species.

References

1. G. Perez, J. Coma, I. Martorell, L.F. Cabeza, Renew. Sust. Energ. Rev. **39**, 139 (2014)
2. P. Sunakorn, C. Yimprayoon, Procedia Eng. **21**, 34 (2011)
3. T. Safikhani, A. Abdullah, D.Ossen, N. Baharvand, Renew. Sust. Energ. Rev. **40**, 450 (2014)
4. N.H. Wong, T.A.Y. Kwang, P.Y. Tan, K. Chiang, N.C. Wong, Build Environ. **45**, 411 (2010)
5. M. Manso, J.Gomes, Renew. Sust. Energ. Rev. **41**, 863 (2014)
6. K. Perini, P. Rosassco, Build Environ. **70**, 110 (2013)
7. S. Sonmez, D. Buyuktas, F. Okturen, S. Citak, Geoderma **144**, 361 (2008)
8. X. Li, J-Z. He, J.M. Hughes, Y-R. Liu, Y-M. G-J. Gao, J.G. Yuan, R.H. Han, G-R. Xin, Z.Y. Yang, Ecol. Eng. **30**, 303 (2007)
9. Zheng, Appl. Soil Ecol. **72**, 58-63 (2014)
10. L. Yang, Y. Yang, Z. Chen, C. Guo, S. Li, Ecol. Eng. **62**, 27 (2014)
11. E.A. Basha, R. Hashim, H.B. Mahmuf, A.S. Muntojar, Constr. Build. Mater. **19**, 448 (2005)
12. D. Zhao, Z. Hao, J. Wang, J. Tao, Sci. Hortic. **154**, 45 (2013)
13. M.J. Zohuriaan-Mehr, K. KAbiri, Iran. Polym. J. **17(6)**, 451 (2008)
14. A.K. Bin, T. Ahmad, T. Bin, R. Mohd, A.L. Bin, *In. Geoenviron. Eng: Integrated Management of Groundwater and Contaminated Land- Fourth Br. Geotech. Assoc. Conf,* 477 (2005)
15. F.L. Buchholz, A.T, Graham, *Modern Superabsorbent Polymer Technology*, (1997)
16. F. Yazdani, I. Allahdadi, G.A. Akbari, Pak. J. Biol. Sci. **10(23)**, 4190 (2007)
17. P. Halamickova, R.J. Detwiler, D.P. Bentz, E.J. Garboczi, Cem. Concr. Res. **25(4)**, 790 (1995)
18. C.A. Anagnostopoulos, Tunn. Underfr. Sp. Tech. **20**, 525 (2005)
19. P.J. Walker, Cem. Concr. Comp. **17**, 301 (1995)
20. P. Zhang, F.H. Wittmann, T.J. Zhao, E. Lehmann, P. Vontobel, S. Hartmann, Restor. Build. Monuments. **15(2)**, 91 (2009)
21. M.Z.Iqbal, M. Shafig, Turk. J. Bot. **25**, 19 (2001)

Water gas shift reaction over Cu catalyst supported by mixed oxide materials for fuel cell application

Pannipa Tepamatr[1], Sumittra Charojrochkul[2] and Navadol Laosiripojana[1]

[1]*The Joint Graduate School of Energy and Environment, King Mongkut's University of Technology Thonburi, Bangkok, Thailand*
[2]*National Metal and Materials Technology Center, NSTDA, Pathumthani, Thailand*

Abstract. The water gas shift activities of Cu on ceria and Gd doped ceria have been studied for the further enhancement of hydrogen purity [1] after the steam reforming of ethanol. The catalytic properties of commercial catalysts were also studied to compare with the as-prepared catalysts. Copper-containing cerium oxide materials are shown in this work to be suitable for the high temperature. Copper-ceria is a stable high-temperature shift catalyst, unlike iron-chrome catalysts that deactivate severely in CO_2-rich gases. We found that 5%Cu/10%GDC(D) has much higher activity than other copper ceria based catalysts. The finely dispersed CuO species is favorable to the higher activity, which explained the activity enhancement of this catalyst. The kinetics of the WGS reaction over Cu catalysts supported by mixed oxide materials were measured in the temperature range 200-400 °C. An independence of the CO conversion rate on CO_2 and H_2 was found.

1 Introduction

The water-gas shift (WGS) reaction ($CO + H_2O \leftrightarrow CO_2 + H_2$) plays an important role in fuel processing. For PEM fuel cell applications, novel low-temperature WGS catalysts are under development to upgrade the hydrogen-rich reformate gas streams where the catalyst possesses improved activity and greatly improved stability over the commercial Cu–ZnO catalysts [1,2]. The latter have been optimized for the production of chemicals, and are not suitable for fuel cell applications as they are pyrophoric and deactivate fast after exposure to air and/or water condensation. Because of its high activity [3] and low price, copper is still an excellent candidate as the active component of a new generation, stable catalyst for the WGS reaction. Many promising such catalyst compositions use ceria carriers, with the latter actively contributing to the making of an active and stable Cu-based catalyst.

As one of the most important rare earth oxides, ceria has been extensively applied in catalysis, electrochemistry, and optics, due to its unique physical and chemical properties. CeO2 retains its defective fluorite-type crystal structure during the oxygen storage and release processes, and thus is an active oxide component of various oxidation catalysts used in diverse redox catalytic reactions [4,5]. Ceria-based catalysts are very good WGS catalysts [6–9], and the Cu–CeO2 system was first reported by Li et al. as a promising low-temperature shift catalyst [6]. The choice of Cu–CeO2 for high-temperature WGS applications was also rationalized [9], because this copper-based system is more stable than the commercial Cu/ZnO. For a practical fuel cell system, operating under frequent shutdown and restart cycles, ceria needs to be modified by addition of zirconia [10] or another dopant to avoid formation of Ce(III) hydroxycarbonate during shutdown to RT in the water-containing reaction gas. Another approach is to regenerate the catalyst frequently by oxidation at 400–450 °C [10,11] or to run the reaction in oxygen-assisted mode [11]. Despite this issue, ceria and doped ceria remains the support of choice when a highly dispersed metal preparation is desired, which must remain stable over a wide temperature range [9]; and when surface oxygen availability to the metal is required, as in the case of the water-gas shift reaction [8].

During the last decade, copper–ceria catalysts have been widely studied for the WGS reaction via different approaches. Synthesis, structural and surface investigations by multiple characterization techniques, activity and stability tests, investigations of the reaction mechanism, and determination of the active sites have been addressed. For example, Wang et al. observed reversible redox properties of copper in $Ce_{1-x}Cu_xO_2$ by in situ XAFS and time-resolved XRD measurements [12], and further found a complex interaction between copper and the oxygen vacancies of ceria [13]. Koryabkina et al. found that the addition of ceria did not increase the WGS rate per unit of Cu surface area, and claimed that ceria is not a promoter for copper [14], while Djinovi´c et al. reported that the activity of $CuO–CeO_2$ catalysts for WGS reaction was related to the extent of surface CeO_2 reduction and the interaction between CuO and CeO_2 [15]. Thus, the basic questions regarding the structure–activity relationship and the interaction of copper and ceria in the WGS reaction have been addressed, but often with divergent conclusions. In our opinion, a consensus must be reached soon to enable better catalyst designs and development of active and stable water-gas shift catalysts for practical fuel cell applications. This will potentially lead to novel, properly designed Cu-containing CeO_2 catalysts of comparable activity to the more expensive choices of Pt- or Au-based WGS catalysts.

In this work, we have studied different preparation methods to obtain the copper–ceria materials, developed a structural model for each sample, and attempted to correlate the catalyst activity to the presence or absence of certain Cu species and their interaction with the surface of ceria

2 Experimental Procedure

2.1 Catalysts Preparation

2.1.1 Preparation of Support

Cerium oxide and gadolinium doped ceria (GDC) were prepared by a combustion synthesis technique [16] using urea as a fuel. Nitrate salts of cerium ($Ce(NO_3)_3.6H_2O$, 99.5%, Alfa Aesar) and gadolinium ($Gd(NO_3)_3.6H_2O$, 99.9%, Alfa Aesar) were used as initial chemical reactants. The stoichiometry between metal nitrate and urea was 2.5:1.

Nitrate salts of cerium ($Ce(NO_3)_3.6H_2O$) and urea were mixed for the preparation of cerium oxide. The preparation of gadolinium doped ceria (GDC), $Ce(NO_3)_3.6H_2O$ and $Gd(NO_3)_3.6H_2O$ were mixed with urea as the desired ratio. A minimum amount of deionized water was added to the mixed reactant to obtain a homogenous solution. The crucible containing the mixed reactant was then heated using a Bunsen burner until an autoignition occurred. The catalytic properties of CeO_2 (AlfaAesar), CeO_2 (Nanophase), 10%gadolinia-doped ceria (Daiichi) and 10%gadolinia-doped ceria (FuelCellMaterials), which denoted as A, B, C and D, were also studied to compare with the as-prepared catalysts.

2.1.2 Preparation of Catalysts

Cu catalyst was prepared by impregnation method. Appropriate amount of $Cu(NO_3)_2.3H_2O$ was dissolved in minimal amount of deionized water. The salts solution was added to CeO_2 (Alfa Aesar), CeO_2 (Nanophase), 10%gadolinium doped ceria (GDC, FuelCellMaterials), 10%gadolinium doped ceria (GDC, Daiichi), CeO_2 (combustion) and 10%gadolinium doped ceria (combustion). All the catalysts were dried at 110°C for 12 hours and then calcined in an oven at 650°C for 8 hours.

2.2 Catalyst characterization

2.2.1 Standard characterization

Specific surface areas (m^2/g) and pore size distribution were determined by N_2 adsorption-desorption isotherms at 77.3 K using Quantachrome NOVA 1200e. The specific surface area of the sample was calculated following the Brunauer Emmett Teller (BET) procedure in the range of 0.05-0.3. Prior to the measurements, the samples were outgassed in vacuum at 300°C for 6 hours.

The XRD diffractograms presented in this study were recorded employing nickel-filtered Cu K_α radiation (λ = 1.5406 Å) operating at 40 kV and the current was 40 mA. The analyses were carried out at 0.02° step and 0.5 s per step over a 2θ range of 20-80°.

2.2.2 Temperature Programmed Reduction

TPR experiments were performed under a flow of $5\%H_2/95\%Ar$ mixture over 0.05 g of catalyst from 40°C to 1000°C using a heating rate of 10°C/min. Prior to the measurement, the catalysts were treated under high purity helium gas at 120°C for 30 minutes. The amount of H_2 uptake during the reduction was measured by using a thermal conductivity detector. The curves in H_2-TPR profile correspond to H_2 consumption.

2.2.3 Scanning electron microscopy (SEM)

SEM images were taken on a FE-SEM)HITACHI SU-8030(with high vacuum mode using secondary electrons and an acceleration tension of 30 kV. Samples were analyzed by spreading them on a carbon tape. Energy dispersive X-ray spectroscopy (EDX) was used in connection with SEM for the elemental analysis. The elemental mapping was also collected with the same spectrophotometer.

2.3 Reaction testing

An amount of 150 mg catalyst was loaded between two layers of quartz wool in a stainless-steel fixed bed flow reactor. The reactor was placed inside a tube furnace. The flow of feed gases was controlled by mass flow controllers)Aalborg(. The initial feed gas of CO and N_2 was mixed with water vapour which was controlled by using a syringe pump. The mixed feed gas composition 5% CO, 10% H_2O, balanced in N_2 and total flow rate was 100 ml/min, was passed through the catalyst bed in the reactor. The temperature of catalysts was increased from 100°C up to 500°C. The column utilized in the chromatograph was a Unibead C, 15 m × 0.53 mm stainless-steel packed column. The outlet gasses were analyzed by using an on-line gas chromatography (Shimadzu GC-14B) equipped with a TCD analyzer. The catalytic activities were calculated by using equation [1]:

$$\%CO conversion = \frac{CO_{in} - CO_{out}}{CO_{in}} \times 100 \quad (1)$$

3. Result and Discussion

3.1 Characterization of monometallic Cu catalysts

XRD patterns of supports and Cu catalysts are presented in Fig. 1. All samples exhibit the diffraction peaks which correspond to the fluorite structure of CeO_2. Copper oxide is not observed in all samples as there are only 5% of Cu in each sample. Based on Bragg's equation, the lattice parameters of ceria and doped catalysts were calculated from (1 1 1) crystallographic planes. For pure CeO_2 support, the lattice parameter is found to be 0.541 nm which is in agreement with other works [17,18]. The lattice parameter of other catalysts (Cu/CeO_2 and Cu/GDC) also displayed similar values in the range of 0.0545–0.0547 nm. From these results, it can be concluded that an addition of Cu did not lead to a significant modification of the ceria bulk structure.

Tables 1 and 2 summarize the results from particle analyses and XRD. Lowering of the surface area and increasing of crystallite size of Cu catalysts are believed to be a consequence of sintering at high temperature. The % metal dispersion data indicate that Cu metal can be better dispersed in GDC (C), GDC (D) and CeO_2 (combustion). However, the dispersion of 5%Cu/CeO_2 (A) cannot be detected. This result was due to agglomeration and the growth of metal–ceria crystallites.

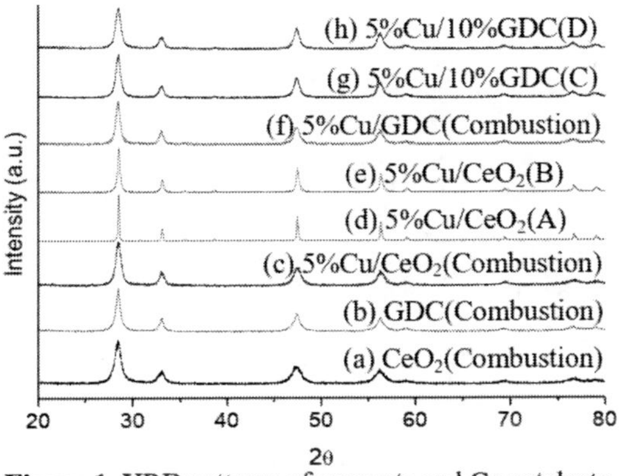

Figure 1. XRD patterns of supports and Cu catalysts.

Table 1. Standard characterization of supports.

Catalysts	Crystallite size (nm)	BET surface area (m^2/g)
CeO_2 (A)	42.27	6.600
CeO_2 (B)	30.60	52.30
GDC (C)	12.26	85.53
GDC (D)	4.540	192.8
CeO_2 (Combustion)	10.15	66.47
GDC (Combustion)	11.02	33.57

Table 2. Standard characterization of Cu catalysts.

Catalysts	Crystallite size (nm)	BET surface area m^2/g	% dispersion
5%Cu/CeO_2 (A)	66.67	1.158	-
5%Cu/CeO_2 (B)	30.04	26.45	11
5%Cu/GDC (C)	14.24	45.61	25
5%Cu/GDC (D)	12.75	41.93	32
5%Cu/CeO_2 (Combustion)	12.52	39.67	31
5%Cu/GDC (Combustion)	15.21	21.01	21

Figure 2 shows H_2-TPR profiles of Cu on various supports. The reduction temperature of all samples is observed at temperature lower than 400 °C. The 5%Cu/10%GDC (C) and 5%Cu/10%GDC (D) catalysts have three reduction peaks at about 150 °C, 180 °C and 200 °C which are denoted as peaks β, γ and δ, respectively. The peaks β, γ and δ assigned to finely dispersed CuO [19,20], the reduction of Cu^{2+} ions in the Cu-O-Ce solid solution [21,22], and bulk CuO particles [23-25], respectively. In case of 5%Cu/10%GDC(D), the peaks β, γ and δ are slightly shifted to lower temperatures compared to 5%Cu/10%GDC(C). The H_2-TPR profiles of 5%Cu/CeO_2(B) and 5%Cu/CeO_2(combustion) exhibit two reduction peaks at about 140 °C, and 200 °C. The first peak appearing at about 140 °C, denoted as peak α, is attributed to the reduction of non-crystalline copper oxide in close contact with ceria. The second sharp peak is assigned to the reduction of Cu^{2+} ions in the Cu-O-Ce solid solution. The reduction of CuO species interacting with CeO_2 of 5%Cu/CeO_2(B) catalyst is easier than that of 5%Cu/CeO_2(combustion). For 5%Cu/10%GDC (combustion), only one reduction peak at high temperature corresponding to Cu^{2+} ions in the Cu-O-Ce solid solution was observed. Moreover, in the case of 5%Cu/CeO2(A), the peak β disappeared while H_2 reduction peak corresponding to the reduction of Cu^{2+} ions in the Cu-O-Ce solution and reduction of bulk CuO particles shift to higher temperatures. Interestingly, the reduction of Cu^{2+} ions in the Cu-O-Ce solid solution of 5%Cu/10%GDC(D) is easier than that of other catalysts. It is expected that the easier reducibility of catalyst resulting in higher CO conversion.

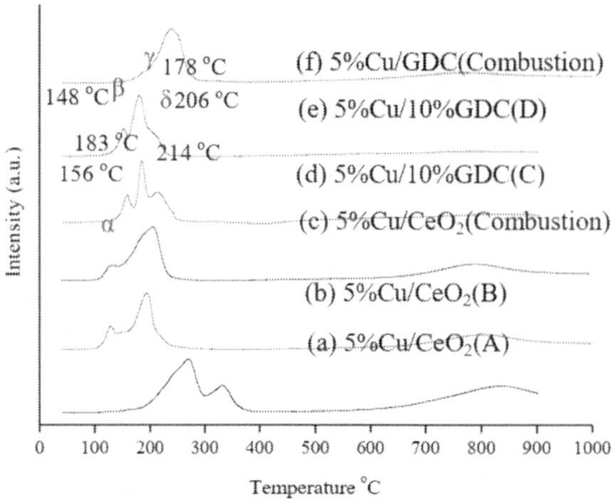

Figure 2. H_2-TPR profiles of Cu on various supports.

3.2 Water gas shift activity of monometallic Cu catalysts

Fig. 3 shows activities of Cu on commercial CeO_2 and GDC mixed oxide. It appears that 5%Cu/CeO_2(A) and 5%Cu/CeO_2(B) have weak WGS activity and start to convert CO to CO_2 at temperature above 150 °C. The activity of 5%Cu/CeO_2(B) is slightly higher than that of 5%Cu/CeO_2(A). Cu catalysts on GDC(C) and GDC(D) were more drastically active than Cu catalysts on ceria. The CO conversion of 5%Cu/10%GDC(D) starts above 100°C and rises up quickly to reach the maximum of 80 %conversion at about 400°C. Above 400°C, the activity reaches equilibrium and slightly declines. The activities of 5%Cu/10%GDC(D) are drastically higher than the activity of 5%Cu/10%GDC(C). For 5%Cu/GDC(C) the conversion started above 100°C and rises up to reach the maximum of 75%conversion at 500°C.

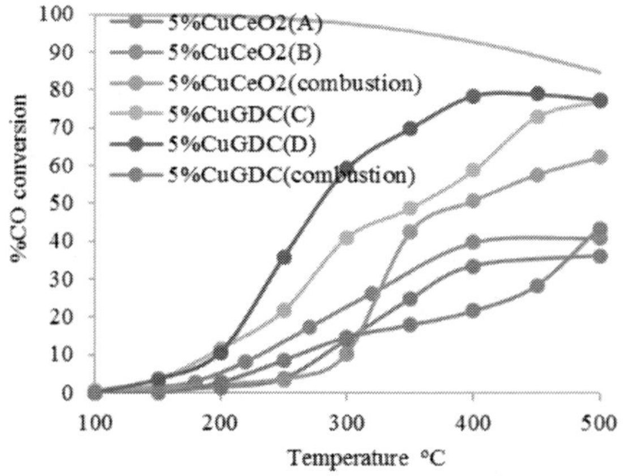

Figure 3. WGS catalytic activity of Cu catalyst on various support.

Catalytic activities of Cu on CeO_2 and GDC prepared by a combustion synthesis technique are also studied. It can be seen that 5%Cu/CeO_2(combustion) is not an active catalyst for WGS reaction. A modification of ceria with Gd drastically decreased the activity for WGS reaction. The order of activity at 500°C is that 5%Cu/10%GDC(D) > 5%Cu/10%GDC(C) > 5%Cu/CeO_2(combustion) > 5%Cu/CeO_2(B) ~ 5%Cu/CeO_2(A) ~ 5%Cu/10%GDC(combustion). For 5%Cu/CeO_2(A), there is a drastic fall in the activity compared with other catalyst due to the reduction peak corresponding to finely dispersed CuO disappear. An additional peak assigned to bulk CuO particles is observed at high temperature. Therefore, the TPR results directly evidenced that the finely dispersed CuO species is more efficient as active sites for WGS reaction. The formation of bulk CuO species is not beneficial for the activity. The finely dispersed CuO species supported on smaller CeO_2 is favorable to the higher activity.

4 Conclusion

The catalytic activities of Cu catalyst on ceria based supports have been studied. It was found that 5%Cu/10%GDC(D) yields the highest catalytic activity. 5%Cu/10%GDC(D) reaches the maximum of 80% conversion at about 400°C. Various evidences seemed to point to the direction that the enhanced catalytic activity of Cu on mixed oxide support was mainly governed by reducibility. The TPR results directly evidenced that the finely dispersed CuO species is more favorable to the higher activity. . The formation of bulk CuO species is not beneficial for the activity.

References

1. D.S. Newsome, Catal. Rev. Sci. Eng. **21**, 275 (1980).
2. C. Rhodes, G.J. Hutchings, A.M. Ward, Catal. Today **23**, 43 (1995).
3. M. Mavrikakis, A.A. Gokhale, J.A. Dumesic, J. Am. Chem. Soc. **130**, 1402 (2008).
4. J.L.G. Fierro, J. Soria, J. Sanz, J.M. Rojo, J. Solid State Chem. **66**, 154 (1987).
5. C. Padeste, N.W. Cant, D.L. Trimm, Catal. Lett. **18**, 305 (1993).
6. Y. Li, Q. Fu, M. Flytzani-Stephanopoulos, Appl. Catal. B **27**, 179 (2000).
7. Q. Fu, A. Weber, M. Flytzani-Stephanopoulos, Catal. Lett. **77**, 187 (2001).
8. Q. Fu, H. Saltsburg, M. Flytzani-Stephanopoulos, Science **301**, 935 (2003).
9. X. Qi, M. Flytzani-Stephanopoulos, Ind. Eng. Chem. Res. **43**, 3055 (2004).
10. W. Ruettinger, X. Liu, R.J. Farrauto, Appl. Catal. B **65**, 135 (2006).
11. W. Deng, M. Flytzani-Stephanopoulos, Angew. Chem. Int. Ed. **45**, 2285 (2006).
12. X. Wang, J.A. Rodriguez, J.C. Hanson, D. Gamarra, M. Fernandez-Garcia, A. Martinez-Arias, J. Phys. Chem. B **109**, 19595 (2005).

13. X. Wang, J.A. Rodriguez, J.C. Hanson, D. Gamarra, A. Martinez-Arias, M. Fernandez-Garcia, J. Phys. Chem. B **110**, 428 (2006).
14. N.A. Koryabkina, A.A. Phatak, W.F. Ruettinger, R.J. Farrauto, F.H. Ribeiro, J. Catal. **217**, 233 (2003).
15. P. Djinovi´ c, J. Batista, A. Pintar, Appl. Catal. A **347**, 23 (2008).
16. S. Charojrochkul, W. Nualpaeng, N. Laosiripojana, S. Assabumrungrat, *Materials Challenge and testing for Supply of Energy and Resources*, Springer.
17. P. Panagiotopoulou, J. Papavasiliou, G. Avgouropoulos, T. Ioannides, D.I. Kondarides, Chem. Eng. J., **134**, 16 (2007).
18. Q. Fu, W. Deng, H. Saltsburg, M. Flytzani-Stephanopoulos, M., Appl. Catal. B: Environ., **56**, 57 (2005).
19. X. Tang, B. Zhang, Y. Li, Y. Xu, Q. Xin, W. Shen, Applied Catalysis A General **190**, 25 (2000).
20. 20. H. Yahiro, K. Murawaki, K. Saiki, T. Yamamoto, H. Yamaura, Catal Today **126**, 436 (2007).
21. M.F. Luo, J.M. Ma, J.Q. Lu, Y.P. Song, Y.J. Wang, Journal of Catalysis **246**, 52 (2007).
22. G. Avgouropoulos, T. Ioannides, Applied Catalysis A General **244**, 155 (2003).
23. J.F. Chen, Y.Y. Zhan, J.J. Zhu, C.Q. Chen, X.Y. Lin, Q. Zheng. Applied Catalysis A General **377**, 121 (2010).
24. X. Zheng, X. Zhang, X. Wang, S. Wang, S. Wu, Applied Catalysis A: General **295**, 142 (2005).
25. L.J. Kundakovic, M.F. Stephanopoulos, Applied Catalysis A: General **171**, 13 (1998).

High-quality AlN films grown on chemical vapor-deposited graphene films

Bin-Hao Chen[1], Hsiu-Hao Hsu [2] and David T.W. Lin[2]

[1]*Department of Materials and Energy Engineering, Far East University, No.49, Chung Hua Rd., Hsin-Shih. Tainan County 744, Taiwan (R.O.C.)*
[2]*Institute of Mechatronic System Engineering, National University of Tainan, No. 33, Sec. 2, Shu-lin St., Tainan City 700, Taiwan (R.O.C.)*

Abstract. We report the growth of high-quality AlN films on graphene. The graphene films were synthesized by CVD and then transferred onto silicon substrates. Epitaxial aluminum nitride films were deposited by DC magnetron sputtering on both graphene as an intermediate layer and silicon as a substrate. The structural characteristics of the AlN films and graphene were investigated. Highly c-axis-oriented AlN crystal structures are investigated based on the XRDpatterns observations.

1 Introduction

The Graphene is single atom thick two-dimensional material, which has the structure of hexagonal carbon atoms that are sp2 bonded, and have high mobility and optical transparency, flexibility, robustness and environmental stability. These characteristics mean that the broader application of graphene has attracted great interest [1-7]. Perfect graphene is a two-dimensional crystal composed of carbon atoms, as shown in Figure 1, in which the atoms are arranged in a honeycomb hexagon, and if they form in pentagons or heptagons then these constitute defects. Current methods of growing graphene are divided into two types, graphene derived from graphite, and that synthesized from other sources. The graphite approach is based on mechanical exfoliation, in which the atoms are chemically cleaved from each other. Graphene synthesis can be carried out using a silicon carbide epitaxial growth method or a chemical vapor deposition method. However, mechanically exfoliating graphene from bulk graphite is not suitable for the synthesis of large surface area devices, nor for use in practical device fabrication processes. In contrast, chemical vapor deposition (CVD) is can be used with large surface area applications and has a relatively low cost [4,7-10]. The low pressure CVD (LP-CVD) method of growing graphene on a Cu substrate has been widely reported in the literature, there are many studies on the mechanism of the resulting materials [3,11,12]. In particular, it has been shown that a large surface area of monolayer graphene with excellent quality can be synthesized by LP-CVD.

Our experiments used graphene as an intermediate layer in epitaxial aluminum nitride films, with aluminum nitride (AlN) being a III-V compound that has a wurtzite hexagonal close-packed structure. Aluminum nitride has a variety of excellent physical properties, such as a high melting point (3273 K), high thermal conductivity (285 W/mK), high direct band gap (6.2 Ev), good dielectric constant (8.5) and high hardness (about ~ 2×103 kgf mm-2) [13,14,15]. Common ways of depositing aluminum nitride films include DC and RF magnetron sputtering [13-18], chemical vapor deposition [19, 20], and molecular beam epitaxy [21]. However, the CVD and MBE methods can only be used at high temperatures, and this limits the base materials that can be used, and if the resulting grains are too big then this will lead to a rough surface, and so it will not be possible to easily manufacture electrodes on the films. In contrast, reactive RF sputtering offers good growth patterns, with the advantages of low temperature deposition, surface smoothness, fast deposition rate, and low complexity.

The use of graphene thin film epitaxial growth of III-v nitride the advantage that graphene has atoms arranged similar to a hexagonal wurtzite structure. In addition, the low migration barrier of III-V metals on graphene allows adatoms to diffuse readily on the surface at reduced growth temperatures. The use of pristine graphene can thus enable large adatom diffusion lengths during nucleation, which may promote the formation of large two-dimensional islands [22]. The graphene films were synthesized in this paper by CVD, and then transferred onto silicon substrates. Epitaxial aluminum nitride films were deposited by DC magnetron sputtering on both graphene as an intermediate layer and a silicon substrate. The structural characteristics of the AlN films and graphene were investigated and analyzed, and the results of this a reported below.

Figure 1. Perfect graphene is a two-dimensional crystal composed of carbon atoms.

2 Experiments

2.1 Graphene synthesis

Synthesis flow chart graphene has three sections, as shown in Figure2.In our experiments the copper substrate was first pre-treated, and then cut into piece of the desired size. Hydrochloric acid (HCL) was used to clean the surface to remove oxide impurities, and then DI water and ethanol were used to remove the HCL liquid residues, as shown in Figure 3.

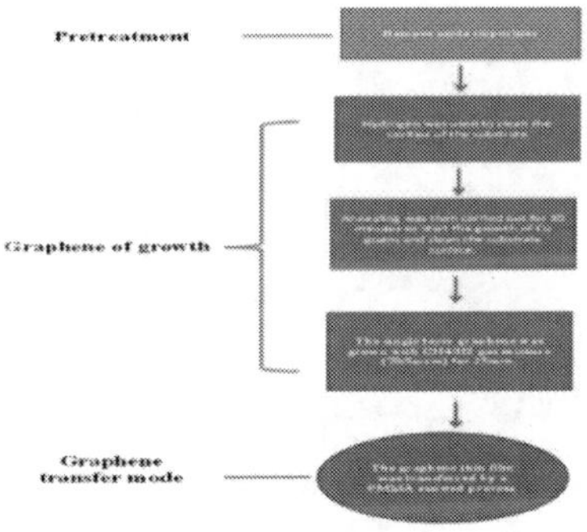

Figure 2. Synthesis of Graphene flow chart shown

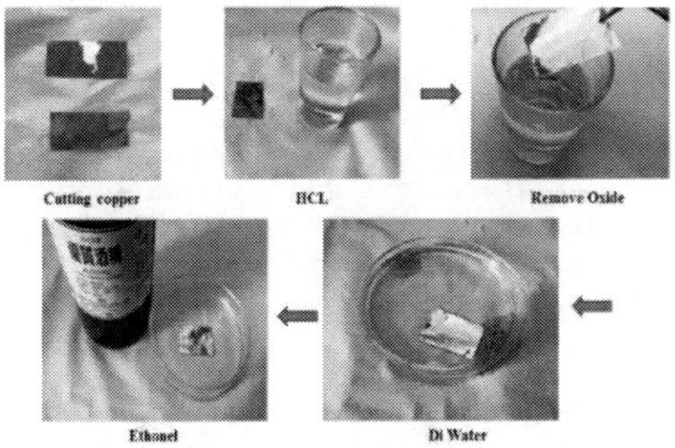

Figure 3. Graphene pre-treated process

Before graphene synthesis a 10sccm flow of hydrogen was used to clean the cavities of the substrate, Annealing was then carried out for 30 minutes to start the growth of Cu grains and clean the substrate surface. Graphene was then deposited at 1050 ℃ for 15 minutes using a precursor mixture of CH4 (5 sccm) and H2 (70 sccm), as shown in Figure4. Finally, the graphene thin film was transferred by a PMMA assisted process in which copper was etched by 5% ferric nitrate, and the graphene thin film was transferred onto a Si wafer. The PMMA was then dissolved in acetone, as shown in Figure 5 and 6.

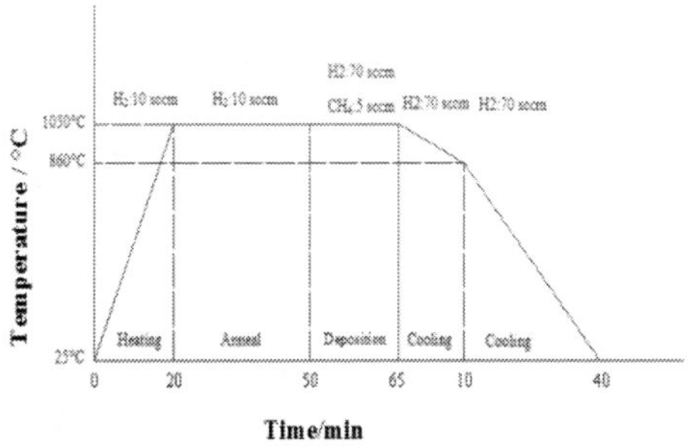

Figure 4. Heating curve diagram of graphene.

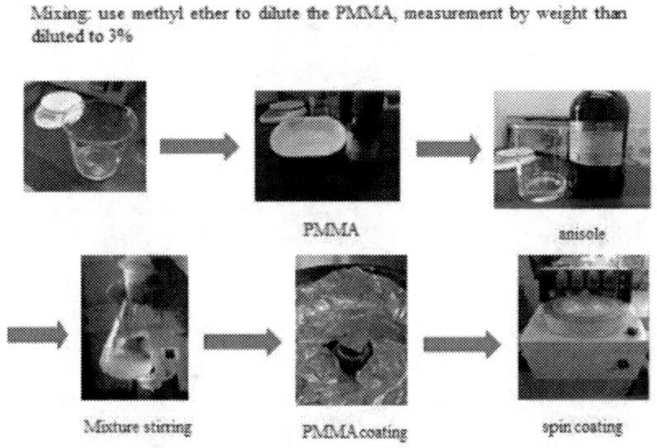

Figure 5.PMMA（Poly(methyl methacrylate)）deployment and coat of sample

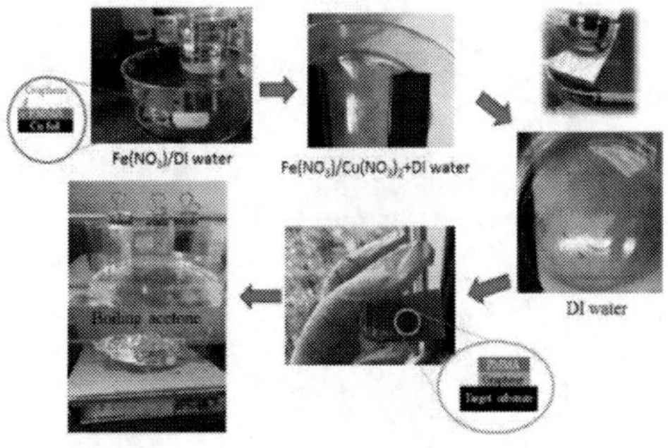

Figure 6. Graphene transfer mode

2.2 Aluminum nitride thin film process

After confirming the presence of graphene on silicon, the epitaxial aluminum nitride films were deposited by DC magnetron sputtering. Aluminum nitride thin films were grown using a high purity 4N 2 inch aluminum target, and high purity argon and nitrogen were used as the reaction gas, with a nitrogen concentration of 40%, temperature of 400 ℃, pressure of 8 mTorr and power 250W at 3 hours. The resulting AlN film was then analyzed by XRD.

3Results and discussion

Figure 7 shows an optical microscopy (OM) image of the graphene film on the silicon substrate. Since graphene is only one atom thick it is difficult for the naked eye to observe. However, the different colors and intensities of the light reflected from the substrate and graphene can be used to distinguish layers, and silicon substrates with oxide layers of different thicknesses also have different colors.

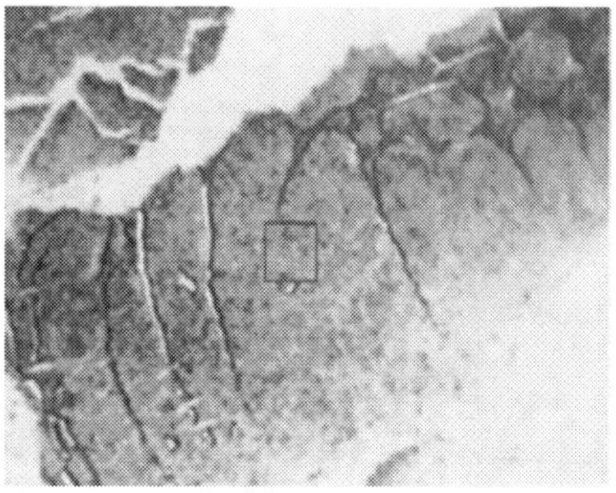

Figure 7. Optical microscope image shows a CVD-graphene film grown on Cu-foil after transfer to a Si wafer.

Scanning electron microscopy (SEM) was used to observe the surface of the thin film, as shown in Figure8, in which the boundary of the copper lattice can be clearly seen. Since the number of graphene layers determines the brightness of different regions, when there is low electron absorption then brighter areas will be seen. The dark areas may be due to of several factors, such as the carbon content or substrate structure.

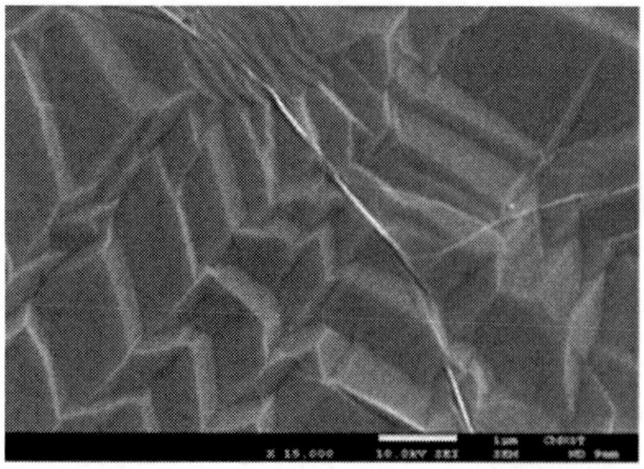

Figure 8.SEM image of a graphene on Cu substrate.

Figure 9 shows there are three peaks, the D peak, G peak and 2D peak at2648 cm-1, 1344cm-1, and 1585cm-1, respectively, due to the different single layer or multilayer graphene structures, with the intensity of the Ramen spectra varying as the number of layers increases. Based on these characteristics, we can determine the number of layers of graphene. D peak is disordered vibration peak graphene, graphene because structural defects, away from the center of the Brillouin zone caused by the movement of the lattice, the defect when the sample is large, the higher its D peak. G Peak, on behalf of graphene good or bad, by stretching vibration between carbon atoms SP2 caused, in the center of the Brillouin zone E2g phonon vibrations. 2D peak double resonance Raman phonon peak use its judgment of graphene layers. Due to the different single or multi-graphene structure, with increasing number of layers, shape, and intensity of seats in the Raman spectrum has not different.The Raman intensity ratio of ID/IG can be used to assess the quality of graphene, and in general the ID/IG value for high quality graphene will not exceed 10%. The results thus show that our graphene is high quality, as it has an ID/IG ratio of about 0.13. A previous study [7] on the synthesis of graphene by CVD reported that when I2D/IG (2.4) > 2 this indicates monolayer graphene, while 1 < I2D/IG (1.1) < 2 indicates bilayer graphene, and I2D/IG (0.4) < 1 is a signature of trilayer graphene. Figure 5 shows the I2D/IG ratio of 1.96, and this, along with the Raman spectra, also shows the existence of graphene on a Si substrate.

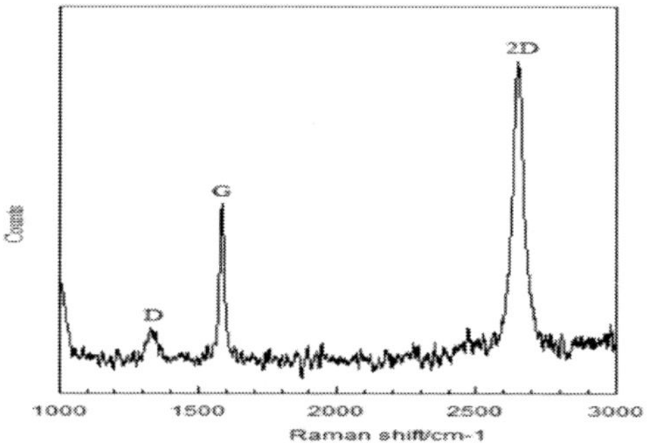

Figure 9. Raman spectra of graphene on Si substrate.

The XRD pattern of as-prepared sample is shown in Figure10 andFigure11. It can be clearly seen that all the peaks can be indexed by hexagonal wurtzite structure AlN. The (002) peak is the highest one in all of the peaks, which suggests that the C-axis growth direction. The measured XRD patterns show prominent diffraction peaks corresponding to the (1 0 0), (0 0 2) and (1 0 1) crystal planes of the hexagonal wurtzite phase of AlN, respectively. These crystal planes value can see the peak value of the graphene substrate intensity all to be higher than the Silicon substrate. Thus can be seen, graphene can let adatoms diffuse easily on the surface, so conducive to the epitaxial growth of aluminum nitride.

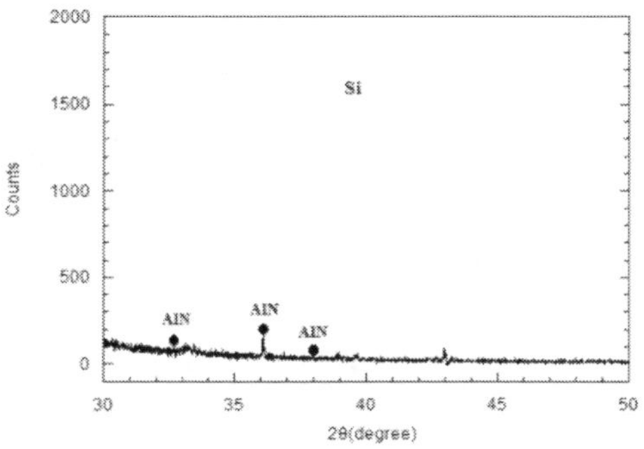

Figure 10.XRD of ALN on Si substrate.

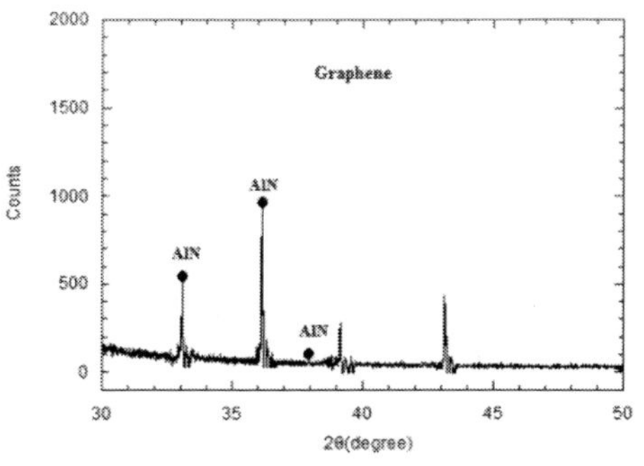

Figure 11.XRD of ALN on grapheme substrate.

4Conclusion

The ID/IG ratio of about 0.13 indicates that the graphene produced in this work is high quality, while the I2D/IG ratio of 1.96) and the Raman spectra results also prove the existence of graphene on the Si substrate in this study. Epitaxial aluminum nitride films were deposited by DC magnetron sputtering on both graphene as an intermediate layer and a silicon substrate. The structural characteristics of the AlN films and graphene were investigated. The XRD patterns of the as-prepared samples show that the peak values of the graphene substrate are all to be higher than those of the silicon substrate. This shows that graphene can let the adatoms diffuse easily on the surface of the substrate, and this is conducive to the epitaxial growth of aluminum nitride.

Acknowledgements

The authors are thankful for the financial support of this research from the National Science Council of Taiwan through the grant no. MOST103-2221-E-269-010.The authors also express their gratitude to the Energy R&D Foundation of the Energy Bureau of the Ministry of Economic Affairs, Taiwan, for providing financial support.

References

1. Y.H. Zhang, Z.Y. Chen, B. Wang, Y.W. Wu, Z. Jin, X.Y. Liu, G.H. Yu,Materials Letters **96,**149(2013)
2. I. Vlassiouk , P. Fulvio, H. Meyer, N. Lavrik, S. Dai, P. Datskos, S. Smirnov, Carbon **54**,58(2013)
3. X. Li, W. Cai, J. An, S. Kim, J. Nah, D. Yang,R. Piner, A. Velamakanni,I. Jung, E. Tutuc,S.K.Banerjee, L. Colombo, R.S. Ruoff, Science **324**,1312(2009)
4. P. Trinsoutrot, C. Rabot, H. Vergnes,A. Delamoreanu,A. Zenasni, B. Caussat,Surface & Coatings Technology **230**,87(2013)
5. Y. Zhang,Y.W. Tan, H.L. Stormer, P. Kim,Nature**438**,201(2005)
6. Y.M. Lin, C. Dimitrakopoulos, K. A. Jenkins,D. B. Farmer, H.Y. Chiu,A. Grill, Ph. Avouris, Science **327**,662(2010)
7. Z. Tu , Z. Liu , Y. Li , F. Yang , L. Zhang , Z. Zhao , C. Xu , S. Wu , H. Liu , H. Yang , P. Richard,Carbon **73**,252(2014)
8. A.K. Geim, P. Kim,Scientific American**298**,90 (2008)
9. Y. Zhang, L. Zhang,C. Zhou, Accounts of Chemical Research**46**,2329(2013)
10. G. Deokar, J. Avila, I.Razado-Colambo , J.L. Codron, C. Boyaval, E. Galopin,M.C. Asensio , D. Vignaud,Carbon **89**,82(2015)
11. P.Y.Huang, C.S.Ruiz-Vargas, A.M.van der Zande, W.S.Whitney, M.P.Levendorf, J.W.Kevek, S.Garg, J.S.Alden, C.J.Hustedt, Y.Zhu,J. Park,P. L. McEuen, D.A.Muller, Nature **469**,389(2011)
12. X. Li, C.W. Magnuson, A. Venugopal, R.M. Tromp, J.B. Hannon, E. M. Vogel,L. Colombo, R.S. Ruoff, J. Am. Chem. Soc.**133**,2816(2011)
13. H.Y. Liu, G.S. Tang, F. Zeng , F. Pan,Journal of Crystal Growth **363**,80(2013).
14. V.I. Dimitrova, D.I. Manova, D.A. Dechev,Vacuum**49**,193(1998)
15. K.A Aissa, A. Achour,J. Camus, L.L. Brizoual, P.Y. Jouan, M.A.Djouadi,Thin Solid Films**550**,264(2014)
16. Z. Vashaei, T. Aikawa, M. Ohtsuka, H. Kobatake, H. Fukuyama, S. Ikeda, K. Takada, Journal of Crystal Growth**311**,459(2009)
17. M.A. Moreira, I. Doi, J.F. Souza, J.A. Diniz, Microelectronic Engineering**88**,802(2011)
18. I.Safi, , Surface and Coatings Technology **127**,203(2000)
19. R. Boichot, A. Claudel, N. Baccar, A. Milet, E. Blanquet, M. Pons,Surface & Coatings Technology **205**,1294(2010)
20. L. Guo, G. Chen , Y. Zhu, X. Duan, H. Ye, Journal of Crystal Growth **426**,49(2015)
21. B. Liu, J. Gao, K.M. Wu, C. Liu,Solid State Communications **149**,715(2009)
22. Z.Y. Al Balushi,T. Miyagi , Y.C. Lin , K. Wangb, L. Calderin,G. Bhimanapati, J.M.Redwinga, J.A. Robinson, Surface Science **634**,81(2015)

Oscillatory reactions in the acetone-glucose-sodium bromate-sulfuric acid system with a tetraazamacrocyclic copper (II) complex as a catalyst

Lin Hu*, BinBin Li, SanChun Wei,YeGui Wang, MiWei Hu,WenYuan Xu

Institute of Applied Chemistry, East China Jiaotong University, 330013,Nanchang , P.R. China

Abstract. The behavior of glycolysis concerning the biochemical reaction was thought to be similar to the oscillatory reaction with glucose being the substrate. The objective of this research is to find a new oscillatory system and study its relevant properties, so new Cu (II)-catalyzed oscillating reaction involved glucose and acetone as the double substrates was investigated systematically. As a function of reaction temperature and concentration of glucose, bromate, and acetone, the features in this novel oscillation system was studied. The results showed that the activation energy reached up to 90.839 KJ/mol and the oxidized product of the double substrates provided the impetus in the course of the new oscillating reaction. A tentative mechanism was considered on the basis of FNK.

1 Introduction

Nature is the best teacher showing us how to effective use energy. Most creatures on the Earth, including animal, plant as well as yeast has shown an example of successful pathway taking energy from glycolysis[1]. Glycolysis that is the ultimate source of energy keeping creatures survive is the metabolic pathway that converts glucose into pyruvate and releases free energy to form the high-energy compounds ATP and NADH .It was first found that glycolysis is a determined sequence of ten enzyme-catalyzed complex reactions resulting in oscillation of NADH concentration in yeast in 1968[2].

The Belousov-Zhabotinskii oscillating reaction was found long ago and studied thoroughly. The classical BZ oscillatory systems consist of malonic acid, Ce(IV)/Ce(III) or Mn(III)/Mn(II) redox catalyst and $KBrO_3$ in solutions of H_2SO_4[3]. With regard to the catalyst in the BZ oscillatory systems, many researchers paid much attention to metallic ion in the past and further research on Ce(IV)/Ce(III) [4-9]. In 1982, the tetraazamacrocyclic complexes of Cu(II) and Ni(II) were first found to act as a catalyst for BZ oscillating reactions. The design and synthesis of macrocycles and their metal complexes attract great attention, resulting from their widespread applications in biology, chemistry and in other applications of science[10]. The features of the oscillating reactions catalyzed by tetraazamacrocyclic complex differ from the classical oscillating reaction in accordance with relevant literature[11-14].The tetraazamacrocyclic complex and organic substrate play a crucial role in the new oscillation reaction because the tetrazamacrocyclic complex is similar to some enzymes, so the newly oscillating reaction is deemed

Corresponding author:hulin21@hotmail.com

to be a man-made biochemical scillation. It is significant to study the chemical oscillation on the basis of glucose as the analogy to glycolysis.

In this research, we have investigated the behavior of glucose and acetone, two important substrates in normal metabolism. As a result, new oscillating reactions in the g-acetone-glucose-sodium bromate-sulfuric acid system using tetraazamacrocyclic copper(II) complex [Cu L]$(ClO_4)_2$ as catalyst are reported. The ligand L in the complex is 5,7,7,12,,14,14 –hexamethyl-1,4,8,11-tetraazacyclotetradeca-4,11-dine.

2 Experiment

In accordance with the relevant dissertation[15-17], [Cu L] $(ClO_4)_2$ was prepared, whose structure is measured by IR . [Cu L] $(ClO_4)_2$ is used as the catalyst in this chemical oscillatory system. The water used in the experiment is twice redistilled to ensure the removal of impurities. The reagents prepared in this system were of analytical grade. The most important thing is that the $NaBrO_3$ needs to be recrystallized from hot water twice to move Br^- and other impurities. This chemical oscillatory experiment was conducted in a closed glass container at 30℃ under constant stirring. A constant N_2 flow of 400 ml/min in 40ml reaction solution was maintained by a capillary flow indicator. The redox potential of the reacting system was measured with a smooth platinum electrode vs. a $Hg(Hg_2Cl_2)K_2SO_4$ reference electrode, and the bromide ions concentration was measured with a bromide selective electrode, then all results were collected on a y-t recorder.

We carried out more than 50 experiments, and the explored ranges of the initial major reactant concentrations are reported in table 1, except for sulfate acid at a fixed temperature. According to the above general regularity, the different behavior of the oscillation in which the major reactant concentration are kept at different temperatures is studied. Typical potentiometric oscillation traces of platinum electrode and bromide selective electrode are shown in Fig.7.

During the oscillations, the solution color changed periodically between orange and faint yellow, indicating the oscillation between $[Cu L]^{2+}$ and $[Cu L]^{3+}$.At the same time ,the relationship between the potential of the platinum electrode and the potential of the bromide ion electrode is also studied in detail, through their potential respectively.

Table 1 the ranges of initial major reactant concentrations in the system

$NaBrO_3$ (mol/L)	Glucose (mol/L)	Acetone (mol/L)	[Cu L]$(ClO_4)_2$ (mol/L)
0.04-0.20	0.06	0.65	0.05
0.08	0-0.06	0.65	0.05
0.08	0.06	0.25-2.05	0.05
0.08	0.06	0.65	0.01-0.21

3 Result and discussions

As a result, this newly oscillatory system catalyzed by [Cu L]$(ClO_4)_2$ displayed classical obvious oscillatory feature. The mean values of the following oscillatory parameters were calculated: oscillatory period(t_p), oscillatory cycles(n) and oscillatory amplitude (A). The characteristics of oscillations is based on the concentration of bromate ions, glucose, acetone and [Cu L]$(ClO_4)_2$. The effects of the double substrate on oscillation parameter is shown in Fig.1.

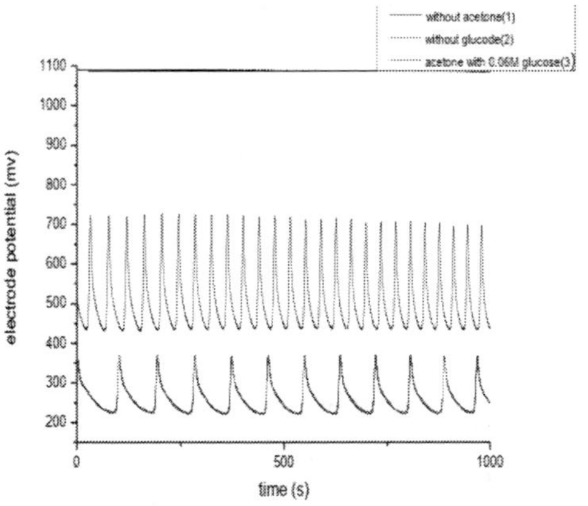

Fig.1. the effects of substrate on the oscillation

As shown from Fig.1, the three conclusion can be made: Firstly,the oscillation can not exist without acetone in the whole oscillatory system. Secondly, when glucose is not added to the system,the oscillatory reaction still happen. Thirdly ,the double substrate can make the oscillatory amplitude bigger and the oscillatory period shorter than without glucose.

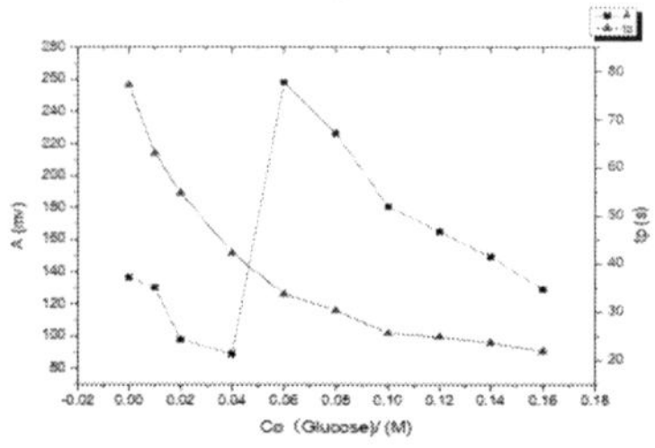

Fig.2. The effects of [Glucose]$_0$ on the oscillation period (tp),the amplitude of oscillation (A) at 30℃:

[Acetone]$_0$=0.65 M, [H_2SO_4]$_0$=1.05 M

[CuL^{2+}]$_0$=0.05 M ,[$NaBrO_3$]$_0$=0.08 M

As shown from the figure.2, oscillation still takes place without the glucose. The amplitude of oscillation decreased with increasing glucose concentration and reached a minimum at [glucose]$_0$ ca. 0.04 M, then rose with further increasing the glucose concentration until the glucose concentration was 0.06 M, finally decreased as the glucose concentration increased. The oscillation period (tp) became increasingly shorter with increasing glucose concentration throughout the experiment.

The effect of the concentration of acetone on oscillation parameter is shown in Fig.3. When the concentration of acetone was 0.85M, the oscillatory amplitude achieves the maximum.when the concentration of acetone increased to 1.05M,the oscillatory period is the longest.The effect of the concentration of $CuL^{2+}$and $NaBrO_3$ oscillation parameter are shown in Fig.4 and Fig.5,respectively. Like the effect of acetone, the trend for the curve of the effect of $CuL^{2+}$and $NaBrO_3$ on the oscillation parameter is similar.With the increasing concentration of reactant in the oscillatory system, no matter the amplitude or period initially, they are boosted and then subsequently lowered.

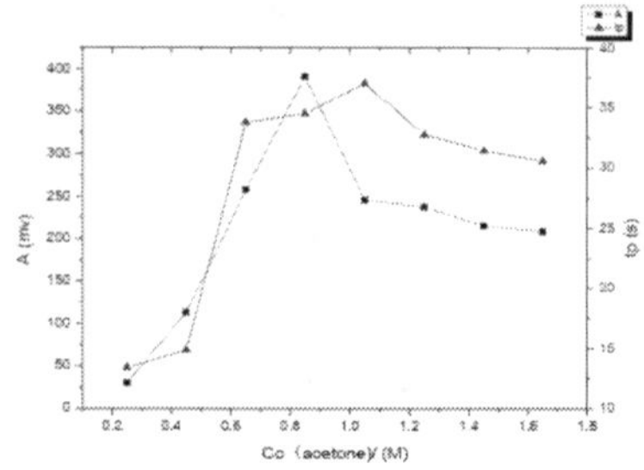

Fig.3. The effects of [Acetone]0 on the oscillation period (tp), the amplitude of oscillation (A) at 30℃:

[Glucose]0=0.06 M, [H_2SO_4]0=1.05 M,

[CuL^{2+}]0=0.05 M, [$NaBrO_3$]0=0.08 M

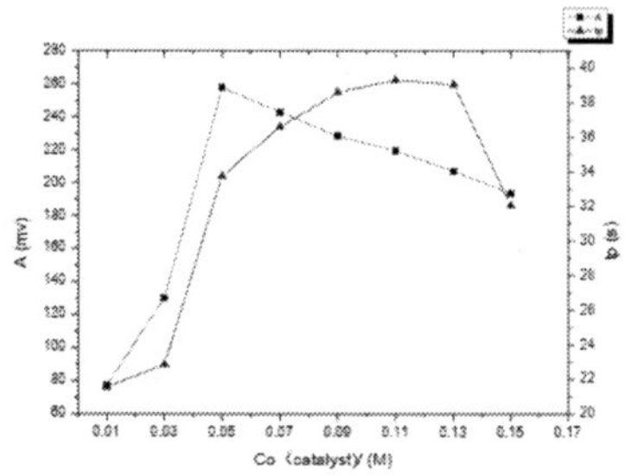

Fig.4. The effects of [CuL^{2+}]$_0$on the oscillation period (tp),the amplitude of oscillation (A) at 30℃:

[Glucose]$_0$=0.06M,[H_2SO_4]$_0$=1.05M

[Acetone]$_0$=0.65M,[$NaBrO_3$]$_0$=0.08M

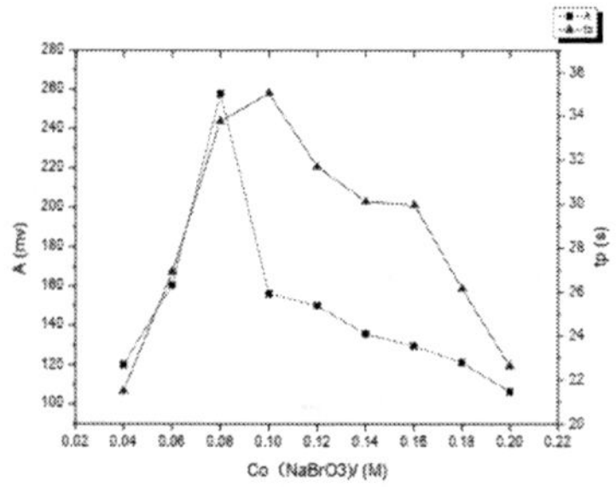

Fig.5. The effects of [$NaBrO_3$]$_0$ on the oscillation Period(t_p),the amplitude of oscillation of oscillation(A) at 30℃:

[Glucose]$_0$=0.06M,[H_2SO_4]$_0$=1.05M

[Acetone]$_0$=0.65M,[CuL^{2+}]$_0$=0.05M

The reaction was also studied at different temperature , and the oscillation period (tp) was found to be inversely related to temperature. As the temperature increased, the oscillation period (tp) decreased. From the Arrhenius plot (Fig.6) of l n ($1/t_p$) v s 1/T, the activation energy was calculated as 90.839 kJ/mol.

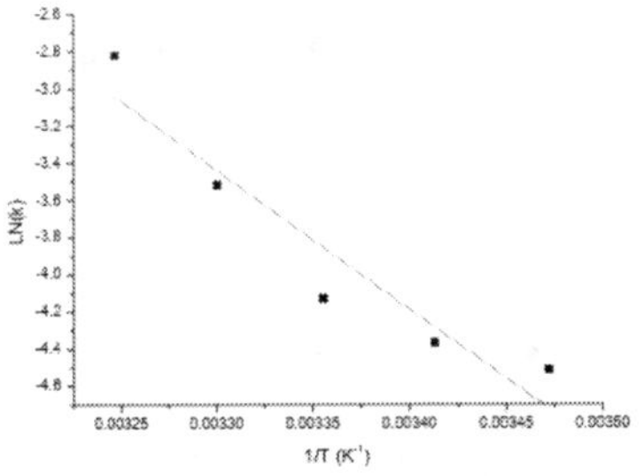

Fig.6. The plot of ln($1/t_p$) vs 1/T: $[Acetone]_0$=0.65 M

$[H_2SO_4]_0$=1.05 M, $[CuL^{2+}]_0$=0.05 M

$[NaBrO_3]_0$=0.08 M,$[Glucose]_0$=0.06 M

The relationship between the potential of the platinum electrode and the potential of the bromide ion electrode is described iconically in the Fig.7.

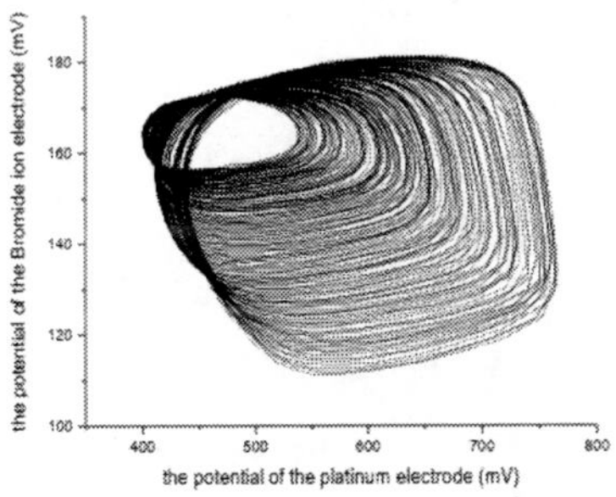

Fig.7. the potential of the platinum electrode vs the potential of the bromide ion electrode for the system at 30℃:

$[NaBrO_3]$ = 0.08 M, $[Acetone]_0$ = 0.65 M,

$[H_2SO_4]_0$=1.05 M

$[Glucose]_0$ = 0.06 M, $[CuL^{2+}]_0$=0.05 M

In order to investigate whether free radical steps were involved, we tested H_2O_2[18-20]. When the oscillations were held up briefly, the amplitude and oscillations decreased as soon as the concentration of H_2O_2 in the oscillation system reached 9.7×10^{-3} M. This indicates that there are free radical steps involved in the mechanism of oscillation.

When the concentration of vitamin C in the oscillation system reached up to 1.0×10^{-2}M, the chemical oscillations could be restrained momentarily, similar to the effect of vitamin C in the Ce(III) catalyzing B-Z oscillation system[21-22] .This phenomenon makes it clear that $HBrO_2$ is a very important intermediate in the system.

By adding Ag^+ to the oscillation system, bromate ions concentration decreased due to the formation of AgBr precipitate, and the potentiometric oscillation of bromide selective electrode could be temporarily suppressed[23]. From this, it can be seen that the concentration of Br^- also acts as an on-off switch in the $NaBrO_3$-lactic acid-H_2SO_4-[Ni L] $(ClO_4)_2$ oscillation system.

It is not easy to clarity the mechanism of the present oscillation system. From everything stated above, this new oscillation can still happen without glucose. At this time, bromine has a substitution reaction with acetone because there are reactive hydrogens in the acetone, which results in bromide ions .There are a series of redox reactions between bromate in the solution and bromide ions ,which render the whole system able to be continuous. This is the reason why oscillation can be found when solely the acetone exists. When acetone and glucose are the double substrates in the system, glucose can have similar reaction with bromine for the reactive hydroxyl radicals in the glycolic chemical reaction. When no additional brominating reagent were supplied, it is necessary to debate on Br_2 consumption. Pyruvic acid is thought to be an intermediate. We assume that the brominating reaction may take place on pyruvic acid, the oxidized product of the organic mixed substrate.

Therefore, the following reactions occur during the oscillation:

$$BrO_3^- + Br^- + 2H^+ \rightleftharpoons HOBr + HBrO_2 \quad (1)$$

$$HBrO_2 + Br^- + H^+ \rightleftharpoons 2HOBr \quad (2)$$

$$HOBr + Br^- + H^+ \rightleftharpoons Br_2 + H_2O \quad (3)$$

$$BrO_3^- + HBrO_2 + H^+ \rightleftharpoons 2BrO_2\cdot + H_2O \quad (4)$$

$$BrO_2\cdot + [CuL]^{2+} + H^+ \longrightarrow [CuL]^{3+} + HBrO_2 \quad (5)$$

$$C_6H_{12}O_6 + Br_2 \longrightarrow C_6H_{11}O_7Br + HBr \quad (6)$$

$$CH_3COCH_3 + Br_2 \longrightarrow BrCH_2COCH_3 + HBr \quad (7)$$

$$BrCH_2COCH_3 + [CuL]^{3+} + H_2O \longrightarrow [CuL]^{2+} + HBr + \text{oxidation product} \quad (8)$$

Glucose is the energy source for human being. The analogical research on the oscillatory reaction with glucose being substrate will help man to make clear the complex and significant biochemical mechanism.

4 Conclusion

New Cu(Ⅱ)-catalyzed oscillating reaction involved glucose and acetone as the double substrates is proved to be effective. This oscillation can not exist without acetone in the whole system but do without glucose. When the concentration of glucose increase to be 0.04M, the amplitude of oscillation is minimum, when the concentration of glucose increase to be 0.06, the amplitude of oscillation is maximum, but with the concentration of glucose increasing ,the period of oscillation reduce constantly. When it comes to the effect of the concentration of acetone , sodium bromate and tetraazamacrocyclic copper(Ⅱ) complex to the oscillatory parameters, there are the maximal of concentration respectively . the activation energy was calculated as 90.839 kJ/mol and this tentative mechanism is proved to be based on FNK through adding H_2O_2, vitamin C and Ag^+ to the system respectively.

Acknowledgment

The authors gratefully acknowledge financial funding of this work by the National Science Foundation of China (31160187).

References

[1]M. Bier, B.Teusink, B.N.Kholodenko,H.V.Westerhoff, Biophys. J. **62** (1996)

[2] J. Wolf, R. Heinrich, Bichem.J. **345** (2000)

[3] Melicherčík M, Treindl Ľ, Chem. Phys. Lett. **196** (1992)

[4] Field RJ, Boyd PM, Phys. Chem. **89** (1985)

[5] Gaspar V, Galambosi P,Phys. Chem. **90** (1986)

[6] Noszticzius Z, Bodiss J,J.Am.Chem.Soc. **101** (1979)

[7] Noszticzius Z, Stirling P, Wittmann M,Phys. Chem. **89** (1985)

[8] Pelle K, Wittmann M, Lovrics K, Noszticzius Z, Phys. Chem. **108** (2004)

[9] Wittmann M, Stirling P, Bódiss J, Chem. Phys. Lett. **141** (1987)

[10] El-Boraey HA, El-Gammal OA,Spectrochim. Acta, Part A. **138** (2015)

[11] Bychkov VY, Tyulenin YP, Slinko MM, Korchak VN, Appl. Catal., A. **321** (2007)

[12] Sun H, Lin H, Zhu S, Zhao G, Su X, ChenY, Polyhedron. **18** (1999)

[13] Zhao G, Sun H, Lin H, Zhu S, Su X, Chen Y,Inorg. Biochem. **72** (1998)

[14] Lin H, Zhu S, Kondiano AB, Su X, Kou F, Chen Y, Polyhedron. **17** (1998)

[15] Choi K-Y, Suh I-H, Inorg. Chim. Acta. **299** (2000)

[16] Choi K-Y, Chun K-M, Suh I-H, Polyhedron. **18** (1999)

[17] Choi K-Y, Lee H-H, Suh I-H, Kim J-G, Shin U-S, Inorg. Chim. Acta. **321** (2001)

[18] Tóthová M, Nagy A, Treindl Ľt, Chem. Phys. Lett. **299** (1999)

[19] Barros WRP, Reis RM, Rocha RS, Lanza MRV,Electrochim. Acta. **104** (2013)

[20]yothi KP, Yesodharan S, Yesodharan EP, Ultrason. Sonochem. **21**(2014)

[21] Tajmir-Riahi HA. J. Inorg,Biochem. **42** (1991)

[22] Gao J, Yang H, Liu X, Ren J, Lu X, Hou J, et al,Talanta. **55**(2001)

[23] Treindl L, Hemmingsen T, Ruoff P,Chem. Phys. Lett. **269** (1997)

NanoCrystalline Cellulose, an environmental friendly nanoparticle for pharmaceutical application – A quick study

Yee Kai Song[1,2], Irene Mei Leng Chew[3], Thomas Shean Yaw Choong[1] and Khang Wei Tan[2]

[1]*Faculty of Engineering, University Putra Malaysia, 43400 UPM, Serdang, Selangor Darul Ehsan, Malaysia*
[2]*Faculty of Engineering, Technology and Built Environment, UCSI University, Cheras, 56000 Kuala Lumpur, Malaysia.*
[3]*School of Engineering, Monash University Sunway Campus, 46150 Bandar Sunway, Selangor Darul Ehsan, Malaysia.*

Abstract. Nanocrystalline cellulose (NCC) is probably the most abundant nano material can be found in nature. It offers not only substantial environmental advantage due to its intrinsic biocompatibility and biodegradability, but also outstanding aspect ratio, tension modulus, bioavailability and permeability. NCC can be isolated from low value biomass through a well-developed acid hydrolysis method, and the structure can be readily manipulated during extraction to obtain desirable size distribution, surface charge and degree of crystallinity. Erratic absorption of drug has been a major issue for years until the emergence of nanoparticle adopted as smart vehicle for drug delivery. However, due to inevitably severe toxicity and the side effect from ordinary nanoparticle, the exploration of green nanoparticles are catching escalating attention is now in urge. Microcrystalline cellulose (MCC) has long been an option, nevertheless, due to their relatively low surface charge compared to total volume, drug release in a control manner was hardly satisfied. It is suggested that the above limitation can be overcome by nanocrystalline cellulose. This article discusses and explores their potential in cosmetic, personal care and pharmaceutical application, and the recent development in term of preparation and characteristic.

1 Introduction

The forest product industry has been consistently searching to unlock potentiality of green innovation, in which stimulated by the fact that industry and consumers are demanding green concept resources. Mohanty *et al.* [1] highlighted that green composite which has high biodegradability, high biocompatibility, high resource availability, and environmental friendly are able to compete with existing products made from petroleum-based material.

Cellulose is an organic component exists in cell walls of plants. It is constructed from repeating β-D-glucopyranose units and having three hydroxyl groups in each anhydroglucose unit (AGU) [2]. Cellulose is present in amorphous state, but it is also exhibits crystalline region depending on the location of hydrogen bonding [3]. Existence of hydrogen bonding gives hydrophilic property to cellulose yet maintaining insolubility in majority of solvents. Cellulose has constantly being a research subject as a green material, by focusing on the high functionality of chemical properties and mechanical properties. Development of nanotechnology over the years is capable of extracting cellulose at nanoscale with minimal defects on cellulose hierarchical structure, producing a new template of cellulose based material for the next generation [4].

Nanocrystalline cellulose (NCC) is a highly potential green and renewable nanomaterial. Ruiz *et al.* [5] reported that NCC is typically elongated rod-shape with dimensions ranging from 10 to 1000 nm in length and 1 to 100 nm in diameter. NCC has appeared in many terms, for instance nanowhisker, nanocrystals, microcrystals, nanofibrils, or cellulose nanocrystals. Different methods had been discovered in preparation of NCC. Acid hydrolysis is the most commonly used method to isolate amorphous region from crystalline region in cellulose [6]. The pretreatment, alkaline treatment, bleaching and acid hydrolysis process to obtain NCC from Oil Palm Empty Fruit Bunch (EFB) was attempted and published in the previous work [7]. As it can be observed from Dynamic Light Scattering (DLS) analysis, NCC particles received from EFB are generally having size much below 1 μm, Figure 1. In comparison, Song Y.K., et al. [8] studied its surface morphology using Field Emission Scanning Electron Microscope (FESEM), and it seems the extracted cellulose particles were slight larger than [7], Figure 2.

However, usage of acid hydrolysis has negative impacts, for example deterioration of NCC, acid corrosiveness, thermal instability, and incompatible to environment friendly concept. Nevertheless, NCC has few supreme properties that gained increasing attentions in recent years, this includes high specific strength, high tension modulus, high aspect ratio, large surface-to-volume ratio, large surface charge, bioavailability, biocompatibility, and competitive cost [6, 9]. Several studies on NCC showed promising industry applications in various optical devices [10], regenerative medicine [11], automotive interior applications [12], and controlled drug delivery [13]. In this paper, only pharmaceutical and personal care applications are focused on.

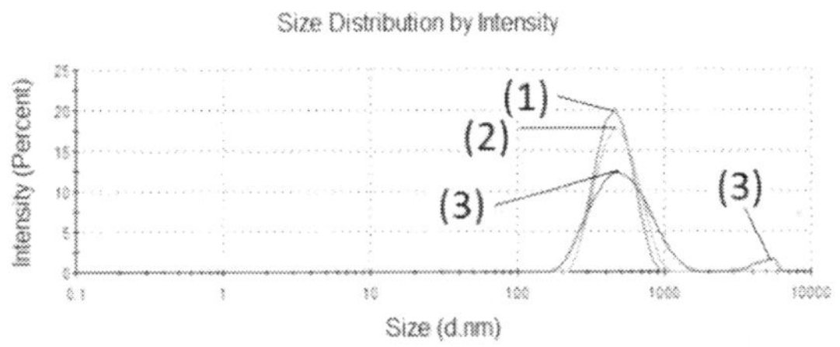

Figure 1. Particle size analysis using Malvern Zetasizer, Nano ZS Series [7]. It was conducted three time, i.e. measurement 1, 2 and 3.

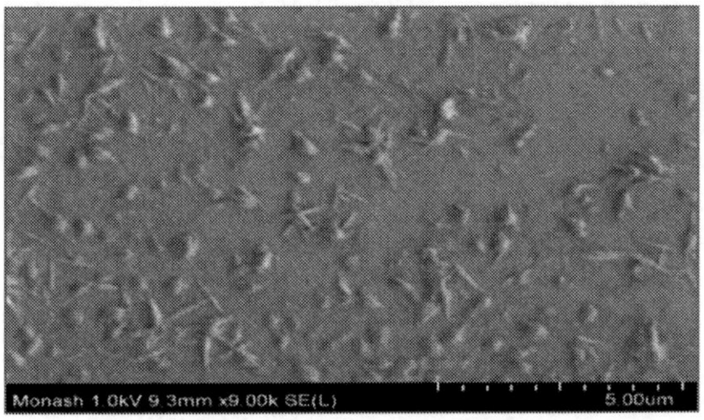

Figure 2. NCC extracted from bleached EFB fibers via ultrasonic cavitation [8]

2 Potential application of NCC

2.1 Pharmaceutical

Drug nanoparticles, or more commonly known as drug nanocrystals had been continuously a research topic in pharmaceutical application over the last two decades. Drug nanocrystals are often prepared in nanosuspension system consists of an aqueous medium and stabilizers. Majority of the pharmaceutical products are developed in the form of most stable solid crystalline. However, this gives rise to the problem of poor solubility, in which leading to low bioavailability [14]. The solution to this problem lies in the hydroxyl groups on the surface of NCC. The hydroxyl group can be altered via specific chemical modification such as surface charge modification, superior hydrophilicity, and enhance stability of matrix of NCC. The surface modification can be utilized to attract drugs to be attached on NCC.

2.1.1 Drug disintergrant

Microcrystalline cellulose (MCC) is widely used partially depolymerized cellulose as one of major component in pharmaceutical industry. MCC is blended with other excipients, for fabricating drug-loaded tablets for oral administration in the field of controlled drug delivery (CDD) technology. Excipient is inert substances that are functioned to initiate, propagate, and even take part in chemical interactions with active drugs compound. This role of excipients contributes directly to the effectiveness of pharmaceutical treatment. Gohel *et al.* [15] stated that MCC is not an excellent drug disintegrant due to its low porosity. Porosity of tablets affects significantly the degree of fluid penetration in aqueous environment, and disintegration of tablets [16]. Another drawback of using MCC in cellulose-based formulation is that it does not have high-accuracy control over the drug content release due to its poor surface charge density caused by relatively higher surface area of MCC. Hence, application of NCC instead of MCC is proposed to overcome the drawbacks mentioned. Jackson *et al.* [17] had performed an experiment that NCC is capable of binding significant quantities of ionizable water antibiotics, namely tetracycline and doxorubicin. Thus, usage of NCC as matrix for drug attachment is justified. Theoretically, NCC has higher porosity because surface area per total volume compared to MCC due to its smaller size. Even though the total surface area per unit total volume is higher in NCC, but the surface area per individual NCC is lower compared to MCC, this directly contribute to higher surface charge density for superior molecular-precision control over drug content release.

2.1.2 Fabrication of drug tablets

Majority of oral drug administration in specific dosage is in the form of tablets. Tablets have been used widely in pharmaceutical due to its chemical and mechanical stability [18]. Fabrication of tablets has been a challenge in pharmaceutical industry as the requirements of a tablet is to have appropriate hardness, desired dissolution characteristic, and optimized drug content. Compression of functional ingredients with excipients is a conventional method of producing tablets. However, increase of pressure might cause higher friction generation between molecules, thus leading to higher temperature [19]. Consequently, thermal degradation occurred for thermal-responsive drug content and other active ingredients. Even though this problem can be overcome by usage of capsule to encapsulate thermal-responsive drugs instead of tablet, but this method is economically unfavorable compared to tablet [20]. Usage of NCC in tablet fabrication is a promising application to provide the desired requirements of tablet, at the same time overcome the negative effect of compression method to fabricate tablet. Robert *et al.* [21] reported that NCC has 7.5-7.7 GPa of tensile strength, 110 – 220 GPa of elastic modulus in axial direction, and 10 – 50 GPa of elastic modulus in transverse direction. Tensile strength of NCC is higher than Kevlar-49 fiber (3.5 GPa), carbon fiber (1.5 – 5.5GPa), and steel wire (4.1 GPa). Elastic moduli in axial and transverse direction are also proven to be superior to Kevlar-49 fiber. These mechanical properties are suggested to have significant improvement on the hardness of tablets to consolidate the excessive frictions generated during compression.

2.1.3 Fabrication of high tensile strength hydrogel

Hydrogels have long been investigated for drug carrier as a biocompatible and swellable polymeric material [22]. The common applications of hydrogel are mostly explored in clinical usage, including tissue engineering, regenerative medicine, barrier materials to regulate biological adhesions, diagnostics, cellular immobilization, and separation of biomolecules or cells. The network structure possess by hydrogels can be non-porous, macroporous, microporous, and potential novel of nanoporous. Their highly porous structure can be manipulated by varying the density of cross-links in the gel matrix and the affinity of the hydrogels for the exposed aqueous environment in which they are subjected to swell. Hoare and Kohane [23] reviewed that the porosity of hydrogel also permits loading of drugs into the gel matrix and subsequent drug release at a rate dependent on the diffusion coefficient of the small molecule or macromolecule through the gel network. The reason why controllable drug release can be achieved in drug delivery is due to the cross-links in the gel matrix that are functioned in preventing the sudden burst of release of drug content. However, there is a major setback for controlled drug release application using hydrogel, which is low tensile strength. This setback does not just limit hydrogel usage use in load-significant applications, but also can resulted in extreme dissolution rate. This property does not favor in controlled drug delivery application. In recent drug delivery vehicle advancement, Zhang *et al.* [24] incorporated NCC into supramolecular hydrogel based on cyclodextrin inclusion to enhance mechanical properties and control drug release. The result showed a staggering increase in modulus as many as 5 times compared to the NCC-less hydrogel. The resulted nanocomposite, hydrogel containing NCC is found to be a potential drug carrier as mentioned previously.

2.1.4 NCC matrix in transdermal drug delivery system

Human skin is a readily route for drug delivery. The technology for non-oral drug delivery is known as transdermal drug delivery system (TDDS) which deliver appropriate amount of specific drug via the skin of a patient. The two biggest driving forces for development in drug delivery via skin are the convenience of patient and able to avoid first pass metabolism [25]. Other reported advantages of TDDS over conventional dosage form are controlled release of drug, more uniform plasma levels, improve bioavailability, control over drug administration duration using patches, and painless [26-28]. The basic components for TDDS are [29]: (i) Polymer matrix/drug reservoir, (ii) Drug, (iii) Permeation enhancers, (iv) Pressure sensitive adhesive (PSA), (v) Backing laminates for patch protection, (vi) Release liner (vii). Other excipients such as plasticizers and solvents.

As mentioned earlier, NCC can be used as matrix former/drug reservoir, which is the main ingredient in formulation of TDDS. NCC matrix, in practical, is biocompatible and has potential compatibility via chemical surface modification with the drug, permeation enhancers, pressure sensitive adhesive (PSA), and other excipients. The outermost layer of skin epidermis is known as stratum corneum. It consists of 20 layers of lipids matrix in average to form a continuous structure. This continuous structure acts as skin barrier to avoid water loss. Each of stratum corneum cell is 40μm in diameter and 0.5μm thick [30]. First obstacle in TDDS is to penetrate stratum corneum layer, which require extremely small particle [31]. Matrix size contributes significantly to overall size of TDDS apart from drug molecular size. NCC is typically from 10 to 1000 nm in length and 1 to 100 nm in diameter, which means an approximate of 40 to 400 times smaller than the size of stratum corneum cell. Logically, usage of NCC should enhance the permeation percentage of TDDS across the skin. In addition, NCC matrix has several advantages over traditional polymer matrix, this include (i) Reduce efforts in formulating biocompatible polymers as drug reservoir. (ii) Avoid usage of health-risk synthetic polymer like polyvinylchloride and synthetic elastomers like rubber silicon. (iii) Smaller in size for ease of skin penetration. (iv) Higher precision in controlled drug release due to exceptional surface charge density.

2.2 Why use NCC in pharmaceutical industry

Potential of NCC in pharmaceutical applications has been an interesting research topic to replace micelles and liposomes in drug delivery system. Researches on NCC in drug delivery increase exponentially after few interesting toxicological assessment on NCC. One of the earliest toxicological assessments of NCC is conducted by Roman *et al.* [32] Human brain microvascular endothelial cells (HBMEC) was used as test substance. HBMEC are highly specialized cells responsible for the functional properties of the Blood-Brain Barrier (BBB). BBB is characterized by cellular tight junctions and active transport systems which allow the selective transport of substances into the brain while acted as a barrier to prevent unwanted substances from entering the brain. Hence, HBMEC is a good candidate to evaluate the toxicity on the drug uptake. NCC was found to be non-toxic to the cells. As for cellular uptake investigations, the results showed minimal uptake of untargeted NCC. Kovacs *et al.* [33] had done a very thorough toxicological assessment by using used two water fleas aquatic organisms, daphnia and ceriodaphnia, to compare the acute lethal toxicity of different nanomaterials. NCC and MCC are practically and virtually non-toxic. It is also found that no evidence of NCC interaction with other tissues, NCC was not passed down to next generation, and NCC is only restricted to digestive tract. Recently, Dong *et al.* [34] conducted a cytotoxicity assessment of nine different cell lines using 3-(4,5-dimethylthiazol-2-yl)-2,5-diphenyltetrazolium bromide (MTT) assay and lactate dehydrogenase (LDH) assay. NCC showed no cytotoxic effects against any of these cell lines in the concentration range of 0 – 50 g/mL and exposure time of 48 hours. Another investigation is conducted involving cellular uptake of fluorescein-5'-isothiocyanate (FITC)-labeled NCC by these cell lines. The result showed that the cellular uptake is physiologically insignificant. These toxicological assessments concluded that lack of toxicity in NCC and untargeted NCC uptake strongly supports the potential of NCC in targeted drug delivery applications because it enables the application of specific uptake mechanism for drug targeting.

3 Conclusion

The applications and practices of NCC in pharmaceutical is still far beyond from its full potential. With the rapid advancement in mentioned industries, the exploration of green nanoparticles are catching escalating attention is now in urge. Nanocrystalline cellulose, or NCC is proposed to be an option, due to their relatively high surface charge density compared to total volume, high porosity, high tensile strength, high stability matrix former for various applications, and most importantly atoxic.

Acknowledgement

The authors are thankful to Ministry of Higher Education (MOHE) for the financial support through ERGS (ERGS/1/2013/TK07/UCSI/03/1), and CERVIE for RGS (PROJ-IN-FETBE-015).

References

1. Mohanty, A. K., Misra, M. & Drzal, L. T. (2002). Sustainable bio-composites from renewable resources: Opportunities and challenges in the green materials world. Journal of Polymers and the Environment, **10(1/2)**, 19–26.
2. B. L. Peng, N. Dhar, H. L. Liu, & K. C. Tam (2011). Chemistry and application of nanocrystalline cellulose and its derivative: A nanotechnology perspective. The Canadian Journal of Chemical Engineering, **89**, 1191-1206.
3. Klemm D, Heublein B, Fink H-P, Bohn A. Angew Chem Int Ed 2005; **44**, 2–37.
4. Robert J. Moon, Ashlie Martini, John Nairn, John Simonsen, & Jeff Youngblood. (2011). Cellulose nanomaterial review: structure, properties, and nanocomposites. Chem. Soc. Rev., **40**, 3941–3994
5. Ruiz M, M., Cavaillé, J. Y., Dufresne, A., Gérard, J. F., & Graillat, C. (2000) Processing and characterization of new thermoset nanocomposites based on cellulose whiskers. Compos. Interfaces **7(2)**, 117-131.
6. Y. Habibi, L.A. Lucia, O.J. Rojas (2010) Cellulose nanocrystals: chemistry, selfassembly, and applications, Chem. Rev. **110**, 3479–3500.
7. Shanmugarajah B., Kiew P. L., Chew M. L., Chong S. Y., Tan K.W. (2015) Isolation of nanocrystalline cellulose (NCC) from palm oil empty fruit bunch (EFB): Preliminary result on FTIR and DLS analysis. Chemical Engineering Transactions, **45**, 1705-1710.
8. Song Y.K., Lim Y. H., Ho Y. L., Chek S.Y., Tan. K.W. (2013) NanoCrystalline Cellulose (NCC) as highly sustainable bio-engineered nanomaterial in solving energy, water and waste crisis at global scale. World Engineers Summits (WES), TUWAE, 9-15 Sept 2013.
9. M.M.D.S. Lima, R. Borsali (2004) Rodlike cellulose microcrystals: structure, properties and applications, Macromol. Rapid Commun. **25**, 771–787.
10. Revol, J. F., Godbout, L., & Gray, D. G. (1998). Solid self-assembled films of cellulose with chiral nematic order and optically variable properties. J. Pulp Pap. Sci. **24(5)**, 146-149.
11. K. Fleming, D.G. Gray, S. Matthews (2001) Cellulose crystallites, Chem. Eur. J. **7**, 1831–1835.
12. Dahlke, B., Larbig, H., Scherzer, H. D., & Poltrock, R. (1998). Natural fiber reinforced foams based on renewable resources for automotive interior applications. J. Cell. Plast. **34**, 361-379.
13. Ruzica Kolakovic, Leena Peltonen, Antti Laukkanen, Jouni Hirvonen, Timo Laaksonen. (2012). Nanofibrillar cellulose films for controlled drug delivery. European Journal of Pharmaceutics and Biopharmaceutics **82**, 308–315.
14. Lipinski C. Poor aqueous solubility – an industry wide problem drug discovery. Am Pharm Rev 2002, **5**, 82-85.
15. Preparation and assessment of novel coprocessed superdisintegrant consisting of crospovidone and sodium starch glycolate: a technical note. AAPS PharmSciTech 8, Article 13.
16. Nogami, H., Nagai, T. and Uchida, H. (1966) Studies on Powdered Preparations XIV. Wetting of Powder Bed and Disintegration Time of Tablet. Chem. Pharm. Bull **14**, 152-158.
17. Jackson J.K., Letchford. K., Wasserman B.Z., et al. (2011). The use nanocrystalline cellulose for the binding and controlled release of drugs. International Journal of Nanomedicine. **6**, 321-330.
18. Joshi, X. Duriez (2004) Added functionality excipients: an answer to challenging formulations, Pharmaceutical Technology, 12–19.
19. H. Abushammala, R. Hashaikeh, C. Cooney (2012) Microcrystalline cellulose powder tableting via networked cellulose-based gel material. Powder Technology **217**, 16–20.
20. C. Shaine, Pharmaceutical manufacturing handbook: Production & process, 2008.
21. Robert J. M., Ashlie M.,John N.,John S.,Jeff Y. (2011) Cellulose nanomaterial review:structure, properties, and nanocomposites. Chem. Soc. Rev. **40**, 3941–3994.
22. Amin, S., Rajabnezhad, S. and Kohli, Kanchan (2009) Hydrogels as potential drug delivery systems. Scientific Research and Essay, **3 (11)**, 1175-1183, 2009.
23. Haore, T. R. and Kohane, D. S. (2008) Hydrogels in drug delivery: Progress and challenges. Polymer **49**, 1993–2007.

24. Zhang, X. L., J. Huang, P. R. Chang, J. L. Li, Y. M. Chen, D. X. Wang, J. H. Yu and J. H. Chen (2010) Structure and Properties of Polysaccharide Nanocrystal-Doped Supramolecular Hydrogels Based on Cyclodextrin Inclusion," Polymer **51**, 4398–4407.
25. Loyd V. Allen Jr, Nicholas G. Popovich, Howard C. Ansel (2005) Pharmaceutical dosage forms and drug delivery systems, 8th Edition., Wolter Kluwer Publishers, New Delhi, 298-299.
26. Robinson J. R., Lee H. L. Controlled drug delivery fundamentals and application. 2nd ed. Marcel Dekker, New York. 524 – 552.
27. Singh J., Tripathi K.P., Sakia T.R. (1993) Effect of penetration enhancers on the in vitro transport of ephedrine through rat skin and human epidermis from matrix based transdermal formulation. Drug Dev. Ind. Pharm, **19**, 1623-1628.
28. Aquil M., Sultana Y., Ali A. (2003) Matrix type transdermal drug delivery systems of metoprolol tartrate: In vitro characterization. Acta Pharm, **53**, 119 – 125.
29. Nikhi S., Bharat P., Shalini S., Uday M. (2012) Blooming pharma industry with transdermal drug delivery system. Indo Global Journal of Pharmaceutical Sciences **2(3),** 262-278.
30. A. W Kenneth, Dermatological and transdermal formulations, New York, USA: Marcel Dekker Inc; 2002
31. Debjit B., Chiranjib, Margret C., Jayakar B., Sampath K.P. (2010) Recent advances in transdermal drug delivery system. International Journal of PharmTech Research. **2,** 68-77.
32. Roman M., S. P. Dong, H. Anjali, Y. W. Lee (2010) Cellulose Nanocrystal for drug delivery. Pollysaccharide Materials: Performance by design. Ametican Chemical Society, Washington DC, 81-91.
33. Kovacs T., V. Naish, B. O'Connor, C. Blaise, F. Gagne, L. Hall, V. Trudeau, P. Martel (2010) An ecotoxicological characterization of nanocrystalline cellulose (NCC). Nanotoxicol. **4(3)**, 255-270 (2010).
34. Dong S.P., Anjali A. H., Katelyn R. C., Lee Y. W., Roman M. (2012) Cytotoxicity and cellular uptake of cellulose nanocrystals. World Scientic Publishing Company. Nano LIFE, **2**, No. 3.

Chicken eggshell as biosorbent: Artificial intelligence as promising approach in optimizing study

Peck Loo Kiew[1], Chun Kit Ang[1], Khang Wei Tan[1] and Shu Xin Yap[1]

[1]*Faculty of Engineering, Technology & Built Environment, UCSI University, 56000 Cheras Kuala Lumpur, Malaysia*

Abstract. Response Surface Methodology (RSM) is the most popular approach for optimization study in various biochemical processes nowadays. Artificial Neural Network (ANN) has emerged as one of the most efficient methods in empirical modeling and optimization, particularly for non-linear systems. In this study, the estimation capability of RSM and ANN models was compared in copper removal from aqueous solution. The experiments were carried out based on a 3-level and 4-variable Central Composite Design (CCD). The RSM results revealed that the relationship between the response and independent variable could be represented by the quadratic polynomial model. In the development of ANN model, the optimal configuration of the model was found to be 4-10-1. Estimated responses from both models were compared with the experimentally determined responses to determine predictive capabilities of both techniques. Comparison of two methodologies showed that the ANN model was more accurate and exhibited better generalization capability than RSM, thus indicated a clear superiority than the latter in capturing the non-linear behaviour of the adsorption process using chicken eggshell as biosorbent.

1 Introduction

Chicken eggshells are waste material from hatcheries, homes and fast food industries which are abundant and readily available worlwide [1]. Without proper management, its disposal contributes to environmental pollution. A recent report described that chicken eggshells contain calcium and trace amounts of other micro elements (boron, copper, iron, manganese, molybdenum, sulphur, magnesium, silicon and zinc). The presence of cellulosic structure and amino acids suggested it a good biosorbent [2]. Calcium carbonate was identified as the main element responsible for metal adsorption [3]. Several studies in recent years concentrating on the evaluation of the possibility of heavy metal removal, particularly copper, fluoride, iron and chromium from aqueous solution had shed lights on the application of discarded chicken eggshell in wastewater treatment [2, 4-6]. In Malaysia, chicken eggshells are discarded from food industry or household as municipal solid waste. They are normally disposed in landfills. Even though these waste materials are available in abundance, its potential in heavy metal removal is still less being investigated.

The efficiency of copper (heavy metal) biosoprtion using chicken eggshell powder can be influenced by many factors such as particle size, biosorption temperature, solution pH, ratio of sorbent to solution and duration of biosorption [2, 7, 8]. Consequently, the biosorption process or design of experiments must be conducted or designed carefully. Various experimental design techniques had been employed for biosorption process which included the conventional One-Factor-At-A-Time (OFAT) method and statical methods such as Central Composite Design (CCD) and Box-Behnken Design (BBD). Under circumstances where multiple process variables may influence the response variables, response surface methodology (RSM) emerges as a favourable technique in optimizing the process [3, 9]. RSM has been extensively employed in various biochemical applications for the purpose of empirical modeling and optimization. It is a collection of statistical techniques for experiment designs, model buildings, evaluation of effects of process factors, as well as identification of optimum conditions, through fitting of the experimental responses to design of experiments (DOEs) into quadratic function [10]. The applications of RSM involve three major steps that start with establishment of experimental design using CCD and BBD to set the experimental level of independent variables. Subsequently, regression analysis will be conducted for the response surface modeling. Finally, the response surface mode will be used as objective function for the process optimization [11]. RSM had been commonly employed in various bioprocesses such as extraction, fermentation and absorption processes. The ultimate objective of the application is to identify the optimum operating conditions for the respective processes.

In most of the RSM problems, the interaction between the response and process variable is not known. Hence, suitable approximation for the true functional relationship between the respond, y, and the set of process variables, x_1, x_2, x_3...x_n must be established where n is the total number of process variables in the process or system. The function can be expressed as:

$$y = f(x_n) + \varepsilon \quad (1)$$

with ε refers to the noise or error observed in the response *y*. The true response function *f* is normally unknown, thus low-order polynomial in some region of the independent variables is preferable for the approximation of *f*. If the response can be defined by a linear function of the independent variables, then the approximation function is a first-order model expressed by:

$$y = \beta_0 + \beta_1 x_1 + \beta_2 x_2 \dots + \beta_k x_k + \varepsilon \quad (2)$$

In the case of curvature in the response surface, a higher degree polynomial can be used (e.g. the second order model) as follow:

$$y = \beta_0 + \sum_{i=1}^{k} \beta_i x_i + \sum_{i=2}^{k} \beta_{ii} x_i^2 + \sum_{1 \le i \le j}^{k} \beta_{ij} x_i x_j + \varepsilon \quad (3)$$

Nevertheless, neural network-based models have recently been shown by several researchers to exhibit superior performance in terms of accuracy and predictive capability [12]. Artificial Neural Network (ANN) has emerged as an attractive tool for non-linear multivariate modeling in the last two decades. It has typically been used as a "black-box" tool, that is, no prior knowledge about the process is assumed but the goal is to develop a process model based only on observations of its input-output behavior [13]. ANN is generic in structure and it exhibits the ability to learn from data gathered by users, thus widely known as a reliable modeling tool for process that is lack of information [10]. ANN is capable of handling multiple independent and dependent variables simultaneously and to do this prior knowledge on the functional relationship does not need to be known. Each neuron receives information through input connections, process the information and produce the output which is distributed through output connections. Each connection is associated with a real number quantity known as weight [14]. In other words, the neurons sum weighted inputs and then applies a linear or non-linear function to the resulting sum to determine the output. These neurons are arranged in layers and are combined through excessive connectivity [14]. A neural network in its basic form is usually composed of several layers of neurons, there being one input layer, one output layer and at least one hidden layer.

The error or bias in prediction is then propagated through the system and the inter-unit connections are changed to minimize the error in prediction [15]. This is a continuous process with multiple training sets until the minimum error is attained. Training an ANN network is therefore accomplished by adjusting these weights by minimizing a non-linear error function [16]. However, a major disadvantage of ANN is the difficulty in explaining the relation between independent and response variables resulting from the ambiguously defined weights, which as mentioned earlier, is a black box [17]. According to Agatonovic-Kustrina et al. [15], 3 types of data sets that are inclusive of training data, testing data and validation or unseen data, are required for network training, neural network performance monitoring during the training, and measurement of the performance of trained network, respectively. The number of the layers and processing elements in layers vary from one process to another. There is no strict rule available to identify the optimal number of layers and neurons. In fact, it is generally defined by trial and error [18].

It is possible to investigate the effect of each independent process variable, alone or in combination in the process through the least number of experiments through the application RSM [18]. Yet, drawbacks associated with it have been pointed out recently. Desai et al. [16] and Lou and Nakai [17] stated that RSM was useful only for quadratic approximations and the sensitivity analysis of input variables was difficult to perform using RSM due to the presence of cross interactions. It is therefore hard to conclude that RSM is applicable to all optimization and modeling studies. Consequently, the main objective of the present work is to investigate the possibility of employing artificial intelligence approach, particularly through the application of artificial neural network to construct a nonlinear relationship among the process parameters and their corresponding responses. The ANN used in this work is the feedforward back-propagation network which is normally applied in analytical applications and is very powerful in function optimization modeling. To our best knowledge, this is the first report on the comparison of RSM and ANN methodologies in predicting the copper removal efficiency from aqueous solution using eggshell as adsorbent. Also, changes on the surface morphology of the eggshell as biosorbent for copper removal was observed and comparison was made before and after the adsorption process.

2 Materials and Methods

2.1 Materials

Chicken eggshells were collected from the restaurants or market around UCSI University Kuala Lumpur (North Wing). The sample was washed with distilled water several times to remove dirt and contaminants, followed by drying in a hot air oven at 110°C for at least 12 hours. The chicken eggshells were then grinded or crushed using pestle and mortar. The sample was stored in an airtight container for future use.

2.2 Chemicals

Pentahydrate copper (II) sulphate was purchased from R&M Chemicals (Malaysia).

2.3 Biosorption Process

The adsorption process was conducted based on the method reported by Nabil et al. [5] with slight modifications. Chicken eggshell was mixed with copper (II) sulphate solution. Agitation was made at a constant stirring speed of 150 – 300 rpm for 60 – 180 minutes. The remaining concentration of Cu(II) after adsorption was measured using UV-Vis spectrophotometer.

2.4 Experimental Design

2.4.1 Response Surface Methodology (RSM)

RSM was employed for experimental design, data analysis, and model building with the aid of the software Design Expert (Version 6.0.6, Stat-Ease Inc., Minneapolis, Minnesota USA). Central Composite Design (CCD) with four variables was used to determine the response pattern and subsequently to establish a response model. According to CCD, the total number of experimental combinations is $2^k + 2k + n_0$, where k is the number of independent variables and n_0 is the number of repetitions of the experiments at the centre point. Four independent variables used in this work were temperature (X_1), agitation speed (X_2), dosage (X_3), and time (X_4), while the dependent variable was the percentage of copper removal. The ranges and center point values of all independent variables were based on the results of preliminary experiments (Table 1). The copper removal percentage was analyzed by multiple regressions to fit into the following polynomial equation:

$$Y = \beta_0 + \sum_{i=1}^{4}\beta_i X_i + \sum_{i=1}^{4}\beta_{ii} X_i^2 + \sum_{i=1}^{3}\sum_{j=i+1}^{4}\beta_{ij} X_i X_j \quad (4)$$

where Y is dependent variable (copper removal percentage), β_0 is intercept, β_i , β_{ii}, βij are regression coefficients, X_i indicates the linear terms, X_i^2 for the quadratic terms for a single variable, and X_iX_j for the interaction terms. Four replicates at the central point of the designed model were used to estimate the pure error sum of squares. 30 experiments were performed in random order to cover all combinations of the factor levels in the experimental design.

Table 1. Experimental range and values of the independent variables used in CCD

Independent Variables	Symbol	Coded Levels		
		-1	0	1
Temperature (°C)	X_1	30	40	50
Agitation speed (rpm)	X_2	150	250	350
Dosage (g)	X_3	0.2	0.6	1.0
Time (min)	X_4	60	120	180

2.4.2 Development of Feedforward Artificial Neural Network (FANN)

The percentage of copper removal was evaluated as the output. Using MATLAB (version 7.8.0.347, Mathworks, Natick, MA), a feedforward artificial neural network (FANN) trained with an error-back-propagation algorithm was designed. All weights and biases were randomly initialized in the range from 0 – 1. Neural network building data were trained and Mean Squared Error (MSE) was plotted against training epoch. When MSE started to rise, the training was terminated and the value of weight that the network had in that previous step was chosen as the result of the training run. The development of ANN (activation functions) adopted in this study was in accordance to the theory and approach suggested by Ang et al. [19].

The FANN model was developed based on the experimental data of RSM (Table 2). To cope with different magnitudes in the input and output, all the data were scaled to zero mean and unit standard deviation. Single hidden layer neural networks with different numbers of hidden neurons were trained on the training data and tested on the testing data. The network with the lowest MSE on the testing data was considered as having the best network topology. In assessing the developed models, MSE and correlation coefficient (R^2) on the unseen validation data were used as the performance criterion.

2.5 Characterization Study

Scanning electron microscopy (SEM) analysis was performed to observe structural and morphology changes in eggshell surfaces both before and after adsorption process. SEM images of the eggshells were obtained using JEOL JSM 6400 LV model scanning electron microscope at 15 kV.

3 Results and Discussion

3.1 Response Surface Methodology

All 30 experimental points were evaluated and the percentages of copper removal results based on the factorial design are shown in Table 2. Response surface optimization is favorable over the conventional single parameter optimization in that it saves time, space, and raw materials [11]. The experiments were run at random in order to minimize errors from the systematic trends in the variables. The percentage of copper removal obtained ranged from 4.55 to 59.1 %.

Table 2. Response surface CCD and results for copper removal percentage

Run	X_1	X_2	X_3	X_4	Copper Removal (%)
1	40	250	0.6	120	50
2	40	250	0.6	120	45.45
3	40	250	0.6	120	50
4	40	250	0.6	60	50
5	40	250	0.6	60	45.45
6	40	250	0.6	60	45.45
7	30	150	0.2	60	13.64
8	50	150	0.2	60	9.1
9	30	350	0.2	60	31.82
10	50	350	0.2	60	31.82
11	30	150	1.0	60	34.54
12	50	150	1.0	60	4.55
13	30	350	1.0	60	45.45
14	50	350	1.0	60	13.63
15	30	150	0.2	180	45.45
16	50	150	0.2	180	59.1
17	30	350	0.2	180	40.9
18	50	350	0.2	180	54.55

19	30	150	1.0	180	59.1
20	50	150	1.0	180	59.1
21	30	350	1.0	180	59.1
22	50	350	1.0	180	45.45
23	30	250	0.6	120	45.45
24	50	250	0.6	120	44.54
25	40	150	0.6	120	50
26	40	350	0.6	120	59.1
27	40	250	0.2	120	20
28	40	250	1.0	120	27.27
29	40	250	0.6	60	27.27
30	40	250	0.6	180	59.1

Exploration and optimization of a fitted response surface might produce poor or misleading results unless the model exhibits a good fit, which makes the checking of the model adequacy essential. Through the analysis of variance (ANOVA), the *P*-value of the model was less than 0.0001. Meanwhile, the lack of fit value of the model was not significant at 0.1417. These values confirmed that the model fitness was good.

In order to check on the adequacy of the predictive model, 3 sets of experiments were repeated at the suggested optimum conditions in order to experimentally obtain the maximum copper removal using eggshell. As shown in Table 3, the percentage error differences between the experimental and predicted values were always less than 1.5 %. Since the differences between the experimental and predicted response were less than 5%, the validity of the model was verified [10].

Table 3 Validation of data and constructed RSM model

Run	Experimental removal (%)	Predicted removal (%)	Error (%)
Trial 1	59.1	58.88	1.31
Trial 2	59.1	58.88	1.31
Trial 3	59.1	58.88	1.31
Average	59.1	58.88	1.31

3.2 Network Topology and Training of Artificial Neural Network (ANN) Model

The copper removal adsorption process using eggshell was conducted in 30 batches contributed to 30 samples data. To create the model, these data were re-sampled

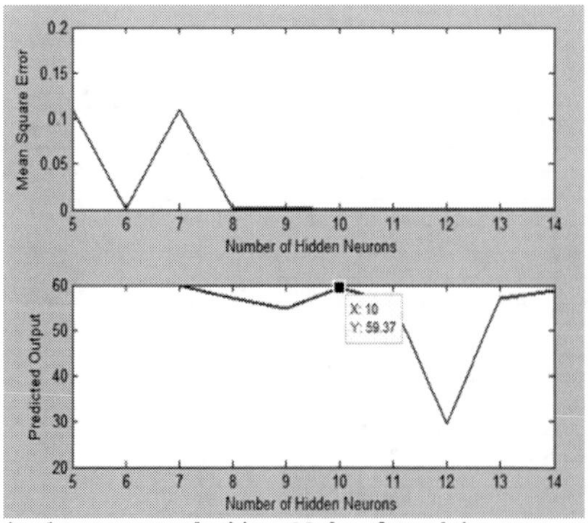

using bootstrap method into 90 data for training purposes.

Figure 1. Selection of number of neurons in the hidden layer.

Several structures were tested to identify the best network topology based on trial and error method. Variation in number of neurons in the hidden layer ranged from 5 – 14. Subsequently, the network was trained and tested after each addition of neuron. There was a drastic drop in the MSE obtained after number of neurons used was more than 7 (Figure 1). With the outputs predicted by ANN trained with different number of neurons, the least error between the predicted and validation result (unseen) was obtained with 10 neurons in the hidden layer. The configuration of the ANN model could therefore be defined as 4-10-1.

The predicted and actual values for the training data matched with each other as shown in Figure 2. In this study, the ANN training resulted in unity R^2 value, proving that the ANN model could fit well with experimental data.

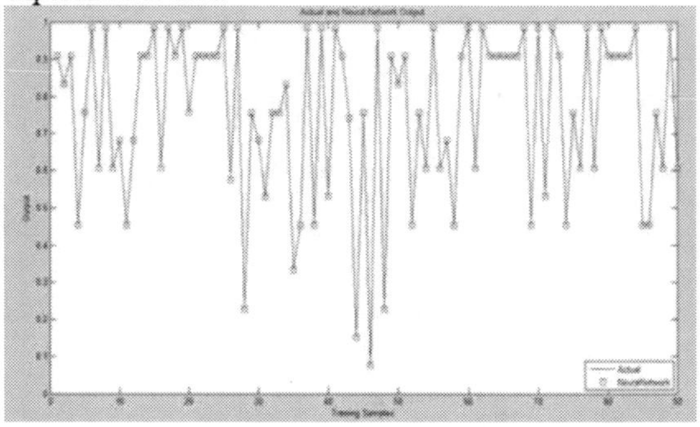

Figure 2. Actual and predicted output for training data.

Validation of the trained model was conducted at the suggested optimum conditions by ANN and it was predicted that 59.37 % of copper removal could be achieved. 3 sets of validation experiments were repeated. The percentage error differences between the experimental and predicted values were less than 1 % (Table 4). Therefore, the validity of the trained ANN model was verified.

Table 4 Validation of data and constructed ANN model.

Run	Experimental removal (%)	Predicted removal (%)	Error (%)
Trial 1	59.10	59.37	0.46
Trial 2	59.10	59.37	0.46
Trial 3	59.10	59.37	0.46
Average	59.10	59.37	0.46

3.3 Validation and Comparison of Both Predictive Models

The estimation capabilities between the statistical- and artificial intelligence-based optimization techniques namely RSM and ANN, respectively, were compared and examined. The comparison was made on the basis of various parameters such as the correlation coefficient (R^2) for both training and validation sets, predicted copper removal percentage by both techniques and the average error of the experimental result based on the optimum conditions suggested by RSM at temperature of 49.26 °C, agitation speed of 233.38 rpm, chicken eggshell dosage of 0.7 g for adsorption process of 178.7 minutes. Desai et al. [16] stated that the generalization ability could best be judged with the unseen (validation) dataset. Therefore, the comparative values are tabulated in Table 5.

Table 5 Comparison of predictive capability between RSM and ANN model

Parameters	RSM	ANN
Training R^2	0.9734	1.000
Predicted Copper Removal Percentage (%)	59.88	59.37
Experimental Copper Removal Percentage (%)	59.10	59.10
Copper Removal Percentage Error (%)	1.31	0.46

RSM was one of the most popular and widely used techniques in optimization studies of chemical and biochemical processes. Nonetheless, it had limitations that modeling of the processes could be represented by only up to second-order equations [14]. However, it could be possible that the relationship between the process variable and response variable is more complex than that. Therefore, in the present study, the R^2 values of training datasets showed that ANN model resulted in significant higher generalization and predictive capability as compared to RSM. Also, it was evidenced by the validation of the predictive results in which prediction suggested by RSM resulted in greater deviation (error) compared to ANN model prediction (Table 5).

3.4 Characterization Result

Scanning electron microscopy (SEM) in Figure 3 (a) shows the morphology of powdered eggshell. It confirmed the crystallinity of eggshell and the existence of pores on the surface of the eggshell. After the adsorption process, success in copper removal was evidenced by the deposition of flake-like copper crystals on the surface of the eggshell. No obvious porous structure was observed on the surface after the adsorption process (Figure 3 (b)).

(a)

(b)

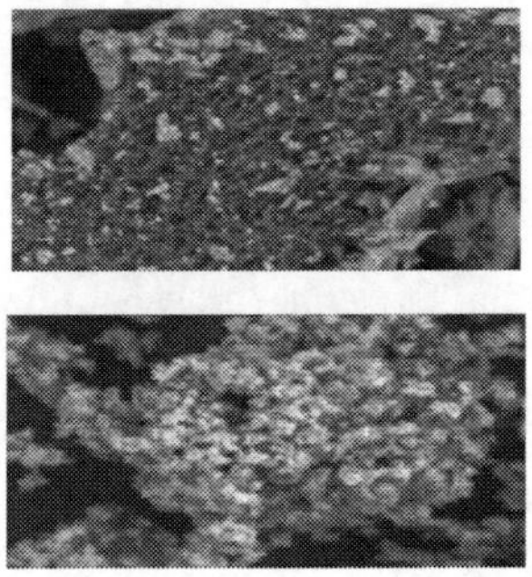

Figure 3. SEM image of eggshell
(a) Before adsorption and (b) After adsorption

4 Conclusion

Two models were developed to predict the percentage of copper removal using response surface methodology (RSM) and artificial neural network (ANN). A feedforward artificial neural network (FANN) was trained by the sets of input-output patterns by implementing the error-back-propagation algorithm. The present study revealed that ANN technique performed better than RSM. Thus, it could be concluded that even though RSM was widely applied in various optimization studies, ANN optimization strategy might present a better alternative to the conventional method. The results of this study also suggested that powdered eggshell, with its low cost and abundant availability has a potential for being used as an adsorbent for contaminants such as copper in wastewater.

References

1. King'or, A.M, Int. J. Poult. Sci. **10**, 11 (2010).
2. Bhaumik, R., N. K. Mondal, B. Das, P. Roy, K. C. Pal, C. Das, A. Baneerjee, and J. K. Datta, E-J. Chem. **9**, 3 (2012).
3. Tsai, W. T., J. M. Yang, C. W. Lai, Y. H. Cheng, C. C. Lin, and C. W. Yeh, Bioresource Technol. **97**, 3 (2006)
4. Nabil, A. J., and A. A. Sameer. J. Envir. Prot. Sci. **3**, (2009).
5. Aimi, R. N., A. H. Norhafizah, and C. S. Wong, Int. J. Eng. Technol. **13**, 1 (2013).
6. Rubcumintara, T., J. Clean Energ. Technol. **2**, 2 (2014).
7. Feng, N. C., X. Y. G. Guo, and S. Liang, Trans. Nonferrous Met. Soc. China. **20**, 1(2010).
8. Liu, C. L., H. H. Ngo, W. Guo, and K. L. Tung. Bioresource Technol. **119** (2012).
9. Singh, R., R. Chadetrik, R. Kumar, K. Bishnoi, D. Bhatia, A. Kumar, N. R. Bishnoi, and N. Singh. J. Hazard Mat. **174**, 1-3 (2010).
10. Kiew, P. L., A. Zainal, and M. D. Mashitah. Biotechnol. Bioprocess Eng. **18** (2013).
11. Zhang, L. L., M. Xu, Y. M. Wang, D. M. Wu, and J. H. Chen. Molecules. **15**, 11 (2010).
12. Chou, C. J., and L. F. Chen. Int. J. Prod. Res. **50**, 7 (2012).
13. Zainal, A., M. D. Mashitah, M. Siti Hatijah, and M. N. Rabiatul Adawish. Bioprocess Biosys. Eng. **5** (2010).
14. Shen, C., L. Wang, and Q. Li. J. Mat. Process. Technol. **183**, 2-3 (2007).
15. Agatonovic-Kustrina, S., M. Zecevic, L. J. Zivanovic, and I. G. Tucker. 1998. J. Pharm. Biomed. Anal. 17, 1 (1998).
16. Desai, K. M., S. A. Survase, S. S. Lele, and R. S. Singhal. Biochem. Eng. J. 41, 3 (2008).
17. Lou, W., and S. Nakai. Food Res. Int. 34, 7 (2001)
18. Baş, D., and İ. H. Boyacı. J. Food Eng. 78, 3 (2007).
19. Ang, C.K., S. H. Tang, S. Mashohor, M.K.A.M. Arrifin. Int. J. Comp. Comm. Contr. **9**, 3 (2014).

Cyanobacteria Dominance in Lakes and Evaluation of Its Predictors: a Study of Southern Appalachians Ecoregion, USA

Sajeela Ghaffar[1], R. Jan Stevenson[2] andZahiruddin Khan[3]

[1] *Institute of Environmental Sciences and Engineering, School of Civil and Environmental Engineering (IESE-SCEE), National University of Sciences and Technology (NUST), sector H-12 campus, Islamabad, Pakistan*

[2] *Center for Water Sciences & Department of Integrated Biology, 203 Natural Science Building Michigan State University, East Lansing, Michigan 48824*

[3] *Institute of Environmental Sciences and Engineering, School of Civil and Environmental Engineering (IESE-SCEE), National University of Sciences and Technology (NUST), sector H-12 campus, Islamabad, Pakistan*

** Corresponding author: sajeela_g@yahoo.com; Tel no. 092-323-5049722*

Abstract: Owing to their ecological and economic implications, the abundance of cyanobacteria has been an important subject in aquatic ecology. There has been a debate about which nutrients are a better predictor of cyanobacteria abundance, total nitrogen (TN), total phosphorus (TP) or nitrogen to phosphorus ratio (N:P). A classic study by Downing and coworkers in 2001 concluded that total N and total P are better predictors than TN:TP ratio. We picked lakes from a different region of the US, the Southern Appalachians (SAP), to test the same hypotheses using the National Lakes Assessment (NLA, 2007) database. This dataset consists of 116 selected sampling sites throughout the ecoregion. In this study, total cyanobacteria abundance was related to three predictors (N, P and N:P) using linear regression, non-parametric statistics and boosted regression trees (BRTs). Total N and total P were more strongly correlated to cyanobacteria abundance than N:P. Partial dependence plots from BRT analysis confirmed results of correlations showing higher relative influence values of TN and TP on cyanobacteria abundance. An important observation from both analyses was that cyanobacteria abundance increased rapidly with increasing TP above 5 µg TP/L. This is a similar response relationship between abundance of cyanobacteria and TP as reported by Downing et al. (2001). These TP thresholds for cyanobacteria dominance are important for TP management targets to control blooms in lakes.

1 Introduction

The algal blooms in lakes, rivers and reservoirs are becoming a major problem worldwide [1]. There are many reasons as to why this is a problem. First is a decline in quality of water, especially freshwater bodies that serve as a source of drinking water for a large population. The second main reason is the ecological disturbance that hinders the normal balance between biotic and abiotic factors of an aquatic ecosystem. Algal blooms make the water body eutrophic which means there is an increase in the primary production which eventually spoils the structure and function of aquatic ecosystems, which has socio-economic implications [2], [3], [4] and [5]. The problem worsens when harmful toxin-producing algal blooms of cyanobacteria or cyano-HABs grow and produce toxins that threaten human and ecosystem health [6] and [7].

Although hydrological and climate factors play a part in proliferation of cyanoHABs [8] and [9], human activities such as urbanization, industrialization and rising agricultural activities due to ever increasing population are the main cause of high wastes. This waste when thrown untreated into water bodies, leads to high levels of nitrogen, phosphorus and other nutrients leading to higher algal growth [10]. More than 50% of phosphorus comes from human waste and 20%–30% from detergents [11]. Animal feedlots are sources of both nitrates and phosphates. As the high levels of nutrients (nitrogen and phosphorus mainly) enter water bodies, they lead to negative ecological costs to aquatic ecosystem; furthering growth of cyanobacteria; high nutrient turnover; loss of ecosystem functions and biodiversity; and decline of water quality and public health [12] and [13].

Since cyanoHABs have and are expected to decrease the water usability for potable and recreational purposes [14], their management is very important from ecological as well as economical point of view. Although various physical and chemical factors play a part in cyanobacteria blooms, identifying the factor contributing the most or the limiting factor is key. This leads to decisive and smart management practices, rather than unsystematically trying to abate nutrients or any physical factor. One of the first steps for identifying the most limiting factor is the finding factor most highly related to cyanobacterial abundance. There has been a debate about the best predictor for cyanobacteria growth, and different studies narrowed the focus to TN, TP and TN:TP (for example [15], [16], [17] and [18]). The conclusions and suggestions vary. A classical 37 year experiment on Canadian lakes by [19] showed phosphorus directly controlling algal blooms. Another extensively cited study by [15] suggested ambient N:P influencing cyanobacteria growth. Many such experiments were later debated for their data collection approach, sample size as well as theoretical basis as more and more studies followed up the classical ones. Most studies have commented and discussed the role of each in theory from surveys and experiments. Few have used an objective survey of nutrient chemistry and cyanobacteria abundance in a large number of lakes as a line of evidence for the limiting factor.

In this study, we aim to identify the most important and efficient predictor for cyanobacteria blooms in freshwater bodies. The rationale being to develop management practices and assigning total maximum daily loads (TMDLs) for water bodies by identifying the nutrient concentration at which risk of high abundances cyanobacteria abundances increase. We aim to further our knowledge and understanding by testing linear regression, non-parametric statistics as well as a relatively new and robust modeling approach called boosted regression trees (BRTs). Boosted regression trees are particularly valuable for characterizing interactions among multiple variables, non-linear relationships and potential thresholds in cyanobacteria abundance along nutrient gradients. Thresholds in ecological responses are particularly valuable for developing stakeholder consensus for management targets [20].

2 Methodology

The data from National Lakes Assessment (NLA) study carried out by US Environmental Protection Agency (EPA) was used for this project [21]. NLA is a large and comprehensive dataset consisting of physical, chemical and biological studies of water bodies across the United States. The US is divided into nine ecoregions in the study. We selected the Southern Appalachians ecoregion for the project as a representation of warm temperate climate region. This subset of data includes 116 lakes.

We tested three predictors with potential influence on cyanobacteria growth, i.e. TN, TP and TN:TP. We chose "abundance" as a measure of cyanobacteria growth as response variable. The selection of variables was based on past studies where these were deemed to be the most important or influential ones for cyanobacteria prediction. Microsoft Access (2013) was used as the tool for extraction of parameters from various datasheets and appending them into a single excel sheet.

Linear models were tested for cyanobacteria abundance as a function of selected predictors. All analyses including the subsequent ones were executed in R ver. 3.0.3 [22]. Residual standard error (RSE) was used as a measure of accuracy i.e. the closeness of fit to the points (for example its value is 0 when model fits to the data perfectly, though this would be most probably due to overfitting). The linear regression analyses were then compared to BRT results.

General trends in relationships of predictors and cyanobacteria abundance were characterized by LOWESS smoothing. LOWESS is a non-parametric regression approach producing a scatter diagram that shows the relationship between predictor and response variables over localized subsets of data, hence the term "local" as opposed to "global" fitting. This is especially useful to visualize patterns in the data as there are no assumptions defining the relationship. The plots were constructed using a tension factor (α) of 0.3. The response variable, cyanobacteria abundance, was log transformed with base 10. The predictor variables were also log transformed to get a better visualization of plots as they expand over a large range and tend to cluster on left hand side of the plot if untransformed.

BRTs are a combination of boosting and regression trees. Boosting is a machine learning (ML) approach, based on the concept that many weak learners make a powerful learner. Regression trees belong to classification and regression tree (CART) assembly of models [23]. This is a relatively novel approach and with strong ML basis that is finding more and more uses in ecological studies due to its high predictive power, better precision and accuracy compared to other modeling approaches [24]. BRTs work by minimizing predictive deviance obtained by fitting successive regression trees. The trees are built by an iterative learning process of modeling interactions between predictors and random selection of observations in each iteration [25] and [23]. The results of BRT models are viewed by partial dependence plots (PDPs) that show relationship as well as relative influence of each predictor on the response variable. The algorithm estimates relative influence by taking into account the number of times the specific variable is used to split tree nodes, given weight by squared improvement at each step, and then averaging it over all trees [26].

BRT analysis was run using the script from [23], a package called gbm – gradient boosting machines. The package also includes 10-fold cross validation in which the data is divided into 10 parts. 9 parts are used for training and 1 part is used for testing. This is an effective way of validating the model by performing analyses on multiple iterations of the dataset.

The analysis was run with setting a bag fraction of 0.5 which means 50% of training data is randomly picked for each iteration; and a tree complexity of 3 that means up to 3 way interactions of predictors can be modelled. A learning rate of 0.001 was set for all analyses, which is a parameter determining contribution from each tree to build the next in making final decision. All variables including the response and predictors were log transformed as convergence could not be achieved over the large range of non-transformed values. 10 fold cross validation was used to validate the model, which is a commonly used approach. PDPs were constructed from the same gbm package, which includes script for the purpose. Predictor selection was then refined by observing PDPs based on relative influence of each predictor.

Model comparison was done by cross-validation (CV) correlation values. The BRT model with combination of predictors giving minimum predictive deviance and maximum explained error variance values was selected. The model was tested for performance on a subset of data withdrawn earlier. The same model parameters were used and cyanobacteria abundance was predicted on new data. The model performance was verified using root mean square error (RMSE) values between observed and estimated values, which is basically the square root of variance of residuals, having same units of response variable and representing accuracy of model for predicting values.

3. Results

The linear regression analysis (summarized Table 1) indicated TN:TP as the weakest predictor of cyanobacteria abundance among selected three (R2 = 0.14, RSE = 0.88). TP came out to be the best predictor with highest R2 (0.28) as well as lowest RSE estimates (0.803) indicating better prediction accuracy. TN also has higher R2 (0.184) and lower RSE (0.857) than TN:TP indicating a relatively stronger relationship. All analyses were significant at $p<0.001$. In summary, TP is related linearly stronger to cyanobacteria abundance than any other predictor, especially TN and TN:TP.

Table 1. Linear regression statistics showing ability of three selected predictors to predict cyanobacteria abundance in Southern Appalachians ecoregion. (RSE on df = 128; all analyses are significant at <0.001)

Variable	Linear model equation	RSE (df 128)	R^2
TP (μg/L)	Abundance = 2.687 x TP + 0.896	0.803	0.28
TN (μg/L)	Abundance = 1.21 x TN + 0.659	0.857	0.184
TN:TP	Abundance = -0.861 x TN:TP + 4.94	0.88	0.14

LOWESS plots (Fig. 1) show little nonlinearity in relationship between cyanobacteria abundance and the selected predictors. The LOWESS plot for TP (Figure 1a) shows that cyanobacteria are abundant even at lowest observed TP concentration (1 μg/L); at around 5 μg/L TP, the abundance of cyanobacteria starts increasing.

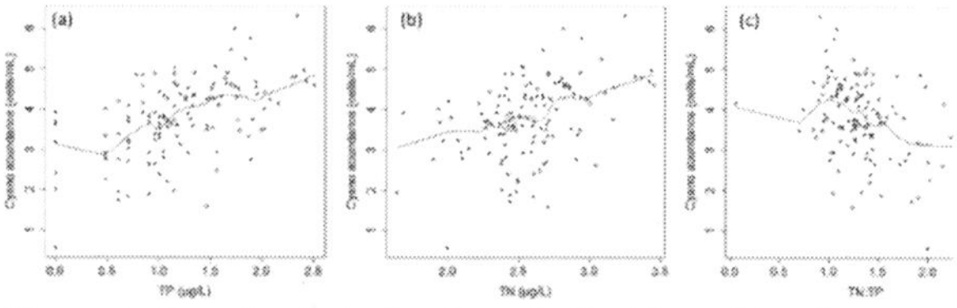

Figure 1: Cyanobacteria abundance in the lakes of Southern Appalachians ecoregion as a function of three selected predictors (a) Total Phosphorus (μg/L) (b) Total Nitrogen (μg/L) (c) Total N:Total P. Scatter represents the observations; solid lines are LOWESS trend lines (α=0.3). All values log-transformed to even out the spread.

The BRT model relating cyanobacteria abundances to TN, TP, and TN:TP was achieved after fitting 1500 trees and explained about 63% and 45% (SE 0.051) of variance in the entire training data set and with cross validation respectively. The partial dependence plots (PDPs) showed that TP had a highest relative influence (41.9%) on cyanobacteria abundance. TN followed TP with a relative influence value of 35.4%. TN:TP had 22.6% influence on cyanobacteria abundance (Figure. 2).

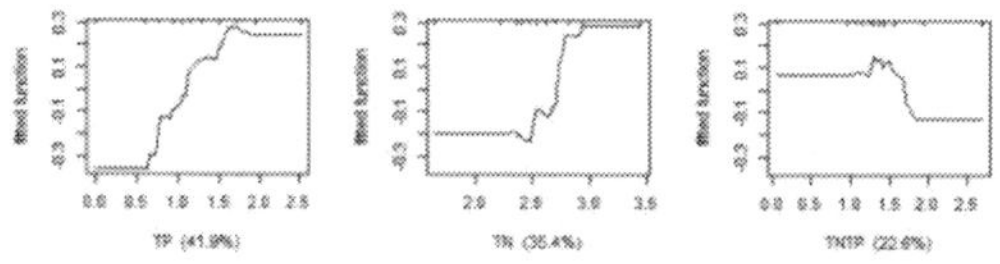

Figure 2: Partial Dependence Plots (PDPs) obtained from boosted regression tree model using 3 predictor variables

RMSE value for training data was 0.73 and for test data was 0.77. If test data RMSE for test data is much larger than training data, it suggests that model validation is poor and it does not have much predictive importance. Since these values came to be quite close, it is indicative of a good predictive performance. The BRT plots show strong nonlinearity between cyanobacteria abundance and nutrient concentrations or ratio. The TP plot shows a sudden rise in abundance at around 5 μg TP/L and saturation around 60 μg TP/L. TN plot shows a relatively sharp rise in cyanobacteria at about 130 μg TN/L. BRT results for TN:TP show that cyanobacteria abundance start to decline at a TN:TP of about 15.

4. Discussion

We aimed to understand the response of cyanobacteria to selected predictors with emphasis on choosing the best one for cyanobacteria prediction from a waterbody management view point. The data included observations over a broad phosphorus and nitrogen range. We expected to see relationships between cyanobacteria abundance and selected predictors and identify the strongest related ones and the most effective nutrient range for cyanobacteria growth. We also expected to see cyanobacteria responses to nutrients similar to those reported by [17]. The results from Southern Appalachians show some interesting trends with a few in-line with [17] findings and a few giving a more thorough insight of Southern Appalachians nutrient dynamics.

Linear models for cyanobacteria and either TN, TP or TN:TP ratio in combination with corresponding LOWESS plots give a few more insights. While all these variables relate to abundance, TP displays the strongest relationship. TN follows TP with an R2 about 35% smaller than that of TP. As for the nitrogen to phosphorus ratio, cyanobacteria are more abundant in lower range of TN:TP; as it goes beyond 15:1, cyanobacteria become less and less abundant. The decline stops around 60:1 and then stays almost constant. This follows resource ratio competition theory that suggests that cyanobacteria have a tendency to thrive well in nitrogen limited conditions [27] and where N:P ratios are lower than 20:1 [28] and [29]. This is attributed to the fact that many cyanobacterial species are capable of utilizing molecular nitrogen [30]. TN:TP however, is the poorest predictor if seen compared to nutrient concentrations (TN and TP) in our results. It shows the lowest R2 value amongst all three which is 50% lower than that of TP and 22% than TN. In addition, TN:TP shows highest RSE; about 8% more than TP and 3% than TN.

The BRT model also shows TP having the higher relative influence on cyanobacteria abundance than both TN and TN:TP. A few points are noteworthy here. First is a long discussed concept about whether the nutrient concentrations (TN and TP) or the ratio TN:TP is a more efficient predictor of cyanobacterial blooms (for example [31], [17], [32], [33],[34] and [35]). Results from this study favor the latter notion based on linear regression, LOWESS and BRT models supported by PDPs. TN and TP clearly outperform TN:TP in predicting cyanobacterial abundance.

An important point from the results is the pattern of response that cyanobacteria abundance shows to total phosphorus concentration. The BRT partial dependence plot (Fig. 2) show that around a TP concentration of 5μg/L, cyanobacteria abundance is high until around 100μg/L TP where almost 90% of maximum cyanobacteria abundance is observed. It is seen to decline afterwards. This range is in line with that reported by [17]. It is suggestive of capability of cyanobacteria to grow well at low TP concentrations. This is most likely owing to several adaptive features of different cyanobacterial species that include nitrogen fixation and capability to store phosphorus in the cells that can be utilized for growth as per need. The practical implication of this observation is related directly to management of water body. It implies that accounting for other predictors, ~5 μg/L TP is a crucial point for increase in probability of cyanobacteria blooms and unless TP is reduce below 60 μg TP/L, little reduction in cyanobacteria abundance should be expected.

It may seem plausible looking at P dependence of cyanobacteria, that adjusting P only can be effective for cyanobacteria control in waterbodies. However, in a watershed where nitrogen influx is high, controlling P alone cannot control non-nitrogen fixing cyanobacteria in the water body. This is especially important considering the dramatic increase in nitrogen influx in water bodies [36]. Reference [33] suggest that in lakes where non-nitrogen fixers are abundant, regulating only P might not be very effective. For example, Microcystis can utilize phosphorus very efficiently by vertical migration, stock phosphorus from sediment-water interface and can reemerge to bloom. This is especially true for shallow lakes where sediment-water interface is easy to get to along with sufficient light penetration and water mixing. Reference [37] also advise that effective control can only ensue by rigorous nutrient management owing to nonlinear nature of association amongst environmental factors and cyanobacteria growth [36]. This study points out that TN and TP are the important predictors that should be given more importance than TN:TP. Although the relationship of the latter is substantiated by resource ratio theory, TP and TN deliver a simpler understanding and a more efficient approach to control cyanobacteria blooms. This is also supported by the fact that not all cyanobacteria have the ability to fix nitrogen nor are they a great competitor for nitrogen [17].

Acknowledgment

The funding to conduct research work at Michigan State University (MSU), East Lansing, Michigan USA was kindly provided by National University of Sciences and Technology (NUST), Islamabad, Pakistan; this study is a part of that research. The authors also acknowledge USEPA whose comprehensive NLA data is openly available and was used for this study.

References

1. X. Yang,X. Wu, H. Hao, Z. He, Journal of Zhejiang University Science B.**9**(3):197-209 (2008)
2. B. Mortazav, R.L. Iverson, W.M. Landing, G.G. Lewis, W. Huang, Marine Ecology Progress Series, **198**:19– 31 (2000)
3. E.J. Phlips, Algae and eutrophication. In Bitton, G. (ed.), Encyclopedia of Environmental Microbiology. Wiley, New York. (2002)
4. E.L. Bledsoe, E.J. Phlips, C.E. Jett, K.A. Donnelly,Ophelia **58**:29–47 (2004)
5. I.T. Webster, N. Rea, A.V. Padovn, P. Dostine, S.A. Townsend, S. Cook, Marine and Freshwater Research **56**:303–316 (2005)
6. J.H. Landsberg, Reviews in Fisheries Science 10: 113– 390 (2002)
7. W.W. Carmichael, A world overview one-hundred, twenty-seven years of research on toxic cyanobacteria—Where do we go from here? In: Hudnell, H.K. (ed.) Cyanobacterial Harmful

Algal Blooms: State of the Science and Research Needs. Advances in Experimental Medicine & Biology, Vol. 619. Springer. 500 pp (2008)
8. Y.S. Li, X. Chen, O.W.H. Wai, B. King, Water Environmental Research,**76**: 2643–2654 (2004)
9. E. Jeppesen, M. Sondergraard, M. Meerhoff, T.L. Lauridsen, J.P. Jensen, Hydrobiologia**584**: 239–252 (2007)
10. M. Schrope, Nature, **452**(7183): 24–26 (2008)
11. F.A. Khan, A.A. Ansari, The Botanical Review, **71**(4):449-482 (2005)
12. W. Liu, R.L.Qiu, Journal of Chemical Technology and Biotechnology, **82**(9):781-786 (2007)
13. M.W. Suplee, V. Watson, M. Teply, H. McKee, Journal of the American Water Resources Association, **45**(1), 123-140 (2009)
14. S. Lyck, Journal of Plankton Research, **26**(7), 727-736 (2004)
15. V.H. Smith, Science, **221**: 669–671 (1983)
16. A.M. Trimbee, E.E. Prepas, Canadian Journal of Fisheries and Aquatic Sciences, **44**(7), 1337-1342 (1987)
17. J.A. Downing, S.B. Watson, E. McCauley, Canadian journal of fisheries and aquatic sciences, **58**(10), 1905-1908 (2001)
18. R. Quirós, The nitrogen to phosphorus ratio for lakes: A cause or a consequence of aquatic biology. Water in Ibero-America: From Limnology to Management in South America. CYTED XVII (Fernandez Cirelli A, ChalarMarquisa G, eds.). Madrid, Spain, 11-26 (2002)
19. Schindler, W. David, R.E. Hecky, D.L. Findlay, M.P. Stainton, B.R. Parker, M.J. Paterson, K.G. Beaty, M. Lyng, S.E.M. Kasian, Proceedings of National Academy of Sciences USA **105**:11254–11258 (2008)
20. R. Muradian, Ecological economics **38**(1), 7-24 (2001)
21. United States. Environmental Protection Agency. Office of Water. & United States. Environmental Protection Agency. Office of Research and Development. National lakes assessment a collaborative survey of the nation's lakes: draft. Washington, DC: U.S. Environmental Protection Agency, Office of Water and Office of Research and Development (2009)
22. R Core Team, R: A language and environment for statistical computing. R Foundation for Statistical Computing, Vienna, Austria. URL http://www.R-project.org/ (2013)
23. J. Elith, J.R. Leathwick, T. Hastie, Journal of Animal Ecology, 77(4), 802-813 (2008)
24. G. De'ath, Ecology 88:243-251 (2007)
25. J.H. Friedman, Annals of statistics, 1189-1232 (2001)
26. J.H. Friedman, J.J. Meulman, Statistics in medicine, 22(9), 1365-1381 (2003)
27. I. Tõnno, The impact of nitrogen and phosphorus concentration and N/P ratio on cyanobacterial dominance and N2 fixation in some Estonian lakes. Tartu University Press (2004)
28. N.G. Bulgakov, A.P. Levich, Archivfürhydrobiologie, 146(1), 3-22 (1999)
29. K.E. Havens, R.T. James, T.L. East, V.H. Smith, Environmental Pollution, 122(3), 379-390 (2003)
30. H. Bothe, O. Schmitz, M.G. Yates, W.E. Newton, Microbiology and molecular biology reviews, 74(4), 529-551 (2010)
31. R.C. Lathrop, S.R. Carpenter, C.A. Stow, P.A. Soranno, J.C. Panuska, Canadian Journal of Fisheries and Aquatic Sciences, 55(5), 1169-1178 (1998)
32. R.W. Howarth, R. Marino, Limnology and Oceanography, 51(1part2), 364-376 (2006)
33. D.J. Conley, H.W. Paerl, R.W. Howarth, D.F. Boesch, S.P. Seitzinger, K.E. Havens,... &G.E. Likens,Science, 323(5917), 1014-1015 (2009)
34. W.M. Lewis Jr, W.A. Wurtsbaugh, H.W. Paerl, Environmental Science and Technology 45(24), 10300-10305 (2011)
35. D.W. Schindler, Proceedings of the Royal Society of London B: Biological Sciences 279(1746), 4322-4333 (2012)
36. H.W. Paerl, H. Xu, M.J. McCarthy, G. Zhu, B. Qin, Y. Li, W.S. Gardner, Water Research 45(5), 1973-1983 (2011)
37. H.W. Paerl, V.J. Paul, Water research, **46**(5), 1349-1363 (2012)

[a]Corresponding author: sajeela_g@yahoo.com

Adsorption of Methylene Blue from Aqueous Solution by AB-8 Macroreticular Resin

Hu Jun-ping[1,2] Liu Yan[1,2] Bi Hui-min[1,2] You Fu-ying[1,2] Xie Peng-tao[1,2]

[1] *School of Chemical and Materials of Handan* College, *Handan, 056005, China*
[2] *Chemistry and Chemical engineering Experiment Teaching Center of Handan College, Handan*

Abstract. Study on adsorption kinetics and thermodynamics of methylene blue by AB-8 macroreticular resin was carried out through static experiments. The results showed that the adsorption rate for the solute appear to be the first order kinetics, the intraparticle diffusion is the main rate-controlling step. In the experimental temperature, thermodynamic studies indicated that methylene blue adsorption onto resin AB-8 conforms to Langmuir and Freundlich adsorption isotherm. Methylene blue adsorption quantity respectively 30 mg/L, 150 mg/L, and 300 mg/L . The adsorption enthalpy change is 40.3 kJ/moL , 39.2 kJ/moL , and 33.6 kJ/moL respectively. The uptake of salicylic acid on AB-8 macroreticular resin is a type of physical adsorption.

1 Introduction

Along with the quick development of national economy, the industrial wastewater has become one of the main sources of water pollution, especially in printing and dyeing wastewater because of its deep color, chemical oxygen demand (COD), biochemical oxygen demand (BOD) value is higher, composition is complicated, wide distribution of surface [1], and is becoming a hot spot of environmental protection workers is becoming more and more attention. The common method of printing and dyeing wastewater treatment have different limitations, such as chemical precipitation treatment of printing and dyeing wastewater will produce a lot of difficult to deal with the sludge; Biological methods while running cost is low, but cannot make dye wastewater decolorizing completely, and toxic side effects on microorganisms. And adsorption method as a kind of important physical and chemical method, in dealing with the organic matter pollution including printing and dyeing wastewater has important applications in [2]

In the paper, we mainly studied the AB-8 macroreticular resin adsorption of methylene blue adsorption behavior, Discussing the different oscillation velocity and initial concentration under the condition of adsorption dynamics, and the influence of different temperature on the adsorption equilibrium, in order to AB-8 macroreticular resin adsorption of methylene blue wastewater process have a further understanding, for the resin in the methylene blue do a preliminary theoretical exploration of the application of wastewater treatment.

2 Experimental

2.1 Experimental materials and equipment

SHA-B thermostatic oscillator (Guohua enterprises); UNICAM uv-vis spectrophotometer (USA); Methylene blue (analysis of pure, Shanghai chemical plant); AB-8 macroreticular resin (Cangzhou treasure well chemical co., LTD); Milky white opaque spherical particles, particle size 0.3-1.25mm, average pore size13~14nm, specific surface area 480~520m^2/g.

2. 2 Experimental method

2.2.1 The pre-treatment of the resin

Take a certain amount of AB-8 resin, 24 h after soaking with ethanol, with deionized water elution clean and dry.

2.2.2 The determination of static adsorption kinetics

AB-8 macroreticular resin was weight 150mg accurate in the iodine flask, dry resin added with a certain volume of deionized water swelling after 24 hours, fully filter to excess water, add a certain amount of methylene blue standard solution, under the temperature of 298 k at 100 r/min constant temperature oscillation, interval time determination of methylene blue solution concentration.

2.2.3 The influence of vibration velocity and initial concentration on adsorption Formatting the title, authors and affiliations

At the first, AB-8 macroreticular resin was weighing out 0.5g after pre-treatments of drying, and dosing to several containing 100 ml methylene blue solution grinding flat bottom flask, preparation of methylene blue solution concentration of 500 mg/L, placed in a thermostatic oscillator, control temperature 28 ℃ , the oscillation speed respectively 100, 150, and 250 r/min, timing sampling analysis. oscillation velocity influence on adsorption was tested; Under the same conditions for the oscillation speed of 200 r/min, initial concentration of methylene blue were 33 mg/L, 152 mg/L, 308 mg/L adsorption experiments.

2.2.4 The influence of temperature on the adsorption equilibrium

Dry resin Said was taking 0.5 g, cast to join respectively containing 50 ml methylene blue solution grinding mouth flasks, concentration of methylene blue of 25 mg/L 50 mg/L, 100 mg/L, 200 mg/L, 300 mg/L, 400 mg/L and 500 mg/L, respectively, at 20 ℃ 、 40 ℃ 、 60 ℃ under three different temperature, speed of 200 r/min oscillations to adsorption equilibrium after sample analysis respectively.

2.2.5 Analysis method

Under 668 nm wavelength, concentration of methylene blue analysis using by UNICAM uv - vis spectrophotometer. Adsorption of methylene blue on the resin based on qt = (C0 - Ct) V/W, C0, Ct of methylene blue respectively when the initial concentration and time t (min) methylene blue solution concentration (mg/L), V (L) for solution volume, W for AB-8 macroreticular resin weight (g).

3 RESULTS AND DISCUSSION

3.1 Adsorption kinetics research

Time, temperature, initial concentration, oscillation speed, pH, and many other external factors influence on adsorption process to some extent[3], this article mainly discusses time, different speed of oscillation and the influence of the initial concentration of the methylene blue adsorption[4]. At the same time, using simulation level of adsorption dynamics model to fitting of adsorption process, having certain prediction for adsorption behaviour. Lagergren equation is solid-liquid adsorption system simulation analysis of one of the commonly used model[4,5,6]. Methylene blue in AB-8 macroreticular resin adsorption equation expression for formula (1)

$$\log C_t = \log C_e + Kt/2.303 \qquad (1)$$

In the formula, K is rate constant, min^{-1}; t as time used for adsorption, min; C_e for salicylic acid solution concentration at the time of adsorption equilibrium, mg/L; C_t was methylene blue on the AB 8 resin adsorption quantity and balance the difference between the adsorption quantity at the t time.

To logCt vs t drawing, if a straight line, are proved to first order kinetics model can describe the adsorption process.

Ion exchange process generally experience three steps:

(1) ion solution by the liquid film diffusion to the resin surface;

(2) by the resin surface to the internal diffusion;

(3) ion exchange reaction active group position within the resin.

The ion exchange process by liquid film diffusion, particle diffusion and chemical reaction rate of 3 steps, The influence of the speed of the slowest step is to control the rate. Generally soon, due to chemical reaction is difficult to become a speed control steps, often using particle diffusion equation (2) to calculate.[7,8]

Particle diffusion rate formula

$$q_t = K_i t^{0.5} \qquad (2)$$

Adsorption rate constant K for min^{-1}; K_i for particle diffusion rate constant in mg/ (g $min^{0.5}$); T, min.

3.1.1 The adsorption kinetics curve

Adsorbent adsorption in liquid phase, is essentially the solvent and the component adsorption of the adsorbent competition, when the adsorption effect of the solvent can be ignored, the adsorption system according to the single component adsorption to deal with.Adsorption curve, namely static exchange kinetics curve, characterization of the adsorbent on the adsorption properties of adsorption process of change over time. As shown in figure 1.

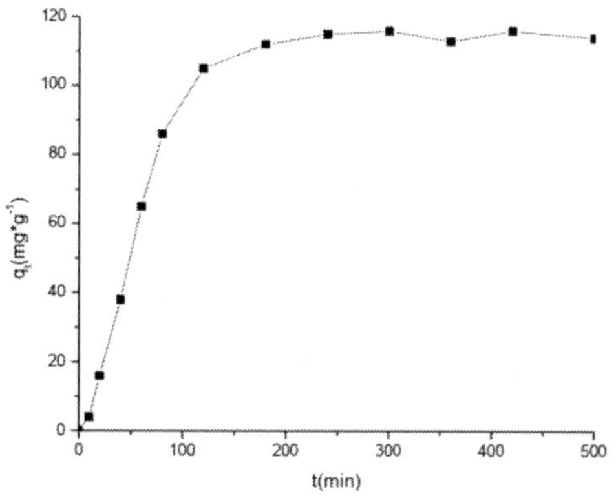

Fig.1 Adsorption kinetics curve of AB-8 resin toward methylene blue

By the figure 1 can be seen AB-8 resin adsorption of methylene blue as time increases, basic equilibrium adsorption at 180 min.

3.1.2 The influence of oscillation speed of adsorption

The figure 2, table 1 shows that logCt on t show a good linear relationship (R^2 more than 0.98), showed that methylene blue on the AB-8 resin adsorption for the first order kinetics. According to Mackay [9], when the type (1) in the q_t with $t^{0.5}$ show good linear relationship and through the origin, the material within the particle diffusion process as adsorption rate control steps.

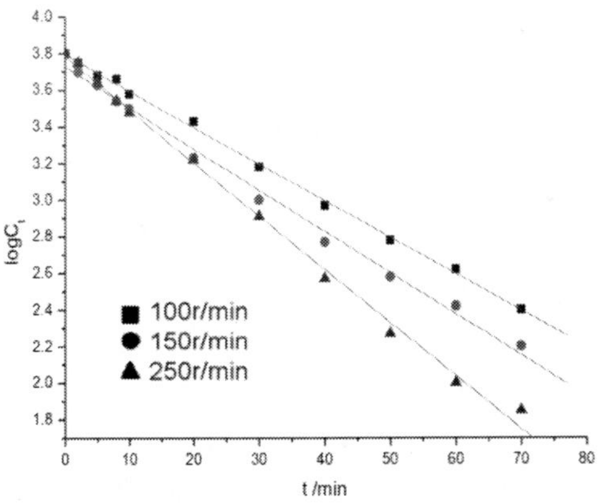

Fig.2 Effect of agitation rate on adsorption rate

Table 1 Adsorptions rates and intrapartical diffusion constants under different agitate speed and initial concentration

	$K,(*10^{-2})min^{-1}$	R^2	K_i, mg/(g min)$^{0.5}$	R^2
Agitate speed,r/min				
100	0.83	0.968	12.88	0.985
150	1.38	0.985	13.54	0.995
250	1.52	0.987	13.67	0.984
C_0,mg/L				
33	0.21	0.978	1.32	0.998
152	0.11	0.972	3.53	0.993
308	0.09	0.968	8.88	0.986

The figure 3, table 1 shows that under the condition of three different oscillation of qt and $t_{0.5}$ show good linear relationship, but as a result of these lines are not through the origin, that methylene blue adsorption on AB-8 resin particle diffusion process inside is not the only control step, the adsorption rate affected by particle diffusion process of adsorption of colleagues.

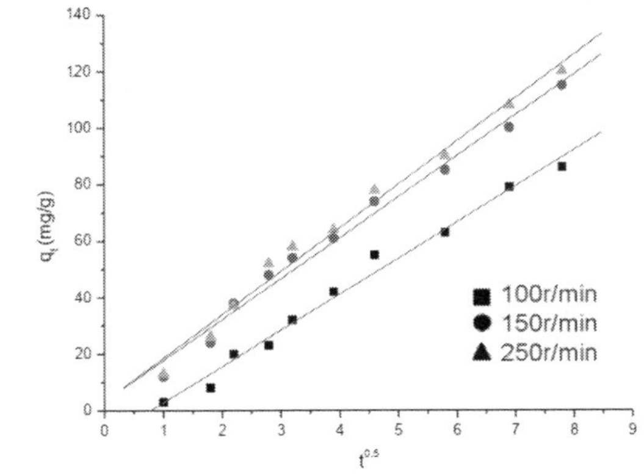

Fig.3 Effect of agitation rate on intrapartical

3.1.3 The influence of initial concentration on adsorption

From table 1 and figure 4, figure 5 shows that changes in the concentration range of methylene blue on the AB-8 resin adsorption rate in line with the first order kinetics equation, in different initial concentration, despite the particle diffusion process has great influence to the adsorption, but still is not the only control the adsorption rate of step.

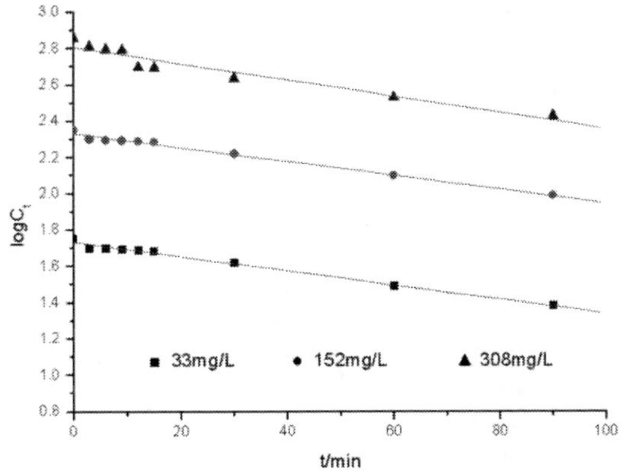

Fig.4 Effect of initial concentration on adsorption rate

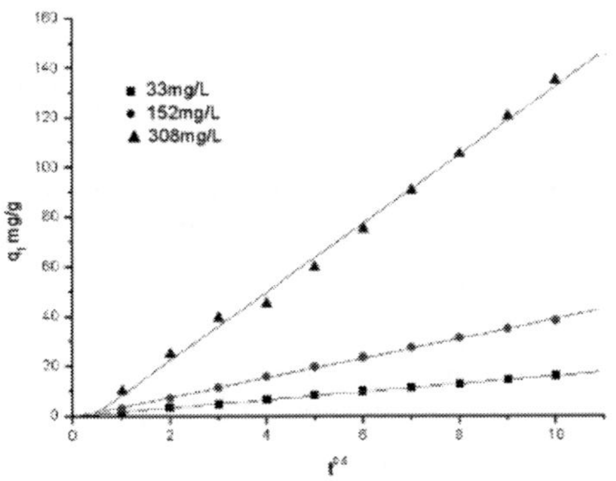

Fig.5 Effect of initial concentration on intrapartical diffusion

3.2 The influence of temperature on the adsorption equilibrium

Figure 6 shows that methylene blue on the AB-8 resin adsorption quantity increases with the increase of equilibrium concentration [10], however, as the adsorption and the rise of temperature, by Langmuir and Freundilich isothermal equation of figure 5 data fitting results are shown in table 2, the results show that the Langmuir and Freundilich isothermal equation can well describe the AB - 8 adsorption of methylene blue

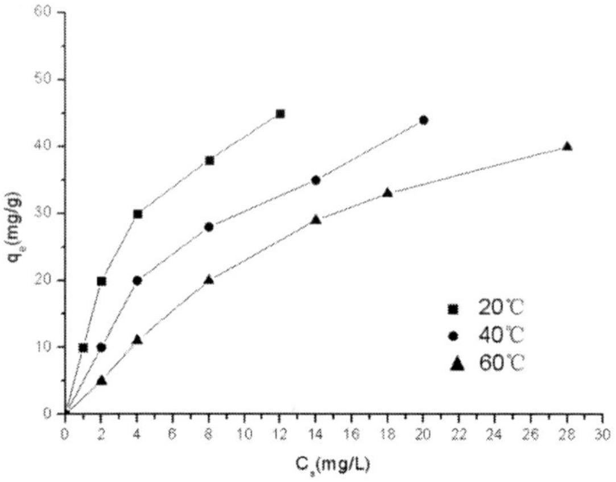

Fig.6 Effect of temperature on adsorption equilibrium

Table 2 Parameters for Langmuir and Freundlich equitions

T, ℃	Xm	a_L	R^2	K_F	b	R^2
20	85.2	0.168	0.995	9.18	0.7755	0.987
40	52.3	0.198	0.996	6.58	0.7152	0.992
60	49.2	0.121	0.981	3.69	0.8205	0.991

3.3 The adsorption enthalpy change of methylene blue on the AB-8 resin

In adsorption, the enthalpy change can be gat by Van 't Hoff equation. Van' t Hoff equation is $\ln(1/C_1) = \ln K + (-\Delta H/RT)$. $-\Delta H^0$ is adsorption enthalpy kJ/moL; T as absolute temperature, K; For the ideal gas constant R is 8.314 kJ/moL, K); C_1 (C_1 = Ce, methylene blue molecular weight 319.86 and Ce values calculated by the Langmuir equation) for the adsorption of methylene blue solution at q equilibrium concentration tendency/ mmoL/L.

Methylene blue adsorption quantity respectively 30 mg/L, 150 mg/L, and 300 mg/L . The adsorption enthalpy change is 40.3 kJ/moL, 39.2 kJ/moL, and 33.6 kJ/moL respectively. Adsorption enthalpy change Δ H negative, indicating that the adsorption process is exothermic process, and the smaller of absolute value of enthalpy change (less than 42 kJ/moL), shows that the adsorption process for the obvious physical adsorption properties. Adsorption enthalpy decreases with the increase of the adsorption quantity, may be due to the inhomogeneity of the surface of the resin adsorption mass adsorption was first to energy of position.

4 CONCLUSION

(1) Kinetics study showed that AB-8 resin adsorption of methylene blue good pseudo-first-order kinetics equation（R2>0.99）。

(2)AB-8 resin adsorption of methylene blue speed steps for particle diffusion process.

(3)Within the scope of the study, methylene blue by AB-8 resin on the equilibrium adsorption data could very well in Langmuir and Freundilich isothermal adsorption equation fitting, adsorption quantity increases with the increase of concentration of methylene blue.

(4)Thermodynamic study, Adsorption enthalpy turned negative, indicating that the adsorption reaction is exothermic physical adsorption process, easier for adsorption and adsorption performance is good, the results show that AB-8 macroreticular resin can be used for methylene blue wastewater pollution control and resource recycling.

References

1. Bhattacharyya K G, Gupta S.Industrial and Engineering Chemistry Research
2. Structure and Fluorescence Emission of Hetero-Nuclear Complexes of Europium-Lanthanum Nitrates with N,N'-Substituted Adipamide - 2002
3. Ho Y S, Mckay G. A kinetic study of dye sorption by biosorbent waste product pith [J]. Resources Conservation and Recycling, 1999 25:171-193
4. The Diels-Alder Reaction on Endohedral Y3N@C-78: The Importance of the Fullerene Strain Energy - 2009
5. Qin Li-hong, Zhang Feng-bao,Zhang Guo-liang. Kinetics and mechanism of the sorption of NTS macroreticular adsorption resin [J]. Chemical Industry and Engineering, 2007,24(3) :245-248
6. E Metwally. Kinetic studies for sorption of some metal ions from aqueous acid solutions onto TDA impregnated resin[J]. Journal of Radioanalytical and Nuclear Chemistry,2006,270(3) :559-566
7. Guo Zhou, Yuan Yue. Kinetics and themodynamics of adsorption of phenol onto mesoporous carbon CMK-3[J]. Chemical Journal of Chinese University, 2007, 28(2) :289-292
8. Dinesh Mohan, Kunwar P Singh. Single-and multicomponent adsorption of cadmium and zinc using activated carbon derived from bagasse – an agricultural waste[J]. Water Research, 2002,9(36) : 2304-2318
9. Mamdouh Mahmoud Nassar. Intraparticle diffusion of basic red and basic yellow dyes on palm fruit bunch [J]. Wat Sci Tech, 1999, 40(7):133-139
10. Enhanced photocatalytic activity of electrosynthesised tungsten trioxide-titanium dioxide bi-layer coatings under ultraviolet and visible light illumination - 2007

Molluscan Fauna in Bang Taboon Mangrove Estuary, Inner Gulf of Thailand: Implications for conservation and sustainable use of coastal resources

Siraprapha Premcharoen[1,a], Soranan Witirawat[1] and Pobporn Tharapoom[1]

[1]Department of Science, Faculty of Liberal Arts and Science, Kasetsart University, Nakhon Pathom 73140, Thailand

Abstract. Molluscan fauna, dealt with diversity and abundance of gastropods and bivalves in Bang Taboon mangrove estuary, Inner Gulf of Thailand was investigated seasonally. Totally 3,337 individuals, representing 18 species (6 bivalves and 12 gastropods) with two dominant species, *Assiminea brevicula* and *Tegillarca granosa*. The mean density of overall molluscs was 11.79±0.47 n/m², recorded in September. The highest value of diversity was 0.78±0.53, recorded in March. The most species-rich and individuals were recorded in the large mudflat within intertidal mangrove zone and formed a distinct community structure in the vegetated area at 15 cm depth. Results from the Analysis of the Similarity (ANOSIM) showed a clearly significant difference of species diversity and abundance at different areas, depths and seasons (P = 0.001). Although Bang Taboon estuary is impacted by human activities, the habitat remains an important nursery ground for numerous molluscs including many commercial species. The findings of this study can be taken as a base data that would help maximizing mangrove and fisheries resources conservation and utilization in a sustainable way in Bang Taboon estuary and adjacent areas in the inner Gulf of Thailand.

1 Introduction

Mangroves are recognized to be among the world's most productive ecosystems which provide important nursery and feeding grounds for juvenile fish and invertebrate species [1-2]. Molluscs are one of the main macroinvertebrates that play major ecological role in nutrient dynamics in the mangrove ecosystem because they form an important link within the food web as predators, herbivores, detritivores and filter feeders [3-4]. They are useful bioindicators of environmental pollution, due to their methods of filter feeding [5]. Many species of which are linked directly, or indirectly, to valuable fisheries, which have a great commercial importance.

Bang Taboon estuary is part of the Ban Laem coastal area, Phetchaburi Province and to be known as one of the most productive fishing areas in the inner Gulf of Thailand [6]. The estuarine ecosystem by influence of the currents that blow the deposition of mud lane. Mangroves occur on the muddy tidal flats along the shoreline on the side of the river mouth. Mangrove forests in the area are currently threatened by exploitation and coastal development, especially converted to shrimp and blood cockle farmings, along with urbanization and reclamation effluents into the Bang Taboon estuary represent environmental concerns. To provide a better understanding of the role of mangrove ecosystem in supporting coastal resources communities, and to date, the limited number of studies have been to observe the seasonal diversity and abundance of molluscs in this area [7]. Therefore, this work is aimed to assess the diversity and abundance of bivalve and gastropod molluscs in Bang Taboon mangrove estuary, Inner Gulf of Thailand and characterized different areas, depths and seasons which affect the patterns of mollusc assemblage in Bang Taboon estuary.

2 Materials and methods

2.1 Study area and sample collections

Bang Taboon estuary covers an area of about 59 km^2. The latitude and longitude is N 13° 16' 7.608" and E 99° 56' 54.708" respectively, located in Bang Taboon Subdistrict, Ban Laem District, Petchaburi Province, Thailand. The area is dominated by intertidal mangrove-fringed, mostly two species of *Avicenia* sp and *Rhizophora* sp., and totally covered by muddy alluvium.

Samplings were performed at low tide from 8 stations (Fig. 1) in November, March, July and September, between November 2012 and September 2013, using a core sample (6 cm in diameter, 15 cm and 30 cm depth). All gastropod and bivalve samples were preserved immediately in 5-7% neutralized formalin,

[a] Corresponding author: faassrp@ku.ac.th

 DOI: 10.1051/matecconf/20166002003

identified to species level according to [8-9], counted and weighted.

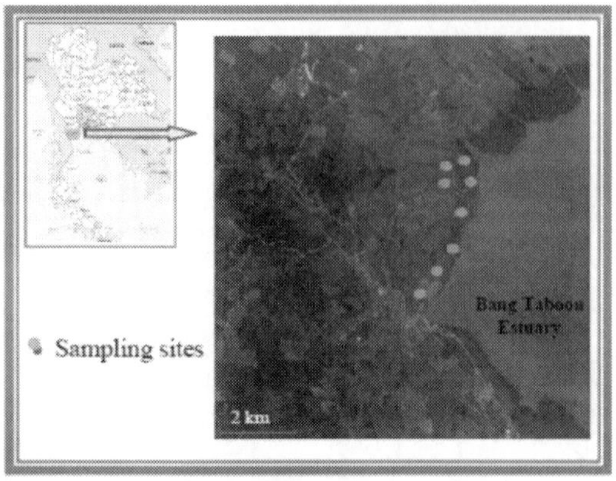

Figure 1. Map of Bang Taboon estuary showing location of sampling sites.

2.2 Assemblage analysis

Species diversity and abundance including diversity index were analysed. Non-parametric multivariate analyses contained in the PRIMER (Plymouth Routines In Multivariate Ecological Research) Version 6.0 [10] were used in the analysis of non-parametric community structure data. Analysis of Similarity (ANOSIM) was tested for comparing and classifying whether bivalve and gastropod assemblages separated a priori into vegetated or non-vegetated areas, depths and seasons differed statistically.

3 Results and discussion

3.1 Species diversity and density

A total of 3,337 individuals representing 16 families and 18 species were recorded during the study period (Fig. 2). Of the recorded species, 12 species (87.24 % individuals) belonged to gastropods, 6 species (12.76% individuals) belonged to bivalves. Among species found, assimineid red snail *Assiminea brevicula* showed the highest individual molluscs (63.08**%**)(Fig. 2), found to occur throughout the study period in all stations and seasons, followed by blood cockle, *Tegillarca granosa* (10.94%) and *Cerithidea cingulata* (8.12%), which found in all seasons and showed consistency in four stations. The 6 species, *viz.*, *Cassidula* sp., *Melampus granifer, Heterocardia gibbosula*, *Solen* sp., *Bedeva* sp. and *Siliqua* sp. were presented by less than 10 individuals. Variation in mean density (number of individuals per m^2) of molluscs showed the highest (11.79±0.47 n/m^2) in the vegetated area at 15 cm depth recorded in September (Fig. 3). There were 10 species found in all seasons, namely, *Assiminea brevicula, Tegillarca granosa*, *Cerithidea cingulata*, *C. obtusa*, *Stenothyra glabra*, *Salinator* sp., *Nassarius* sp., *Tellina* sp., *Littoraria articulata* and *L. Melanostoma*. The study was reported earlier by many research workers who pointed out that the predominantly mangrove-associated bivalves and gastropods of localities within the Indo-Pacific consists predominantly of the families Arcidae, Tellinidae, Amphibolidae, Ellobiidae, Nassariidae, Assimineidae, Stenothyridae, Littorinidae, and Potamididae [11-15].

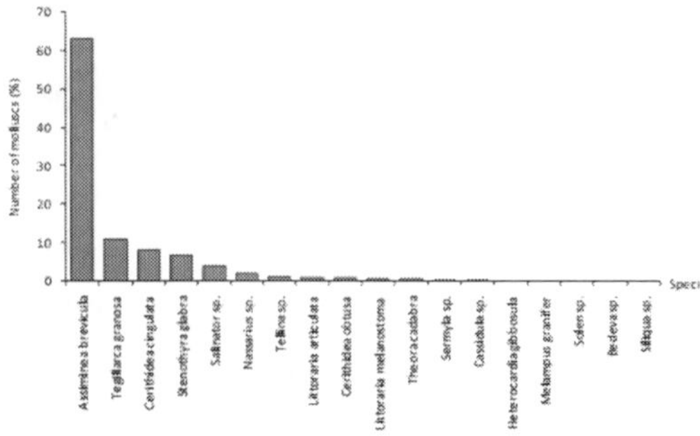

Figure 2. Number of mollusc species (%) recorded in Bang Taboon estuary.

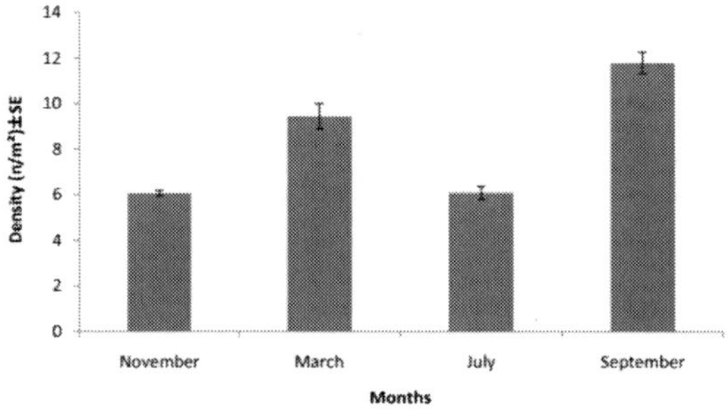

Figure 3. Mean density (number of individuals per m^2) ±SE of molluscs between November 2012 and September 2013 in Bang Taboon estuary.

The study showed numbers of bivalve species that were less than of gastropods, similar were the findings of Iwahig River-Estuary, Palawan, the Philippines [14] and Raigad district, Maharashtra, West coast of India [15]. This was due to bivalves which mostly live in the mud, confined to a narrow seaward zone, unlike gastropods which occupy a wide range of ecological niches and are known to climb mangrove trees, usually inhabit the trunk of trees and stilt roots [16-17]. Therefore, they are more adapted to the very changing and harsh mangrove environment.

Diversity index is commonly used to characterized species abundance relationship in a biological system. From the study, diversity index was calculated per site and among seasons (Table 1). The highest value of diversity was 0.78±0.53, recorded in March. The results showed the mollusc species fluctuated seasonally from a high between March and July to low during September and November. The diversity index value in November is lower than in March might be that July to October dominated by southwest monsoon in

DOI: 10.1051/matecconf/20166002003

Thailand [18] that caused temperature and salinity changed. The salinity acts as a limited factor which made the condition opposed to the molluscs and probably exceeded an optimal intermediate level of disturbance, reflecting absence of species using estuaries as nursery areas [19]. However, the mollusc density was at its peak in September during late monsoon period. It is clearly that the post monsoon is the most favourable time for the new inflow of mollusc species [14].

Table 1. Average diversity index (±SE) in various seasons in Bang Taboon estuary, S1-8= station 1-8.

M	BS-1	BS-2	BS-3	BS-4	BS-5	BS-6	BS-7	BS-8	Ave
Nov 2012	00.25	00.39	00.47	00.22	11.07	00.47	00.35	00.00	0.40 ±0.10
Mar 2013	00.22	00.00	00.00	00.47	11.76	11.01	11.78	11.01	0.78 ±0.53
Jul 2013	00.00	00.28	00.00	00.11	11.10	00.92	11.96	00.00	0.55 ±0.51
Sep 2013	00.86	00.25	00.50	00.05	11.47	00.23	11.58	00.48	0.68 ±0.33

3.2 Variabilities of mollusc assemblages

A classification carried out using Bray-Curtis similarity (Fig. 4) and non-metric multi dimensional scaling (MDS) (Fig. 5) revealed the same groupings as recognized in the cluster within the vegetated and non-vagetated areas, depths and seasons, given by the presence and abundance of common species. The 2D stress value (0.14) indicated that the results were credible. Results of the analysis of similarity (ANOSIM) determined that there was significant difference in species numerical abundance of mollusc in different areas, depths and seasons (R=0.959, P = 0.001). There were generally 2 groupings on the dendrogram. The largest was comprised of three dominant molluscs (*Assiminea brevicula, Cerithidea cingulata* and *Tegillarca granosa*) in the Bang Taboon estuary. This group mostly comprised molluscs those were spreaded all over mangrove region at 15 cm depth samplings. Results based on species numerical abundance of the similarlity percentage (SIMPER) analysis indicated that Bang Taboon estuary was dominated with *Assiminea brevicula*. The average dissimilarity in different areas, depths and seasons was 72.72%, and the taxa which together contributed of *Assiminea brevicula* (35.55%), *Cerithidea cingulata* (12.86%) and *Tegillarca granosa* (10.66%).

The dominant gastropod species, assimineid red snail, *Assiminea brevicula,* which accounts for 26.31, 13.84, 12.1 and 10.81 percents of total individuals in September, November, March and July respectively. This species has a wider range of distribution and can be found on the surface of muddy sediment. *Assiminea brevicula* can be used as an indicator of the mangrove ecosystem health because its sensitivity to the physical-chemicals environment, especially to the proportional of organic content in the mangrove forest [20]. A large number of blood cockle, *Tegillarca granosa* can be found in all seasons inhabiting in mudflats on mangrove strands, where it lies partially buried in the sediment. This species plays a vital role in local people those are living on the coastal regions and is being cultured for commercialization [14]. *Cerithidea* sp, was observed to the mud banks, mud flats, mud forest, sandy muddy, mud swamps, tree trunks, fallen wood, prop-roots and pneumatophores [14, 19, 21] and also known to be abundant in eulittoral zone and shallow-water [16]. Cerithiids were among the most abundant gastropod molluscs in the seaward part and seaward edge of a mangrove forest in a study in Phuket Island, Thailand [20, 22].

There have been several reports described earlier that mollusc diversity and abundance varies seasonally and also depends on locality and depth, and is related to sediment rate, tidal inundation, flowing current, evaporation, rainfall, groundwater input, temperature, salinity, dissolved oxygen, organic carbon, nutrients, pollution, habitat destruction, over- exploitation and predation [1, 19, 20, 22-23]. In the present study, salinity seemed to be one of a limiting factors which determine the diversity and abundance of molluscs in the estuary during monsoon season. The present study did not show characteristic relationship between salinity and mollusc distribution. The variation of salinity at study sites was probably due to freshwater runoff, evaporation and rain that influenced the diversity and abundance of molluscs since these factors may vary with seasons.

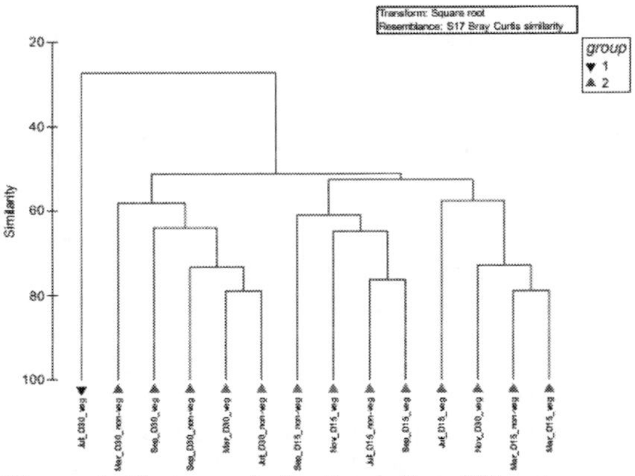

Figure 4. Cluster analysis of sampling different areas, depths and seasons based on species numerical abundance. veg=vegetated area, non-veg= non-vegetated area; D15= 15 cm depth, D30=30 cm depth.

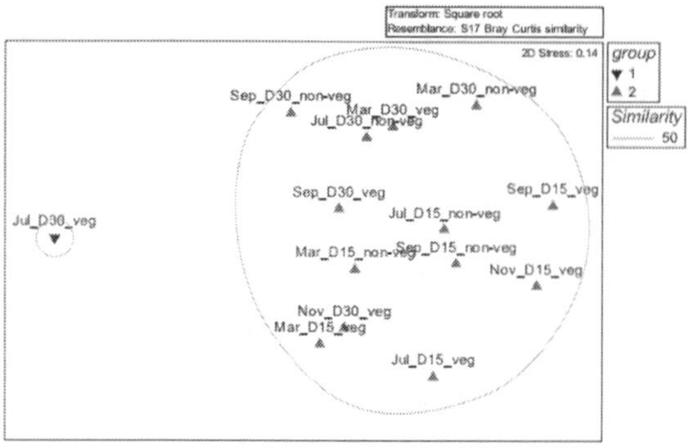

Figure 5. Ordination (nMDS) of sampling different areas, depths and seasons based on species numerical abundance. veg=vegetated area, non-veg= non-vegetated area; D15= 15 cm depth, D30=30 cm depth.

The results showed a large number of molluscs at 15 cm depth more than those at 30 cm depth in the sediment of vegetated mud habitats (mangrove zone). As it is the zone at the land water and marine-estuarine interface zone, true estuarine species may find this zone a better shelter from the open estuary. Availability of decomposed material of the plant litter and also the adequate safety can be the factors that govern a large number of diverse epifauna species in this mangrove zone. Some snails, *Bedeva* sp., *Cassidula* sp., *Melampus* sp., *Littoraria* sp. and *Cerithidea* sp. were recorded in the present study as well and found restricted to the mangrove zone. It was known that these species are instantly climb down from the tree trunks or prop roots to the mangrove floor for feeding when the ebb tide starts and again climb up during the high tide [24].

Conclusions

The molluscs have a significant role to play in the mangrove ecosystems in term of ecosystem health and comprise an important trophic component of detrital food webs. Although, Bang Taboon estuary has been threatened by anthropogenic activities which result in the loss of mangrove and contribute to the reduction of mollusc communities, the results from this study support moderately high species diversity with high numbers of individuals, indicate fairly good health of the sediment in the area. The dominant species, assimineid red snail, *Assiminea brevicula* shows the integrity bioindicator of the mangrove ecosystem and blood cockle, *Tegillarca granosa*, which has a great commercial importance. Therefore, the data presented in this study can be taken as a base line data that would help maximizing mangrove and fisheries resources conservation and utilization in a sustainable manner in Bang Taboon estuary and adjacent areas in the inner Gulf of Thailand.

Acknowledgement

The authors are grateful to everyone involved in the sampling and data collections. Thanks are also due to the Faculty of Liberal Arts and Science and Graduate School, Kasetsart University for financial support for participation in the IEEA 2016 conference at which this paper was presented.

References

1. C.R. Beasley, C.M. Fernandes, C.P. Gomes, B.A. Brito, S.M. Lima dos Samtos, C.H. Tagliaro, Molluscan diversity and abundance among coastal habitats of northern Brazil. Ecotropica, **11**, 9-20 (2005).
2. K. Kathiresan, B.L. Bingham, Biology of mangrove and mangrove ecosystem. Adv. Mar. Biol. **40**, 81-251 (2001).
3. P.R. Pawar, Molluscan diversity in mangrove ecosystem of Uran (Raigad), Navi Munbau, Maharashtra, west coast of India. Bull. Environ. Pharmacol. Life Sci. **1**, 55-59 (2012).
4. M. Suresh, S. Arularasan, K. Ponnusamy, Distribution of molluscan fauna in the artificial mangroves of Pazhayar backwater canal, southeast coast of India. Adv. Appl. Sci. Res. **3**, 1795-1798 (2012).
5. D.W. Frith, A preliminary list of macrofaunafrom a mangrove forest and adjacent biotopes at Surin Island, western Peninsular Thailand. Phuk. Mar. Biol. Centre Res. Bull. **17**, 1-14 (1977).
6. Department of Marine and Coastal Resources, Mangrove forest n Phetchaburi Province, (Department of Marine and Coastal Resources, Ministry of Natural Resources and Environment, 2012).
7. C. Printrakoon, F.E. Wells, Y. Chitramvong, Distribution of mollusks in mangrove at six sites in the upper gulf of Thailand. Raffles Bull. Zool. **18**, 247-257 (2008) .
8. R.T. Abbott, S.P. Dance,*Compendium of Seashells*, (Odyssey Publishing, 1998.
9. T. Kira, *Shell of the western Pacific in color*, (Hoikusha, Osaka, 1995).
10. K.R. Clarke, R. N. Gorley, *PRIMER-e v6*: Usual *manual/Tutorial*. (Promer-E, Plymouth, 2006).
11. D.W. Frith, R. Tantanasiriwong, O. Bhatia, Zonation and abundance of macrofauna on a mangrove Shore, Phuket Island. Phuk. Mar. Biol. Center Res. Bull. 10, 1-37 (1976).
12. G. Ghasemi, M. Zakaria, N.M. Hoveizen, Abundance of molluscs (gastropods) at mangrove forests of Iran. J. Am Sci, **7**, 660-668 (2011).
13. D.J. Macintosh, E.C. Ashton, S. Havanon, Mangrove rehabilitation and intertidal biodiversity: a study in the Ranong mangrove ecosystem. Thailand. Estuar. Coast. Shelf Sci. **55**, 331-345 (2002).
14. R.G. Dolorosa, F. Dangan-Galon, Species richness of bivalves and gastropods in Iwahig river-estuary, Palawan, the Philippines. Int. J. Fish Aquat. Sci. **2**, 207-215 (2014) .
15. S.N. Khade, U.H. Mane, Diversity of Bivalve and Gastropod Molluscs in Mangrove ecosystem from

selected sites of Raigad district, Maharashtra, West coast of India. Recent res. sci. technol. **4**, 16-20 (2012).

16. A. Appadoo, N.B. Roomaldawo, A preliminary study on benthic and arboreal fauna associated with mangrove *Rhizophora mucronata* on the east coast of Mauritius (Indian ocean) with special reference to the molluca. J. Coast. Dev. **8**, 107-121 (2005).
17. M. Vannini, C. Coffa, E. Lori, S. Fratini, Vertical migrations of the mangrove snail *Cerithidea decollata* (L.) (Potamididae) through a synodic month. Estuar Coast. Shelf Sci. **78,** 644-648 (2008).
18. A. Limsakul, S. Limjirakan, B. Suttamanuswong, Asian summer monsoon and its associated rainfall availability in Thailand. EnvironmentAsia **3**, 79-89 (2010).
19. S. Barkati, S. Rahman, Species composition and faunal diversity at three sites of Sindh mangroves. Pakistan J. Zool. **37**, 17-31 (2005).
20. N. Dissanayake, U. Chandrasekara, Effects of mangrove zonationand the physiochemical parameters of soil on the distribution of macrobenthic fauna in Kadolkele mangrove forest, a tropical mangrove. Adv. Ecol. Article ID 564056,1-13 (2014).
21. B. Hookham, A.T. Shau-Hwai, B. Dayrat, W. Hintz, A baseline measure of tree and gastropod biodiversity in replanted and natural mangrove stands in Malaysia: Langkawi Island and Sungai Merbok. Trop Life Sci Res. **25,** 1-12 (2014).
22. T. Suzuki, M. Nishihira, N. Paphavasit, Size structure and distribution of *Ovassimines brevicula* (Gastropoda) in a Thai mangrove swamp. Wetlands Ecol. Manage. **10**, 265-271 (2002).
23. M.V.A Salem, M. van der Geest, T. Piersmab Y. Saoud, J.A. van Gils, Seasonal changes in mollusc abundance in a tropical intertidal ecosystem, Banc d'Arguin (Mauritania): Testing the 'depletion by shorebirds' hypothesis. Estuar. Coast. Shelf Sci. **36**, 26-34 (2014).
24. M. Włodarska-Kowalczuk, E. Jankowska, L. Kotwicki, P. Balazy, Evidence of season-dependency in vegetation effects on macrofauna in temperate seagrass meadows (Baltic sea). PLoS One. **9**, e100788 (2014).

Water quality assessment and analysis for rehabilitate and management of wetlands: a case study in Nanhai wetland of Baotou, China

Jing tian Gao, Xiao yun Wang, Wei ping Li*, Ling hong Yu, Wen huan Yang and Zhen yu, Yin

School of Energy and Environment, Inner Mongolia University of Science &Technology, No.7 Arden Street, Baotou 014010,China

Abstract. Wetland plays an irreplaceable role in many aspects and waters are important part of wetland, the water quality can easily reflect the situation of Wetland. In this study, water quality was assessed on the basis of 5 parameters (DO, NH_4^+-N, TN, TP and CODcr) that were monitored monthly at 5 sites (N_1,N_2,N_3,N_4 and N_5)from April, 2014 to March, 2015 of the Nanhai Lake in Baotou, China by water pollution index method and comprehensive water quality identification index method. The twelve monitoring months were divided into wet season (Mar., Aug. and Sep.), normal season (Jan., Feb., Apr., Nov. and Dec.) and dry season (May., Jun. and Jul.). The assessment results determined using the water pollution index method showed that the water quality of all the five monitoring sites were inferior Ⅴ, the main contamination was COD. The comprehensive water quality identification index showed that the water quality of the Nanhai Lake were classes Ⅴ, except for the N_2 in wet season and dry season, the N_1 in dry season and the N_5 in normal season, which were classes inferior Ⅴ. All the five monitoring sites don't achieving the desired water quality standard. According to the analysis, domestic discharge, industrial activities and developed recreation were major threats to water quality of Nanhai Lake.

1 Introduction

Wetlands have huge environmental functions in maintaining ecological balance and regional stability due to their particular characteristics in hydrology, soil, plant, as well as landscape ecological structures. Baotou Yellow River wetland is a vital wetland resource in Inner Mongolia, Nanhai wetland is a part of it, which well known as northern west Lake. Managing the limited water resource and dealing with the relationship between water resources and ecological environment well are the foundation for the protection, management and utilization of wetlands. In recent years, due to the development of tourism and the increase of the surrounding human activities, water quality of the Nanhai Lake is gradually turns bad. It's essential to assess the water quality of Nanhai Lake. As to the method of water quality assessment, such as single-factor assessment method, water quality grading method, nemerow pollution index, comprehensive pollution index, principle component analysis, fuzzy comprehensive evaluation method, water pollution index method and comprehensive water quality identification index[1,2] each have their own characteristics and apply to different waters.

Water pollution index method (WPI) has been widely used to assess water quality in America[3], China [4], and Spain [5]. It overcomes the problem of single factor evaluation and comprehensive water evaluation method. In this method, the water quality is quantified, and can reflect the main pollution parameters. The comprehensive water quality identification index is a relatively new tool for general water quality assessment of surface waters, It evaluates general water quality both qualitatively and quantitatively, and also evaluates water quality inferior to V, identifying whether the waters malodorous and black[6]. In recently years, this method was applied widely to assessed the water quality lake and river [7,8].In this study, Water pollution index method and comprehensive water quality identification index method were selected to assess water quality of Nanhai Lake.

2 Materials and methods

2.1 Study area

The Nanhai wetland (109°59'02″～110°02'26″E, 40°30'08″～40°33'32″N) is one of the largest lake scenic areas located in the South of Baotou, China, the north shore of the Yellow River, with a total area of 2104.6hm^2(Fig.1), in which protected waters area 333 hm^2 and floodplain wetlands area 1,000 hm^2. The main geomorphology of Nanhai wetlands is Yellow River alluvial plain, with an average altitude of 1020m, including the waters, marshes, shrub and grassland. The region is in a semiarid continental monsoon climate with a large diurnal and seasonal range of temperature, an uneven distribution rainfall seasonally and a different wind direction between winter and

*

Innovation fund of Inner Mongolia University of Science & Technology (No.2015QDL16)

summer. The annual average rainfall is 307.4mm with most precipitation occurring between Jun. and Aug., but the annual evaporation is 2343mm. The annual average temperature is 8.5℃, with the mean lowest temperature -12.7℃ and the highest 22.2℃ come from the coldest month Jan. and the warmest Jul., respectively. In 2008, Nanhai wetland was named the national grade 4A tourist scenic spot, and the same year was authorized to be a provincial nature reserve of the Yellow River Wetland by the Inner Mongolia Autonomous Region. In recent years, due to the development of tourism and the increase of the surrounding human activities, the water quality of the Nanhai Lake is gradually turns bad. It is essential to evaluate the water quality of the Nanhai Lake for the quality of water environment directly affects the growth status of each species.According to the location and function of wetland, 5 monitoring sites are arranged in the study area.

2.2 Field sampling and Laboratory analysis

Water samples were collected every month from Apr., 2014 to Mar., 2015 (the twelve months are divided into wet season(Mar., Aug. and Sep.), normal season(Jan., Feb., Apr., Nov. and Dec.) and dry season(May., Jun. and Jul.)) at five sampling sites on Nanhai Lake, in the condition that without extreme weather, such as large scale rain and sand dust storm within seven days and no sudden pollution events. The sampling sites concrete spot position locates with GPS. All water samples were taken at 0.5m below the surface. Preservation, transportation and storage were performed according to the Chinese National Standard Methods (SEPA,2002). Six water quality parameters were selected due to their continuity in measurement and importance for assessing the water quality conditions of the Nanhai Lake: pH,dissolved oxygen (DO), biochemical oxygen demand index(COD), ammon-ium nitrogen(NH_4^+-N), total nitrogen(TN) and total phosphorus(TP).The specific methods are summarized in Table1.

2.3 Water quality assessment methods

2.3.1 Water pollution index method

In this method, the seriously polluted parameter is taken as water quality classification criteria like single factor evaluation. It can quantify the water quality, according to which intuitively identify the water quality grades of river and directly reflect the change of water quality in time and in space. The assessment was performed as follows: (1)calculating the WPI value of each parameter using interpolation method based on water quality grades and its corresponding WPI value (Table 2); (2)selecting the highest one as the WPI value of monitoring sites. I. If the classification of water quality is between class I and class V according to GB3838-2002, as to each variable except DO,

$$WPI(i) = WPI_{ij-1}(i) + \frac{WPI_{ij}(i) - WPI_{ij-1}(i)}{C_{ij}(i) - C_{ij-1}(i)}(C(i) - C_{ij-1}(i)) \quad (1)$$

$$C_{ij-1}(i) < C(i) \leq C_{ij}(i)$$

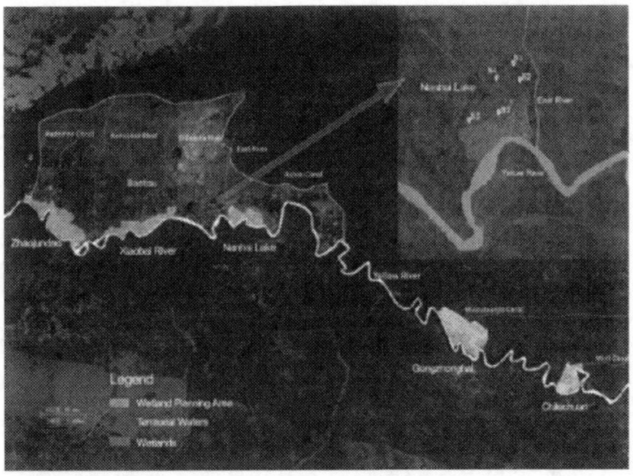

Figure 1. Baotou Yellow River wetland and monitoring sites of the Nanhai Lake

where $C(i)$ is the measured value of parameter i, $C_{ij-1}(i)$ and $C_{ij}(i)$ is the standard value of parameter i for class j-1and j, respectively, $WPI_{ij-1}(i)$ and $WPI_{ij}(i)$ is the standard WPI value of parameter i for class j-1and j, respectively.

In addition, if the standard value of two class is equal, the lower value interval is selected[9].

As to DO≥7.5mg/L, WPI=20; (2)

As to 2≤DO <7.5mg/L,

$$WPI(i) = WPI_{ij-1}(i) + \frac{WPI_{ij}(i) - WPI_{ij-1}(i)}{C_{ij-1}(i) - C_{ij}(i)}(C_{ij-1}(i) - C(i)) \quad (3)$$

II. If the classification of water quality is inferior to class V according to GB3838-2002, as to each variable except DO,

$$WPI(i) = 100 + \frac{C(i) - C_5(i)}{C_5(i)} \times 40 \quad (4)$$

As to DO<2mg/L,

$$WPI(DO) = 100 + \frac{2.0 - C(DO)}{2.0} \times 40 \quad (5)$$

III.As to W value of a monitoring site,

WPI=max(WPI(i)) (6)

2.3.2 Comprehensive water quality identification index method

The comprehensive water quality identification index is a relatively new method, which derives from Single factor water quality identification index. It evaluates general water quality both qualitatively and quantitatively, and also evaluates water quality inferior to V. it classifies general water quality conditions into five classes: type I , type II, type III, type IV, type V, inferior to class V and inferior to V and malodorous and black[6]. The assessment criteria (Table 2).

The water quality identification index comprises integral digits and three fractional digits. The structure is as follows:

$$I_{wq}=X_1.\ X_2X_3X_4 \quad (7)$$

The meaning and calculation method for each digit are as follows:

I. X_1 indicates the comprehensive classification of water quality, generally compared the monitor data with national standard data to determine the location of the

Table 1 measurement methods of parameters

parameters	measurement methods	apparatus
pH	Precise pH instrument	handy Precise pH instrument(AZ8601)
DO	Water analysis apparatus	multiparameter Water analysis apparatus(WTW Multi 340i, Germany)
TP	Ammonium molybdate spectrophotometric	UV rays visible spectrophotometer UV-3200(meipuda, Shanghai)
TN	Alkaline potassium persulfate method	UV rays visible spectrophotometer UV-3200(meipuda, Shanghai)
NH_4^+-N	Pay reagent luminosity law	UV rays visible spectrophotometer UV-3200(meipuda, Shanghai)
COD_{cr}	Potassium dichromate method	UV rays visible spectrophotometer UV-3200(meipuda,Shanghai)

Table 2 the classification standards for water quality

Comprehensive water quality identification index					Water pollution index	
Comprehensive water quality assessment standard		qualitatively assessment standard				
classification	$X_1.X_2$	$X_2\neq0$	$X_2=0$	conclusion	classification	WPI
I	(1.0, 2.0]	f- $X_1\geqslant1$	f- X_1-1$\geqslant$1	excellent	I	$\leqslant$20
II	(2.0, 3.0]	X_1=f	X_1-1=f	favorable	II	20-40
III	(3.0, 4.0]	X_1-f=1	X_1-f-1=1	Slightly pollution	III	40-60
IV	(4.0, 5.0]	X_1-f=2	X_1-f-1=2	Moderately pollution	IV	60-80
V	(5.0, 6.0]	X_1-$f\geqslant$3	X_1-f-1$\geqslant$3	Heavily pollution	V	80-100
Inferior V	(6.0, 7.0]				Inferior V	>100
Inferior V,malodorous black	>7.0					

monitoring value, thus obtaining the value of X_1. X_2 is the location of water quality within the variation range of Class X_1, so as to evaluate the water quality that in the same Class more specifically. $X_1.X_2$ can be calculated as follows:

(1)If the classification of water quality is between class I and class V according to GB383-2002,as to each variable except DO,

$$(X_1.X_2)_i=j+(C_i-C_{ij-1})/(C_{ij}-C_{ij-1}) \quad (8)$$

As to DO,

$$(X_1.X_2)_i=j+(C_{ij}-C_i)/(C_{ij}-C_{ij-1}) \quad (9)$$

(2) If the classification of water quality is inferior to class V according to GB3838 -2002, as to each variable except DO,

$$(X_1.X_2)_i=6+(C_i-C_{i5})/C_{i5} \quad (10)$$

As to DO,

$$(X_1.X_2)_i=6+(C_{i5}-C_i)/C_{i5} \quad (11)$$

where C_i is the measured value of parameter i, C_{ij} is the standard value of parameter i for class j, and $(X_1.X_2)_i$ is the single-factor water quality identification index of j =1,2,3,4,and 5 for classes I, II, III, IV,and V, respectively.

Finally, calculate the comprehensive water quality index $X_1.X_2$, The weight distribution of each single-factor water quality identification index influences the comprehensive water quality index, affecting the results and the accuracy of the assessment. Through improving and comparing methods previously used, the following three methods are initially selected [8].

$$\text{(A)} \quad X_1.X_2=\frac{1}{n}\sum_{i=1}^{n}(X_1.X_2)_i \quad (12)$$

Where n is the number of the single-factor water quality identification index.

$$\text{(B)}\ X_1.X_2=\frac{1}{2}((X_1.X_2)_a+(X_1.X_2)_{max}) \quad (13)$$

$$(X_1.X_2)_a=\frac{1}{n}\sum_{i=1}^{n}(X_1.X_2)_i \quad (14)$$

$$\text{(C)}\ X_1.X_2=\frac{1}{m+1}(\sum_{i=1}^{m}(X_1.X_2)_i+\frac{1}{n-m}\sum_{j=1}^{n-m}(X_1.X_2)_j) \quad (15)$$

where m is the total number of parameters that are inferior to the water quality standard for the region, and absolutely n-m is the total number of the rest parameters.

I. X_3 is the total number of parameters that do not meet the water quality standard for the region.

II. X_4 is used to identify whether the comprehensive water quality classification is inferior to the desired water quality standard for the region, usually is one or two significant digits. There are three kinds of situations:

(1) If the comprehensive water quality is inferior to the desired water quality standard and $X_2\neq0$

$$X_4=X_1-f \quad (16)$$

(2) If the comprehensive water quality is inferior to the desired water quality standard and X_2 =0

$$X_4=X_1-f-1 \quad (17)$$

(3) If the comprehensive water quality is superior or equal to the desired water quality standard

$$X_4=0 \quad (18)$$

where f is the desired water quality standard for the region.

3. Results and discussion

The assessment results determined using the water pollution index method showed that the water quality of all the five monitoring sites were inferior V(Table 3), the main contamination was COD in both the five monitoring sites because of high WPI (COD) which ranged from 176.8 to 244.2 as presented in fig.2. TN was another main contamination of the Nanhai Lake, varied from 109.6 to 153.4 over the year. Owing to more domestic sewage and industrial wastewater from surrounding residents and industry branches directly discharged into the Nanhai Lake[10,11].

From the time point of view, the degree of pollution of NH_4^+-N in wet season was slightly higher.This perhaps results from the rainfall in Aug. and Sep., and the pollution of agricultural area and other nonpoint source pollutants washed into the water body under the action of the rain[12]. However, the pollution of TN, TP and DO were higher in dry season because of the high demand for nitrogen and phosphorus in wet season, the high growth period of aquatic plants, the nutrient elements and DO are fully utilized, and the water was purified to some extends.

The comprehensive water quality index X_1. X_2 was calculated as listed in table3 in three methods. It is observed that X_1. X_2 calculated by method1 was smallest because of the same weight of all the parameters. Calculated by method2 was biggest due to the too much emphasis on parameters inferior to the water quality standard.So the method3 was selected to determine the comprehensive water quality index, but if all the parameters were inferior to the water quality standard, select the method 1 instead.

The comprehensive water quality identification index showed that the water quality of the Nanhai Lake were classes V, except for the N_2 in wet season and dry season , the N_1 in dry season and the N_5 in normal season, which were classes inferior V. The five monitoring sites were't achieving the desired water quality standard. But single-factor index of some parameters reached, so the standard achieving rate of each monitoring sites in each flow season and the comprehensive standard achieving rate of the Nanhai Lake in each flow season were calculated, it's easy to find that the comprehensive standard achieving rate of normal season>wet season>dry season. In wet season, the Yellow river replenishing water to Nanhai Lake,contaminations were diluted and the water quality was better than dry season[10], but the rainfall in wet season made pollutants washed into the water body result in the water quality worse than normal season. In addition, qualitatively assessment of water quality was basically consistent with the quantitatively assessment.

It is clear that the domestic discharge, industrial activities and developed recreation were major threats to water quality of Nanhai Lake. It's essential to take some measures to protect the Nanhai Lake from domestic sewage, industrial wastewater and solid waste from tourists. In addition, plants can consume nitrogen, phosphorus and organic matter, result in the water quality improved, so rehabilitate of wetlands is a vital measures.

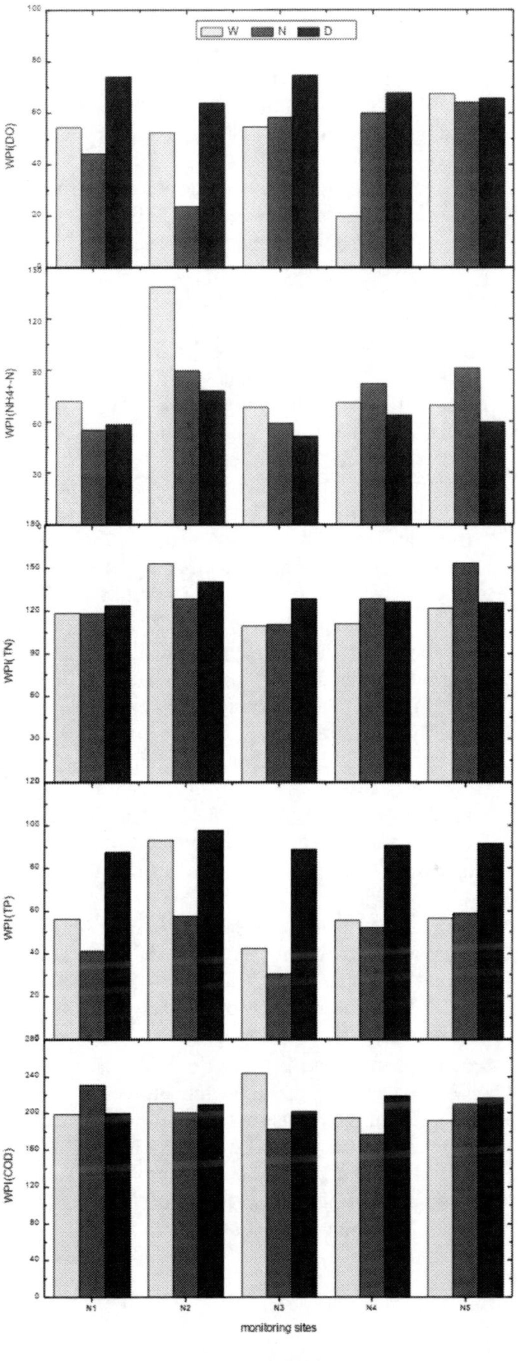

Figure 2. The WPI values of DO, NH_4^+-N, TN, TP and CODcr

4. Conclusions

(1) COD and TN were the main contaminations of the Nanhai Lake, result from domestic discharge, industrial activities and developed recreation.

(2)The water quality classes of the five monitoring sites of the Nanhai Lake were inferior Vbut not malodorous and black, neither of them achieved the desired water quality standard. Single-factor index of some parameters reached, the comprehensive standard achieving rate of normal season>wet season>dry season.

(3) Water pollution index method is simply calculated but can't assesses the water quality inferior V specifically. Comprehensive water quality identification index method can be applied to judge whether the water quality meets environmental water functions and suitable for assessment of the overall water quality when the water is seriously polluted.

References

1. XL Ji , Randy A. et al.Comparison of seven water quality assessment methods for thecharacterization and management of highly impaired river systems. Envi- ron Monit Assess. **188**:15 (2016)
2. W Hou, et al. Assessing water quality of five typical reservoirs in lower reaches of Yellow River, China:Using a water quality index method. Eco Indi, **61**, 309- 316 (2016)
3. Chow-Fraser, P.,Development of the Water Quality Index to assess effects of basin-wide land-use alteration on coastal marshes of the Laurentian Great Lakes. In: Coastal Wetlands of the Laurentian Great Lakes: Health, Habi Indi, pp. 137–185.(2006)
4. Liu, Z.,Sun, G.,et al, Water quality index as a simple indicator of drinking water source in the Dongjiang River, China. Int. J.Environ.Prot. **2**, 16–21.(2012).
5. Bordalo,A.A.,et.al,A water quality index applied to an international shared river basin: the case of the Douro River. Environ.Manag.**38**, 910–920. (2006).
6. ZX，Xu. Comprehensive Water Quality Identificati- on Index for Environmental Quality Assessment of Surface Water. Jour. of TongJi uni..**33** 4(2005)
7. C Hu,et al. Application of comprehensive water quality identification index in water quality assessment of Hun River. Eco. Environ. Sci. **20** 1 186-192 (2011)
8. H Zhang, et al. Application of comprehensive water quality identification index method to water quality assessment of park lakes. Jour. of Water Res & Water Engin. **23** 2 (2012)
9. Yan Liu et al. Application of Water Pollution Index in Water Quality Assess-ment of Rivers. Environ Monit in China..**29** 3 (2013)
10. CL Miao, The research of Nanhai wetland of Inner Mongolica Baotou city protection and exploitati- on.[D] Northwest Agric. & Fore. Uni. （2008）
11. Wei Yu et.al. Strengthen the protection and man- agement of the Nanhai wetland to play the ecological function. Journ. of Inner Mongolia Fore.(2007)
12. CL Miao, Biodiversity conservation and sustainable utilization of Nanhai wetla-nd of Baotou. Journ. of Inner Mongolia Fore.(2007)

Table 3 Water quality assessment results comparing the two assessment methods

flow sea-sons	moni-toring sites	desired water quality classes	Comprehensive water quality identification index													Water pollution index method		
			$(X_1.X_2)_i$					$X_1.X_2$			I_{wq}	Water quality classes	Standard achieving rates (%)		qualitatively assessment	WPI	Water quality classes	main pollution parameter
			DO	NH_4^+-N	TN	TP	COD_{cr}	Method 1	Method 2	Method 3								
W	N_1	IV	3.7	4.6	6.5	3.8	8.5	5.4	6.9	5.8	5.831	V	40	32	Slightly polluted	199.3	Inferior V	COD
	N_2	IV	3.6	7.0	7.3	5.6	8.8	6.5	7.6	6.5	6.542	Inferior V	20		Moderately polluted	211.6	Inferior V	COD
	N_3	IV	3.7	4.4	6.2	3.1	9.6	5.4	7.5	5.9	5.931	V	40		Slightly polluted	244.1	Inferior V	COD
	N_4	IV	1.0	4.6	6.3	3.8	8.4	4.8	6.6	5.4	5.431	V	40		Slightly polluted	195.4	Inferior V	COD
	N_5	IV	4.4	4.5	6.5	3.8	8.3	5.5	6.9	5.5	5.541	V	20		Slightly polluted	192.5	Inferior V	COD
N	N_1	IV	3.2	3.8	6.5	3.1	9.3	5.2	7.2	6.4	6.422	Inferior V	60	44	Moderately polluted	230.8	Inferior V	COD
	N_2	IV	2.2	5.5	6.7	3.9	8.5	5.4	6.9	5.9	5.931	V	40		Slightly polluted	201.2	Inferior V	COD
	N_3	IV	3.9	3.9	6.3	2.5	8.1	4.9	6.5	5.9	5.921	V	60		Slightly polluted	182.8	Inferior V	COD
	N_4	IV	4.0	5.1	6.7	3.6	7.9	5.5	6.7	5.9	5.931	V	40		Slightly polluted	176.8	Inferior V	COD
	N_5	IV	4.2	5.6	7.3	3.9	8.8	6.0	7.4	6.0	6.041	Inferior V	20		Slightly polluted	210.4	Inferior V	COD
D	N_1	IV	4.7	3.9	6.6	5.4	8.5	5.8	7.2	5.8	5.841	V	20	12	Slightly polluted	200.4	Inferior V	COD
	N_2	IV	4.2	4.9	7.0	5.9	8.8	6.1	7.4	5.1	6.152	Inferior V	0		Moderately polluted	210.1	Inferior V	COD
	N_3	IV	4.7	3.6	6.7	5.4	8.6	5.8	7.2	5.8	5.841	V	20		Slightly polluted	202.8	Inferior V	COD
	N_4	IV	4.4	4.2	6.7	5.5	9.0	5.9	7.5	5.0	5.951	V	0		Slightly polluted	219.4	Inferior V	COD
	N_5	IV	4.3	4.0	6.6	5.6	8.9	5.9	7.4	5.9	5.941	V	20		Slightly polluted	217.0	Inferior V	COD

Analysis of the origin of soluble COD in Lake Sanaru

Qian Yu[1], AyakaOomae[1],Hiroyuki Futamata[1]and Satoshi Matsuda[1]

[1]*Faculty of Engineering, Shizuoka University, 3-5-1 Johoku, Naka-ku, Hamamatsu 432-8561 Japan*

Abstract. This study intends to investigate the origin of Sol-COD by analyzing the water quality data and the amount of detergent consumption in the Lake Sanaru region,located at Hamamatsu City in Japan. It was found that there is a correlation between the total COD in the lake and the sewer coverage in the upstream region of Lake Sanaru. This fact implies that the water quality of upstream rivers receiving untreated wastewater from homes would greatly affect the water quality of the lake. Since the value of Sol-COD in the rivers is close to that of Lake Sanaru, the main origin of the Sol-COD was supposed to be the organic components in the river water especially those of detergents, because the Sol-CODof detergent didn't decompose completely.Thus, the amount of detergent emitted in the upstreamlakeregion was determined using the public data of the pollution from the sewer system and the national average of detergent consumption. The calculated value of the detergent concentration agreed very well with the one measured in the upstream rivers, meaning that detergents strongly affected water quality in upstream rivers and also the majority of Sol-COD in Lake Sanaru.

1 Introduction

The Total-COD can be subdivided into several parts according to wastewater characteristic because in many instances wastewatersare very different[1,2].Theclassification of CODhas often been used inthe design of technology forsewage treatment plants[3,4] butnot much in regards to natural bodies of water.We consider that total COD consists of two parts in general:suspended solid COD (SS-COD) and water-soluble COD (Sol-COD). This classification can give us a much clearer conception of water quality variation including seasonal change and the origin of effluents.

Lake Sanaru is located in the western area of Hamamatsu City, Shizuoka Prefecture in Japan.Since 1999, the annual average of Total-COD value reached 12mg/L and it was the worst ranking value of national public water quality in Japan from 2001 to 2006. In recent years, along with the continuously increasing sewer coverage area in the lake region, the amount of nutrients in upstream rivers has decreased. Also, the Total-COD value is no longer in the top five for worst public water quality in Japan[5].We want to clearly define how the Total-COD decreased and use this information as guidance for the governance of other lakes.First and foremost, the origin of pollution in the lake as well as the condition of SS-COD and Sol-COD in the upstream rivers need to be clarified.

Based on Japan's experience of tackling pollution since the 1970s,strict standard setting and financial support for environmental protection have had a more positive effect[6] than in other countries.We can collect a large quantity of data due to strong monitoring capacities.In this study we focus on sewer systems,population distribution, basin environment, point sourcesand nonpoint sources to discover the primary originof pollution in Lake Sanaru.

2 Method

2.1 Study area

2.1.1Basin environment

The average water depth of Lake Sanaruis 2 m. The lake area is 1.2 km^2, and the volume is 2.4 million m^3. There are three main upstream rivers flowing into the lake. Lake Sanaru is a brackish tidal lake, although it is located far from the seashore, and is affected by organic matter and salt water from one 13 km downstream river which flows backward. There are 2 exits, which flow from the lake;refer to the map in Figure 1.

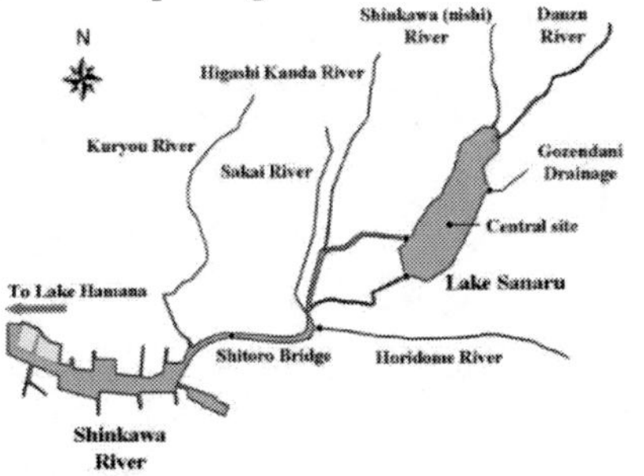

Figure 1.A map around Lake Sanaru

With an average depth of just 2 meters, Lake Sanaru is considered to be in an aerobic condition on the lake bottom. The nutrient release from sediment is not thought to be very large. As a characteristic of a brackish lake, the lake level due to tides has ± 30~40 cm variations, and the amount of water volume fluctuates by 3.6×10^5-5.0×10^5 m^3 per day[7]. However, nutrients like soluble phosphorus in the downstream river are almost the same as in the lake and the average concentration from the three upstream rivers. That means the nutrient transfer and water exchange between Lake Sanaru and the downstream river is not so large. So the main pollution load contributing to the amount of nutrients comes from the upstream rivers. In addition to the weak self-purification water environment, once the pollution flows into the lake, it can promote internal production, like the abnormal growth of phytoplankton. All of these conditions are not conducive to water purification in Lake Sanaru.

2.1.2 Analysis of pollution origin

The catchment zone of Lake Sanaru is higher on the north side than that the south, and the area of the upstream rivers region is 17.67 km^2. Farmland area covers only 14% of that area. After the comprehensivesewerage program was completed in the lake region, the foul sewer coverage rate increased from 77% in 2001 to 93% in 2012.It was found that there was a rough correlation between the COD in the lake and the sewer coverage in the upstream region of Lake Sanaru (Figure2).With the increase of foul sewer coverage rate, both total COD and Sol-COD decreased in the lake. Contrastingly, Sol-COD has not decreased at the same level as the Total-COD. The most direct and knowable effect relates to inflow from the river. Although we wantto compare Sol-COD directly in rivers and in the lake, the water quality monitoring data only tested for Total-COD in the three rivers. After working to improve water quality in the rivers for many years, Total-COD is at 2 to 4mg/L in three of the rivers, so they are not so dirty. So we compared the Total-COD in the three upstream rivers and Sol-COD in the lake. It was observed that both of them have a similar fluctuation (Figure3). This fact implies that the main cause of the COD value decrease canbe attributed to the increased coverage of the sewer system.In particular, Sol-COD in the lake is greatly affected by the upstream rivers' pollution load.Therefore, the origins

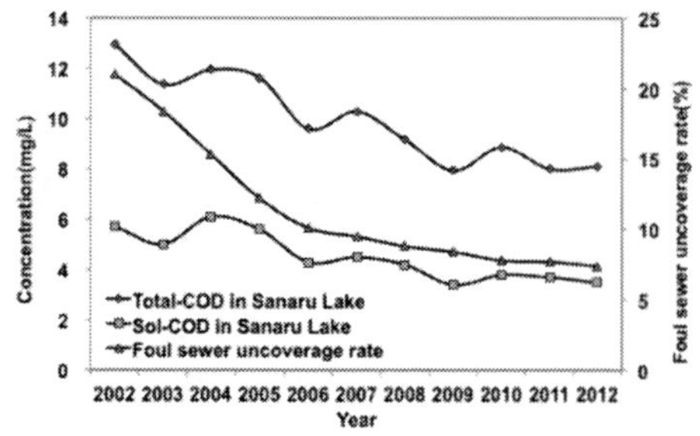

Figure 2. The change of COD and foul sewer coverage rate in the Lake Sanaru region

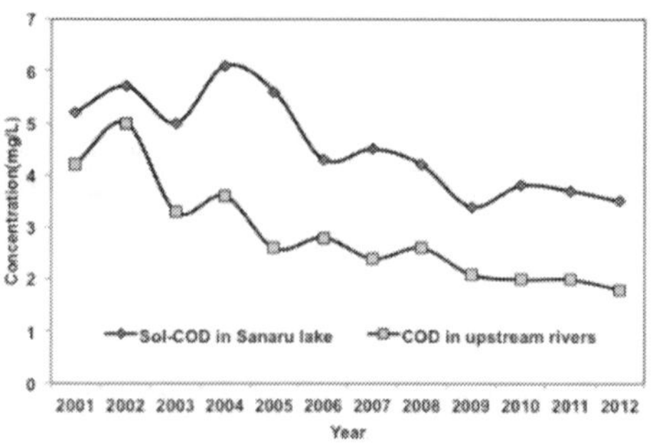

Figure 3. The change of Sol-COD in Lake Sanaru and COD in the upstream rivers

of Sol-COD inthe upstream rivers need to be clarified.We focus on the point resource aboutdomestic sewage.In addition, because Sol-COD in Lake Sanaru is larger than in the upstream rivers by almost 2 mg/L, whether or not the difference between these two numbers comes from the decomposition of algae in Lake Sanaru also needs to be clarified.

Generally, in daily life, the domestic sewage discharge from a kitchen is 17%, the bath is 25%, laundry is 30%, hands and face washing is 8%, and other activities is 10%. Detergent isthe most important pollution contained in domestic sewage. Although the load of domestic wastewater has been reduced due to the development of the sewage system, the pollution load comes from theuncovered sewer area still flowing into the river. These untreated runoffsfoundin the lake can finally be assumed to be difficult-to-decompose organic substances. In order to verify this assumption, we need to check the decomposability of detergent, and to validate whether or not it contributes to the real COD valuein the water.

2.2Detergent consumption and emission

Figure 4 is the percentage figure of detergent consumption in Japan[8]. Detergent used for laundry (liquid and powder) accounts for 47% of the total detergent consumption. Combined with the next highest seller, kitchen detergent, kitchen and laundry detergent account for 60% of all detergent consumption. In recent years, people like to use liquid laundry detergent rather than powder laundry detergent, but the total laundry

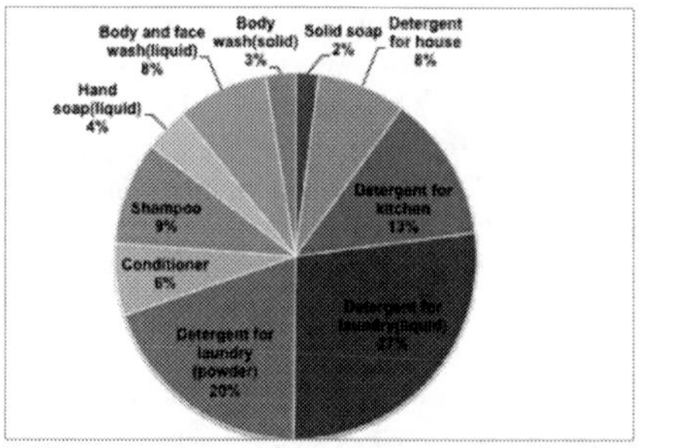

Figure 4. The percentages of detergent consumption in Japan

DOI: 10.1051/matecconf/20166002005

detergent market in Japan has remained at almost the same level since 2002 in terms of sales. Also, other detergent consumption has mostly kept steady; only hand-soap sales showed an increase. According to the condition of Japanese detergent consumption, we can choose the most widely used detergent as our detergent solution samples.

Fromformer analysis of the Lake Sanaruregion, we know that the main cause of decrease in COD value is attributable to the increased coverage of sewer systems. The water quality of upstream rivers receiving untreated wastewater from homes would greatly affect the water quality of Sanaru Lake. Thus, in order to estimate the concentration of the detergent in the lake, we need to know how many people are outside of the sewer system and how many detergents are used and flow into the river. Eq. (1) calculates the detergent concentration with public data on the population outside of the sewer system in Sanaru Lake upstream region and the national average of detergent consumption[6,8].

$$C=\frac{Dc \times Po}{T_f} \quad (1)$$

Where C is the detergent concentration in the upstream rivers (g/L).*Dc* is the detergent consumption per head (g/day · person). P*o* is the population outside of the foul sewer coverage (person). T_f is the total influx of the upstream rivers (m^3/day).

2.3Detergent solution in rivers

2.3.1Biodegradation of detergent

In order to clarify whether the detergent will decompose in river water easily over time, a certain amount of detergent solution was put into river water and kept in a static condition until it reached a steady state. Water quality tests were completed at regular intervals. The experiment was carried out in open air conditions. The detergent solution in the experiment was prepared by the calculation of Eq. (1) for several detergents.Ordinarily, COD_{Mn} is used as the indicator for water quality in Japanbecause dichromate, used in Western countries, is itself a heavy metal pollutant[9]. However, owning to the weak oxidation ability of potassium permanganate and the poor additive, TOC is mainly used in our experiment to express the value of dissolved organic matter more exactly[10], especially for the majority of organic substances in detergent that are difficult to decompose.

2.3.2 Field survey

Samples were taken twice from six sites along the three upstream rivers (sites A-F, Figure5) on January 10 and January 19, 2012. Two days were sunny. Filled black circles represent sites A-F. The A and B sites are in Gozendani Drainage. The C and D sites are in Danzukawa River. The E and F sites are in Shinkawa River. Samplings were taken by bottle, and then immediately filtered and determined for TOC and anionic surfactants using standard measurement method.

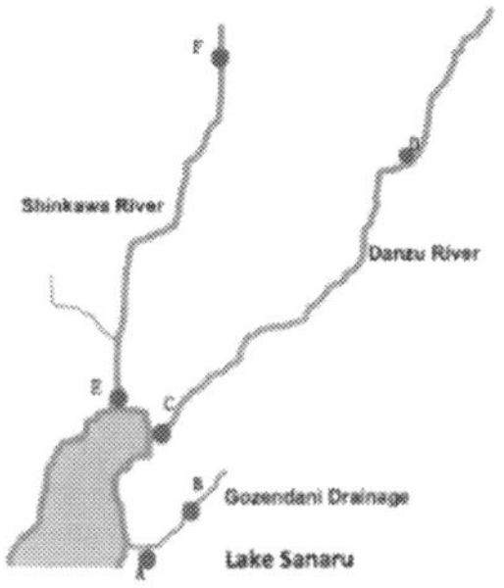

Figure 5. Sampling sites of Lake Sanaru

3Results

3.1Detergent consumption

By the calculation of Eq. (1), we get the detergent concentration from different years. The result is shown in Figure6, as the foul sewer coverage rate increased in recent years, more and more discharge from population was treated. With the type of laundry change, only the liquid laundry detergent increased, other detergentsconcentrations were significantly reduced. The total concentration of the eight detergents in the upstream rivers was 2.7 mg/L in 2012 (Table 1). Because the detergent and soap included large amounts of organic

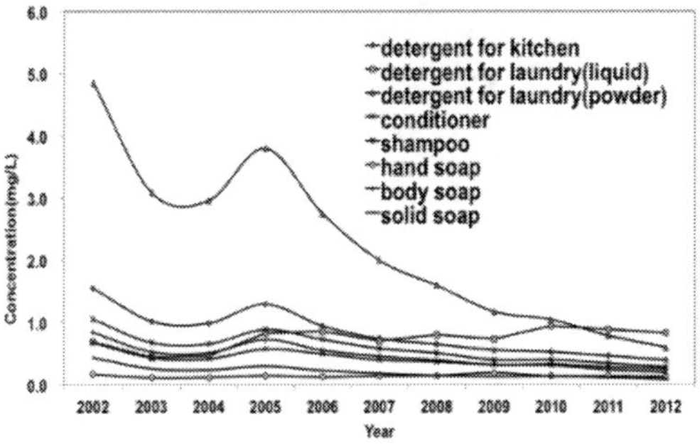

Figure6. Detergent concentrations in the rivers, which are calculated by Eq. (1)

Table 1. Concentration of detergent in solution by calculattion

Kind of detergent	Concentration (mg/L)
Detergent for kitchen (liquid)	0.39
Detergent for laundry (liquid)	0.82
Detergent for laundry (powder)	0.59
Conditioner	0.19
Shampoo	0.28
Hand soap	0.12
Body soap	0.24
Solid soap	0.08
Total	2.70

matter, such as surfactants and fatty acids, in detergent is 40% percent carbon, TOC of the detergent solution can be estimated to be 1.1mg/L. When we tested the TOC of the detergent solution that was made according to the concentration of Table 1, the measured value was 0.94 mg/L, which showed a good agreement.

3.2Detergent solution in rivers

3.2.1Biodegradation of detergent

As shown in Table 1, eight kinds of detergent were accurately weighed and dissolved in 1 L of water. The detergent and water were mixed, and 1 ml of the resulting detergent solution was dissolved in 1 L of river water and 1 L water. For a blank test, 1 L river water was used. These three solutions were kept in a static condition until they reached a steady state; TOC of the three solutions was tested every 12 hours. The results are shown in Figure 7. River water was taken from the upper reaches of Danzukawa River and the total TOC was 6.08 mg/L. TOC of 2.7 mg/L detergent solution is 0.94 mg/L. For the detergent solution added to the river water, total TOC was 6.85 mg/L. A small part of detergent may have decomposed immediately after addition to the river water. At first, the detergent concentration decreased quickly. After three days, nearly the same value remained. The detergent did not decompose completely. There was 0.32 mg/L left. This part of the detergent would eventually flow into the lake.

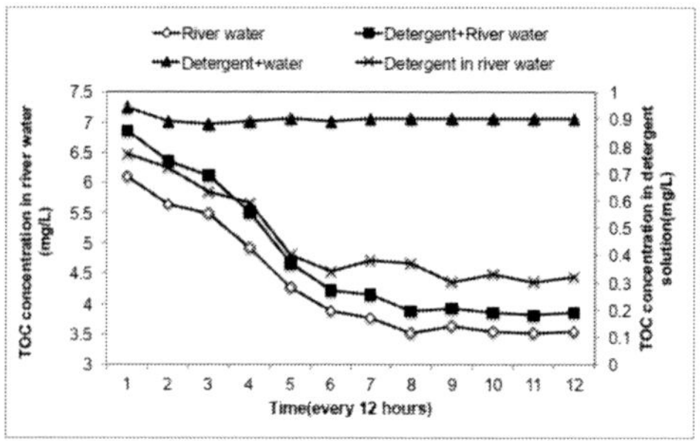

Figure 7. The variation ofdetergent concentration

3.2.2 Field survey

The values of TOC, COD_{Mn}, nitrogen and surfactants in each site were shown in Table 2. For all water samples, values in the downstream sites A and E were larger than in the upstream sites B and F. Along with the river flow, more pollution substances flow into the river. However, values in the Danzukawa River at the D site were higher than at the C site, because the D site samples were taken in an irrigation channel, not in a natural river. The nitrogen pollution is serious in Shinkawa(nishi)River. There is a positive correlation between TOC and the anionic surfactant concentration (Figure 8). Therefore, detergents strongly affect TOC in Lake Sanaru upstream rivers. We calculated the average TOC in the three rivers, and then obtained a value of 1.2mg/L, which was almost the same as the experiment result. The detailed calculated course is shown in Table 3. The value of TOC was the average value of two tests in the downstream sites.

Table 2. Test results in 3 rivers (mg/L)

Site	TOC	Surfactant	COD_{Mn}	NO^{2-}	NO^{3-}
A	1.1	0.10	3.5	0.02	8.3
B	0.2	0.05	3.5	0	8.5
C	1.7	0.15	3.5	0.02	8.3
D	3.5	0.45	5.0	0.35	14.0
E	0.9	0.10	3.0	0	20.0
F	0.1	0.05	3.0	0	20.0
(First time)					
Site	TOC	Surfactant	COD_{Mn}	NO^{2-}	NO^{3-}
A	0.8	0.25	2.5	0.02	7.3
B	0.1	0.10	2.5	0	7.5
C	1.2	0.15	3.5	0.03	7.2
D	5.3	0.50	5.5	0.70	13.0
E	0.5	0.10	2.5	0	25.0
F	0.0	0.05	2.0	0	25.0
(Second time)					

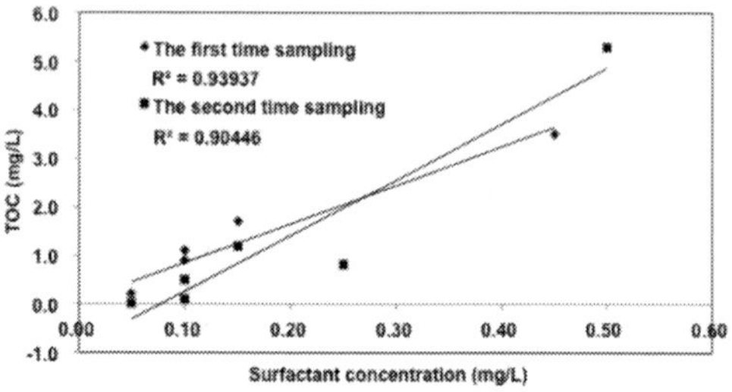

Figure 8 Relationship with TOC and surfactant concentration in the upstream rivers

Table 3. Flow rate and TOC in the three rivers

River	Average annual flow (m^3/s)	TOC (mg/L)
Gozendani Drainage	0.01	1.0
Danzu River	0.29	1.5
Shinkawa River	0.19	0.7

$$\text{Average TOC}=\frac{0.01\times1.0}{0.49}+\frac{0.29\times1.5}{0.49}+\frac{0.19\times0.7}{0.49}=1.2$$

4 Discussion

4.1Compare the result

In our research, we obtained three TOC values through different methods (Table 4). In the first method, the TOC value was estimated by the detergent concentration that was calculated by Eq.(1), and the result of TOC was 1.1mg/L. In the second method, we weighted eight kinds of detergent and diluted them to the concentrations that were calculated by Eq.(1) for the needed detergent solution, and its TOC was 0.94 mg/L. In the third method, according to the calculation with flow rate and field survey of TOC values in the samples from three rivers, average TOC concentration was determined to be 1.2 mg/L. We compared the results of the three methods and

Table 4. TOC results compared in different way (mg/L)

TOC by estimate from detergent concentration	TOC test in detergent solution	TOC sampling test and calculation in all rivers
1.1	0.94	1.2

all of them were roughly the same. The result suggested that TOC in the upstream rivers mostly comes from detergent.

4.2The origin of other Sol-COD

We have known that there is a similar fluctuation change of Sol-COD in the rivers and the lake, but they have about a 2 mg/L difference between them. In order to prove the origin of this part, water from Lake Sanaru was treated with 40 kHz ultrasonic waves to disrupt algae cells.The color of the lake water (pale yellow) did not change much after the use of the ultrasonic wave machine, but the Sol-COD value changed from 4 mg/L to 6 mg/L. Thus, after applying the ultrasonic wave to the water, the Sol-COD components of about 2 mg/L eluted from the carcasses of the algae (Table 5). This is the part of Lake Sanaru which internally generated content.

Table 5.COD internally generated (mg/L)

From the difference between rivers and lake	From algae carcass in the experiment
1~2	1~2

5Conclusions

In this study, we clarified the origin of Soluble COD in Lake Sanaru.In the upstream rivers, there is a positive correlation between the anionic surfactant concentration and the TOC in the water. In addition, the TOC of detergent solution was the same as the actual measurement in the upstream rivers. This indicated that the origin of Sol-COD in Lake Sanaru came from detergent in domestic sewage. The same approachcan also be applied to other lakes'governance and estimationof pollution concentration especially detergents.Although we found good agreement between the experiment and field survey, twice samplings were obtained on sunny days; the nonpoint pollution effect from farmland and streets didn't show clearly in this research. Therefore, future studyshould be done on a rainy day. In the future, the way to reduce Sol-COD is improve the sewer connection rate in the Lake Sanaru region. If the sewer connection rate is completely improved, organic matter concentration in the river could be greatly reduced. In addition, with the sewer connection completed, the phosphorus in the upstream rivers also will be greatly reduced, so that phytoplankton amount in the lake would also be inhibited. Thus, Total-COD in Lake Sanaru could possibly be below 5 mg/L. But one problem to which attention must be paid is that the total river flux will also be decreased, owning to a sewer connection rate increase. This may influence the water exchange as well as the phytoplankton growth and fish catch. Therefore, it is necessary to discuss whether or not the value of Total-COD in Lake Sanaru can achieve expectations. Lake Sanaru is a brackish tidal lake affected by organic matter from one downstream river which flows backward. It can be assumed that the tidal water exchange between Lake Sanaru and Lake Hamana must control or affect internal generation by being mixed in the dilution process in the lake and downstream Shinkawa River. Attention needs to be paid to this part of the research in the future.

References

1. H. Melcer, P. L. Dold, R. M. Jones,C. M. Bye, I.Takacs, H. D.Stensel, A. Warren Wilson, Paul Sun, Scott Bury*Methods for Wastewater Characterization in Activated Sludge Modeling*(WERF, 2003)
2. C. Köhler, *COD fraction dynamics: Respirometric analysis &modeling sewer processes* (Diplomarbeit to reach the degree of a Diplomingenieur of Water Management Academic year 2007-2008)
3. F. Raposo, M.A. de la Rubia, R. Borja, M. Alaiz, J. Talanta**76**, 448–453 (2008)
4. N. H. Tran, H. H. Ngo, T. Urase, K. Y.H. Gin, J. Bior Tech**193**, 523–533 (2015).
5. Hamamstsu, Shizuoka Prefecture public waters water quality measurement results (2001-2014). (http://www.hamamatsu-kankyo.jp/suishitsu/district/naka_situation.htm)
6. K. Otsuka, K. Fujita, Y. Isono, and M. Mizuochi, *Governance for Water Environment Conservation: Implications from Japanese Experiences*(2005)
7. K. Yasuda, M. Iga and S.Sawamoto,J. The School of Marine Science and Technology, Vol.8,No.1, 13–25(2010)
8. Japan Soap and Detergent Association statistics (2002-2014) (http://jsda.org/w/index.html)
9. S. Araki, M. Numata, O. Wada, *Kankyokagakuziten (A Dictionary of Environmental Sciences)Kagakudozin, Tokyo* (1985)
10. G. Visco, L. Campanella, V. Nobili, J. Microc**79**, 185–191(2005)

MATEC Web of Conferences 60, 02006 (2016) DOI: 10.1051/matecconf/20166002006
ICCBS 2016

Ensemble Trajectory Simulation of Large Jellyfish in the Yellow and Bohai Sea

Lingjuan Wu , Jiangling Xu

North China Sea Marine Forecast Center of State Oceanic Administration,266061,Qingdao, China,

Abstract. Combining Lagrange particle tracking method with the ensemble forecasting concept, one ensemble trajectory model is developed for large jellyfish with vertical movement of jellyfish taken into account in the Yellow and Bohai Sea. Besides, this ensemble model can give quick simulation of drifting trajectories and potential affected range. The comparison between observation and simulation of jellyfish trajectories in the coastal waters of Qingdao and Qinhuangdao as the typical place of the Yellow Sea and Bohai Sea has been made. And the results indicate that the ensemble trajectory model is more reliable than single one, and could provide more useful information for emergency response of jellyfish bloom, especially under the circumstance that autonomic (vertical) movement of of jellyfish and its mechanism is not clear.

1 Introduction

Due to the global change and human activities, the occurrence frequency and disaster category of large jellyfish increase continuously [1-3]. In recent years, jellyfish has been blooming in the coastal waters of Qingdao and Qinhuangdao including dominant species such as Nemopilema nomurai, Aurelia aurita and so on. According to their occurrence mechanism, these jellyfish could be divided into two types: one is foreign type and the other is local type. The recent observation shows that Aurelia aurita may belong to the local type since they mainly distribute in Jiaozhou Bay and Qinhuangdao coastal sea, replenish themselves and complete their life history; Nemopilema nomurai and Cyanea nozakii may belong to the foreign type, because their larvae were not found in these places during the past winter and spring. That agrees with the observation and study of Wang et al. and Sun et al.[4-5]. Here, only large jellyfish of foreign type is discussed in the following section.

Scientists analyze the key dynamic mechanism of jellyfish drift and accumulation with numerical models. Barz et al.[6] found the advection process played an important role in the distribution of jellyfish in the central Baltic Sea. Without considering tidal movement in ROMS (Regional Ocean Model System), Moon et al.[7] showed the jellyfish distribution was affected by the coastal current and wind stress between East China Sea and Japan Sea. However, Luo et al.[8] opposed Moon's view with considering the tidal movement in ocean model. Johnson et al.[9] indicated that the seasonal variation of circulation played an important role in the distribution and abundance of jellyfish.
Graham et al.[10] showed the 'Diurnal Vertical Migration' (DVM) phenomena was common in jellyfish of hydromedusae, Scyphozoa and Siphonophorae. However, in 2006 Yellow Sea survey cruises, Zhang et al. [11] found that the number changes of jellyfish in 24 hours is not caused by DVM. During 2006~2007 bottom trawl survey, large jellyfish appeared in many samples of surface or subsurface layers. In the jellyfish survey of Jiaozhou Bay, DVM is not obvious from August to September. In southern Japan Sea, Nemopilema nomurai mainly distributed in the waters shallower than 40m, and its distributed depth at night is often greater than that at day. According to the observation of 2011~2013 near the coastal waters of Qingdao and Qinhuangdao, jellyfish are found on the surface at day. Therefore, the autonomic movement of different or the same kind of jellyfish under different conditions is not the same, which may be related to light or not. The mechanism of jellyfish autonomic movement is vague. The main distribution depth of jellyfish is not the same. Hence, with reference to the ensemble forecasting concept, an independent ensemble model of large jellyfish would be developed to simulate its trajectory quickly.

2 Model introduction and configuration, data source

ROMS is a three-dimensional, free-surface, terrain-following numerical model that developed as a community model by Rutgers University and has been widely used by the scientific community for a diverse range of applications. According to the location of large jellyfish and the forecasting demand for emergency response, three sub-region dynamic ROMS models are established by using the multi-nested technique (Fig.1).

Big, middle and small region models cover northwestern Pacific, Yellow and Bohai Sea, and the coastal waters of Qingdao and Qinhuangdao with increasing resolution, respectively. The surface forcing parameters are supplied by operational Weather Research and Forecasting Model (WRF). The initial and boundary fields of big-region model such as elevation, current, temperature and salinity are obtained from the operational forecasting dataset (HYCOM+NCODA Global 1/12° Analysis). The tidal influence is definitely included in our small-region model by adding the tidal elevation to the sea level height at the open boundaries.

The observation of large jellyfish mainly applies the following methods: fisheries survey, trawl monitoring, visual observation underway period, monitoring by airplane, underwater camera monitoring, sonar monitoring.

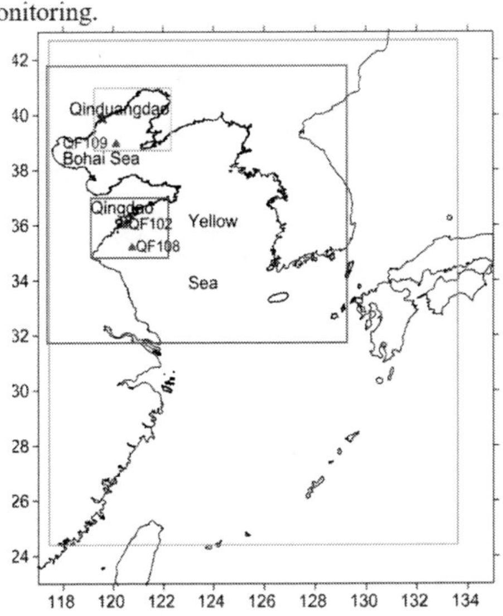

Figure 1. Nested domain of model (*purple, red and pink boxes* represent the domain of the Yellow and Bohai Sea, the coastal waters of Qingdao and Qinhuangdao, respectively; red triangles represent QF102, QF108 and QF109 buoy).

3 Ensemble trajectory schemes for large jellyfish

Because of the characteristics of jellyfish autonomic movement, independent ensemble trajectory model of large jellyfish is developed to simulate its trajectory based on Lagrange particle tracking method and WRF and ROMS results. The WRF and ROMS results is the operational forecast results of marine and atmosphere environment and could be obtained at fixed time every day. Once large jellyfish was found, this model would simulate the ensemble trajectories in short time.

With the vertical movement of large jellyfish taken into account, seven ensemble schemes are designed for Qingdao. Since the water depth near the coastal waters of Qingdao is about 20m, and the jellyfish maybe distribute between surface and middle layer, the vertical motion amplitude of jellyfish is set to half of the depth and vertical depth from the seabed is not more than 5m.

1. Without vertical diurnal movement, particles initial released in 0.5m
2. Without vertical diurnal movement, particles initial released in 1.5m
3. Particles released in 0.5m, sinked and located at about half of depth during day (6:00~18:00) and floated and located at surface during night (18:00~6:00)
4. Same as scheme 3 except with particles released in 1.5m
5. Particles released in 0.5m, floated and located at about half of the depth during day (6:00~18:00) and sinked and located at surface during night (18:00~6:00)
6. Same as scheme 5 except with particles release in 1.5m
7. Same as scheme 3 except with located at deeper than half of the depth

Since the water depth there is relatively shallow, only the first five identical schemes are implemented for Qinhuangdao.

4 Meteorological and marine model validation

The observation of QF102 in the Yellow and Bohai Sea such as wind at 10 meter and ocean current agrees well with the model results (Fig. 2~3). So does the QF108 and QF109 buoy (figure omitted). For QF102 and QF108 buoy near Qingdao, the forecasting errors of wind speed (direction) of 48h is less than 1.77m/s (35.2°), the (relative) forecasting errors of ocean current speed (direction) of 48h is less than 19% (28°) (Tab. omitted). For QF109 buoy near Qinhuangdao, the 48h forecasting errors of wind speed (direction) is less than 1.61m/s (19.2°), the (relative) 48h forecasting errors of ocean current speed (direction) is less than 13% (17°) (Tab. omitted). The results indicate that the meteorological and marine simulated results are reasonable and reliable in the Yellow and Bohai Sea.

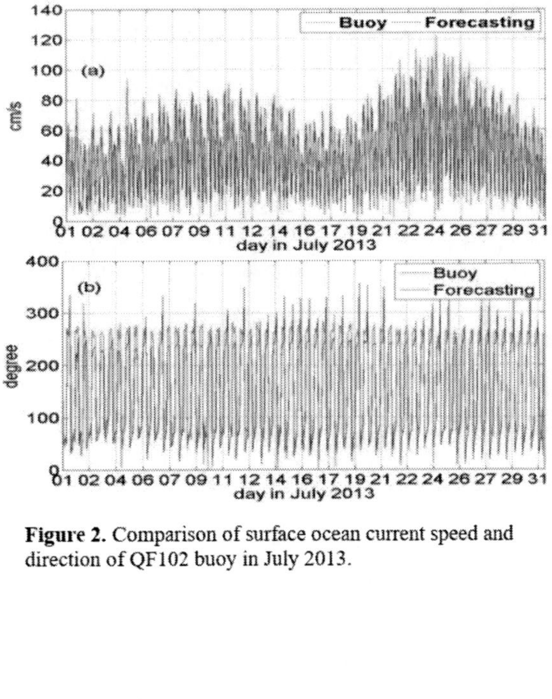

Figure 2. Comparison of surface ocean current speed and direction of QF102 buoy in July 2013.

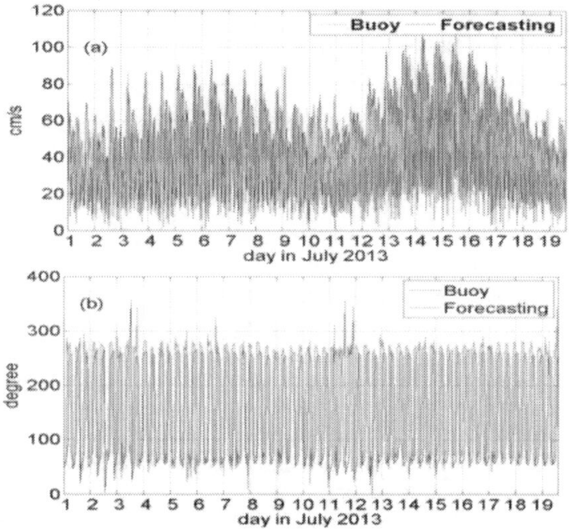

Figure 3. Same as Fig. 2 except with ocean current at 10m.

5 Ensemble Trajectory Simulation Results and Validation

5.1 Simulation in the Coastal Waters of Qingdao

With the observation of large jellyfish (Nemopilema nomurai) on July 5th 2012 as the initial state (Fig. 4, green triangle) , the ensemble model of jellyfish could simulate the trajectory (Fig.4, blue line), velocity, trend, as well as influenced range (Fig.4, purple line) in about 15min.

In order to the factors that influence the jellyfish net displacement, 24 hours of low pass filter on the ocean current at surface and middle layers is conducted to filter out the periodical tidal current. As you can see from Fig. 5, from 5th to 12th July, 2012, the mean wind is southeastward with 2~3m/s residual ocean current is northward with 0.2~0.3m/s; ocean current speed at mid-layer is less than surface layer with counterclockwise deflection. With the influence of strong southeastward wind (figure omitted) and northward surface mean circulation (Fig. 5), Nemopilema nomurai observed in July 5th 2012 moved northward with 5km/d and may affect the waters near Grand Island in future days. Then, the simulated results at 7:50 on July 2th were compared with observation (Fig. 4, blue triangle), the relative errors of jellyfish drift path ranged from 5.2% to 24.1%, which indicated that jellyfish observed near Dagong Island might drift from where jellyfish observed on July 5th.

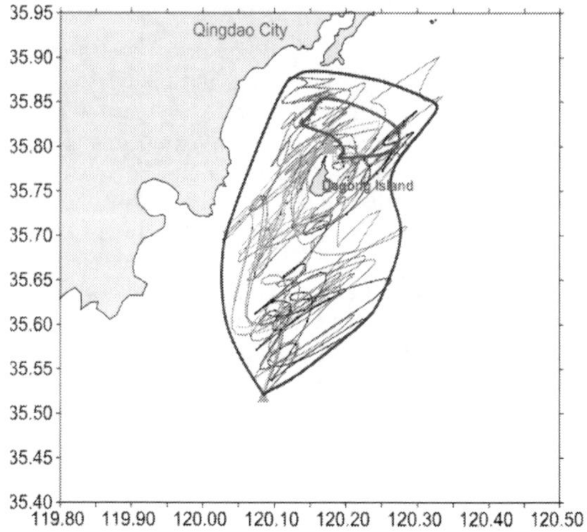

Figure 4. Ensemble trajectories (black lines) from 5th (green triangle) to 12th July 2012 and potential affecting domain (blue line), green(blue) triangle represents nemopilema nomurai monitoring on 11:16 5th (7:50 12th) July, respectively. The region surrounded by purple lines represents the ensemble results on 7:50 July12th.

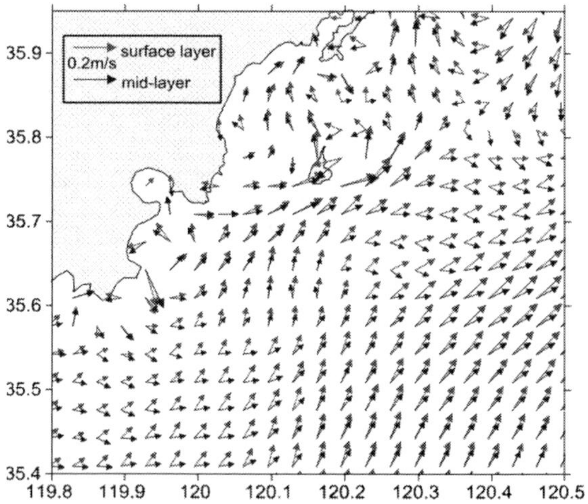

Figure 5. Averaged wind at 10m(unit :m/s) and residual ocean current (unit :m/s) at surface and middle layerfrom 5th to 12th July of 2012

On July 18th of 2013, a large number of nemopilema nomurai were observed in the coastal waters from Huangdao to Jiaonan, Qingdao. Based on seven ensemble schemes, the ensemble model was implemented to simulate the drift trajectories. The results showed that due to the strong southerly winds (figure omitted) and northward ocean current (Fig.6), large jellyfish moved northward, influenced the coastal waters of XueJia Island, passed the mouth of Jiaozhou Bay, and then entered the Jiaozhou Bay. Large jellyfish moved with velocity about 8km/d. The ensemble trajectories from 18th to 21st July and influenced range at different times are shown in Fig. 7.

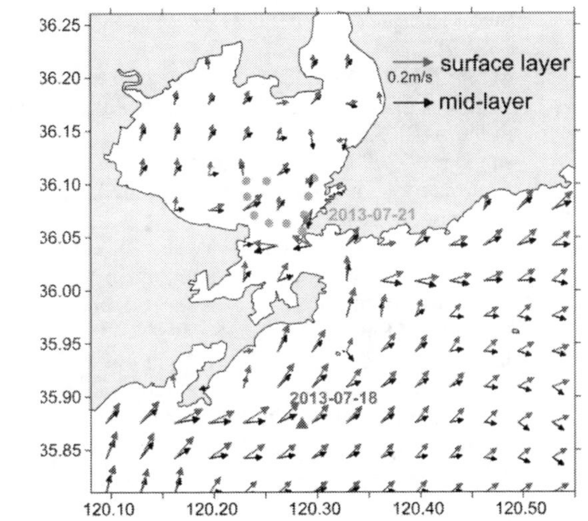

Figure 6. Same as Fig.5 except from 18th to 21st July 2013 (red triangle represents the observation on 18th July while green dots represent the observation at different times of 21st July)

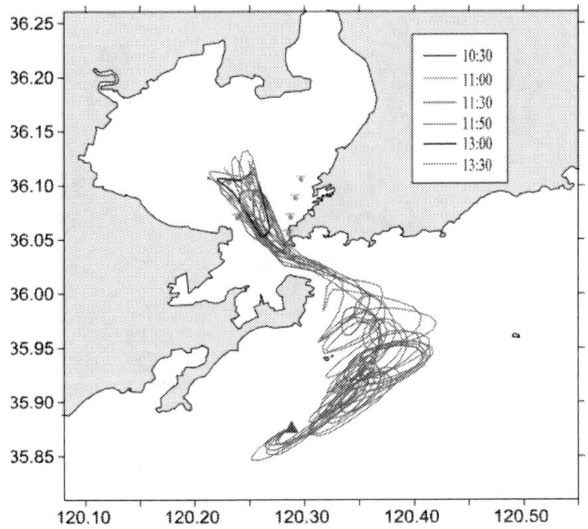

Figure 7. Ensemble trajectories (black lines) from 18th to 21st July 2013 and influenced ranges at different times

As we can see, influenced ranges characterized by different shapes at different times, while moves northeastward with influence of northward-northeastward current in Jiaozhou bay. The relative errors of drift distance are less than 28% (Tab. 1).

Table 1. Relative error of drift distance of large jellyfish at different times

Date	Time	Minimum relative error(%)	Maximum relative error(%)
21th July, 2013	10:30-10:35	13.9	27.5
	11:00-11:05	10.7	27
	11:30-11:35	5.1	16.4
	11:50-11:55	4.1	16.9
	13:00-13:05	5	19.3
	13:30-13:35	11.7	18.2

5.2 Ensemble simulation in the coastal Waters of Qinhuangdao

The ensemble trajectories from 21st to 25th July and its potential influenced ranges (Fig. 8b, blue line) are simulated. The results show that with the effect of weak easterly (figure omitted) and ocean current (Fig.8a), jellyfish moved westward, then moved northwestward and consequently influenced Zhongzhi beach near Beidaihe (Fig.8b). The potential influenced range is almost as the same as the observation of 25th July. The relative errors of drift distance range from13.5% to16.7%.

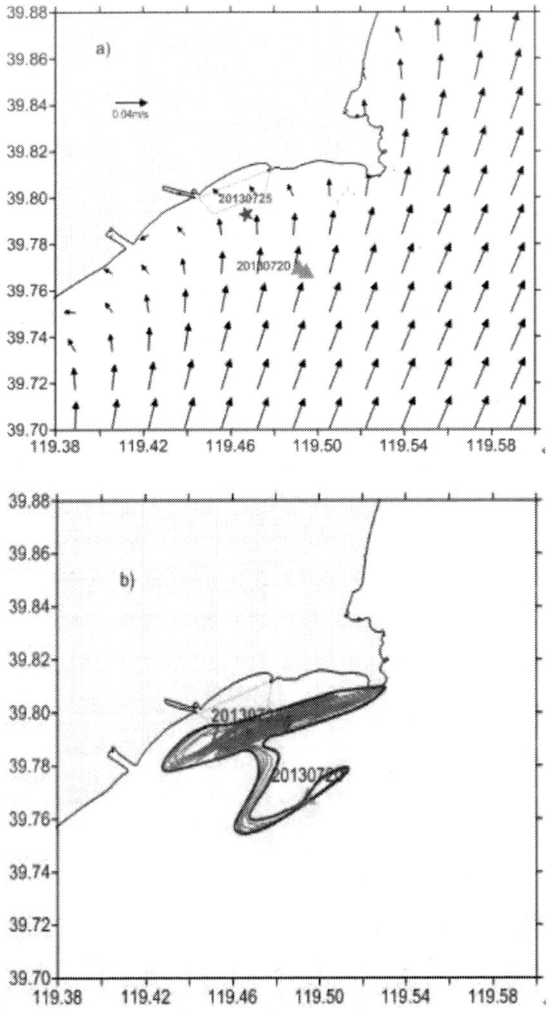

Figure 8. Depth-averaged residual ocean current from 21st to 25th July 2013(a), ensemble drift trajectory (black line) from 21st (green triangle) to 25th July 2013 of jellyfish and potential affecting domain (blue line)(b).

6 Summary and Discussion

In recent years, in addition to Red Tide and Green Tide, one ecological disaster induced by mass propagation of jellyfish starts to get attention in global. Since the vertical movement vary largely, and its mechanism is vague.

Hence, we apply the ensemble forecasting concept to simulate the trajectory of jellyfish, with its living habits taken into account. The coastal waters of Qingdao and Qinhuangdao are chosen as the typical jellyfish disaster places of the Bohai and Yellow Sea. Different schemes are designed according to the different environment parameters related to the vertical movement of jellyfish, such as ocean topography or different habits of vertical movements.

Besides, this ensemble model for jellyfish trajectories is independently developed, based on operational forecast results of marine and atmosphere environment, in which seven (five) different ensemble schemes are adopted. And we could simulate the ensemble trajectory in short time because the oceanic dynamical results could be operationally obtained at fixed time every day. Once large jellyfish was found, multi-source observation would be integrated, and this model would simulate the ensemble trajectories in short time.

This model can deliver ensemble simulation of trajectory, direction and potential influenced ranges. The comparisons between monitoring and numerical simulation results in 2012 and 2013 indicate that our ensemble model is suitable for description of jellyfish movements and can behave very well. Compared to single trajectory model, our ensemble model can present more accurate information of potential influenced range, which provides scientific decision support for emergency disposal and management of jellyfish disaster.

Acknowledgements

This work is supported by the Marine Public Foundation of State Oceanic Administration (No. 201005018) and open found of Key Laboratory of Ocean Circulation, Institute of Oceanology, Chinese Academy of Science (KLOCAW1403). Many thanks for the reviewers' helpful comments.

References

1. Brodeur R D, Mills C E, Overland J E, et al. Evidence for a substantial increase in gelatinous zooplankton in the Bering Sea, with possible links to climate change[J]. Fish Oceanogr., 1999, 8(4): 296-306.
2. Mills C E. Jellyfish blooms: are populations increasing globally in response to changing ocean conditions? [J]. Hydrobiologia, 2001, 451: 55-68.
3. Nagai T. Recovery of fish stocks in the Seto Inland Sea[J]. Mar. Pollut. Bull., 2003, **47**: 126-131.
4. Sun, S., 2012. New perception of jellyfish bloom in the east china sea and yellow sea[J], Oceanologia et Limnologia Sinica, **43**(3), 406-410.
5. Wang S W, Zhang G T and Sun S. et al., 2012. Population dynamics of three scyphozoan jellyfish species during summer of 2011 in jiaozhou bay[J], Oceanologia et Limnologia Sinica, **43**(3), 471-479.
6. Barz K, Hinrichsen H H, and Hirche H J, 2006. Scyphozoa in the Bornholm basin, (central Baltic Sea)--The role of advection[J]. J. Mar. Syst., **60**(1-2), 167-176.
7. Moon, J. H., Pang, I. C., and Yang, J. Y., et al., 2010. Behavior of the giant jellyfish Nemopilema nomurai in the East China Sea and East/Japan Sea during the summer of 2005: A numerical model approach using a particle-tracking experiment [J]. Journal of Marine Systems, **80**: 101-114.
8. Luo, X. F., Wei, H., Wang, Y. H., 2012. Processes influencing jellyfish particle tracking in the Yellow Sea and East China Sea[J], Oceanologia et Limnologia Sinica, 43(3): 635-642.
9. Johnson, D. R., Perry, H. M., and Burke, W. D., 2001. Developing jellyfish strategy hypotheses using circulation models[J]. Hydrobiologia, **451**: 213-221.
10. Graham, W. M., Pagès, F., and Hamner, W. M., 2001. A physical context for gelatinous zooplankton aggregations: a review[J]. Hydrobiologia, **451**: 199-212.
11. Zhang F, Zooplanktivorous Gelatinous Taxa: Medusas in the Yellow Sea and East China Sea[D], Qingdao, doctoral dissertation, 2008,1-130.

Study Added of Waste Chicken Egg Shell in Soils

Irwan Lie Keng Wong

Universitas *Kristen* Indonesia Paulus/Civil Engineering, Makassar, South Sulawesi, Indonesia

Abstract. Soil is the foundation of structure or construction that will receive the load transfer through to foundation. If the soil has a carrying capacity of small and cannot withstand the load transfer can result in the failure of construction. If the soil has a carrying capacity of small ground it is necessary to stabilize or improve the soil so that an increase in the carrying capacity of the land so that it can be used for construction. One material is commonly used for soil stabilization with the addition of lime. Waste chicken egg shell is waste that is still rarely used, the results of research [1], states that composition egg shell broadly consists of water (1,6%) and dry material (98,4%). The total dry ingredients are there, in shell eggs contained mineral elements (95,1%) and protein (3,3%). Based on the existing mineral composition, then the egg shells are composed of crystalline $CaCO_3$ (98,43%), $MgCO_3$ (0,84%) and $Ca_3(PO_4)_2$ (0,75%). This research was done by adding powdered chicken egg shell waste in clay with a composition of 5%, 7,5%, 10% and 14% with physical properties test and soil compaction test.

This research is an experimental study dan the results showed classification of soil with system USCS (*Unified Soil Classification System*) shows the percentage of grains of soil sieve No. 200 amounted to 60, 24% (> 50%), then USCS ground based system can be classified as fine-grained soil, liquid limit is 31, 92% and plasticity index is 5, 44 so that the soil is classified as silt organic and clay organic with low plasticity (OL). Unit weight dry (γ_{dry}) value in the addition of 10% waste chicken egg shell has increased a maximum of 1.58 gr/cm^3 into 1,630 gr/cm^3 at 14 days of aging (increase occurred at 3,164%). Water Content Optimun (w_{opt})in the addition of 10% waste chicken egg shell has decreased a maximum of 22,18% into 20,33% at 21 days of aging (decrease occurred at 1,85%).

1 Introduction

Land is an essential element in the development of the construction as a beachhead foundation that bears the entire burden of the construction that was built on it. If bearing capacity of soil is low and soil properties are sensitive to water maybe cause damage to the construction, examples of the damage that may occur landslip, cracks in the pavement (cracking), the structure of plates uplifted, the soil expands, etc.

Soil stabilization can be done to increase the carrying capacity of the land The first method is the mechanical stabilization for example, mixing several types of soils with the aim of producing well graded soil, the installation of other materials as reinforcement on the ground, use of geotextiles as well as the improvement of the soil surface with a drainage system. The second method is soil improvement with the addition of chemical stabilization of added materials (additive) to the soil to improve soil physical properties, such as strength, texture, workability and plasticity. Materials that can be used as an added ingredient in the improvement of the soil include lime, portland cement, asphalt (bitumen), fly ash, etc.

Soil stability is mixing the soil with certain materials to improve the engineering properties of soil stabilization. Process covering soil mixing with other soil or added material to obtain the desired gradation so that the engineering properties of the soil such as: strong support, compressibility, permeability, potential for development and sensitivity becomes better [1].

2 Chiken eggshell

Egg shell is the outermost part of the egg that serves to give protection to the component of the egg content from damage,either physically, chemically and micro- biologically, the rest of the hatchery is meant here is everything waste generated from industries such hatching eggs that did not hatch (sterile), eggshell of chicks that have hatched and eggshells in it still contains the embryo dead. When observed through a microscope, it can be seen that the eggshell is composed of four layers, namely:

2.1 Cuticle layer.

Cuticle is a transparent layer of proteins that coat the surface of the eggshell. This layer lining the pores in the eggshell but nature can still be passed gas so that the discharge of water vapor and CO_2 gas can still occur.

2.2 Foam layer.

This layer is the biggest part of the skin layer of the egg. This layer consist of protein and limestone layers consisting of calcium carbonat, calcium phosphate, magnesium carbonat and magnesium phosphate.

2.3 Mammillary layer.

This layer is the third layer of the egg shell consists of a cone-shaped layers with a round or oval cross-section. This layer is very thin and consists of woven protein and minerals.

2.4 Membrane layer.

This is part of the deepest layer of the egg shell. Composed of two layers of the membrane that envelops the entire contents of the egg.

Based on this research, chicken egg shell powder contains calcium of 401 ± 7.2 grams or about 39% of calcium, in the form of calcium carbonate. There is also a strontium at 372 ± 161µg, toxic substances such as Pb, Al, Cd, and Hg in small quantities, as well as V, B, Fe, Zn, P, Mg, N, F, Se, Cu, and Cr [2].

The composition of egg shell broadly consists of water (1,6%) and dry matter (98,4%). The total dry ingredients are there, in shell eggs contained mineral elements (95,1%) and protein (3,3%) [3], the existing mineral composition, then the egg shells are composed of crystal $CaCO_3$ (98,43%), $MgCO_3$ (0,84%) and $Ca_3(PO_4)_2$ (0,75%).

3 Calcium carbonate (CaCO3)

Calsium carbonate is the calcium salts contained in chalk, limestone, marble and is a major component contained in egg shells [4]. Calcium carbonate powder form, white, odorless, tasteless, stable in air. Partically insoluble in water, solubility in water increases with little ammonium salts or carbon dioxide. Insoluble in nitric acid to form gas bubles [5].

One of the chemical properties of calcium carbinate that can neutralized the acid. The use of calcium carbonate in the pharmaceutical field is as an antacid because of its ability to neutralize the acid but calcium carbonate can cause constipation.

If calcium carbonate is heated to a temperature of 900°C, the calcium carbonate will form kalsium oxide (CaO) due to release of carbon dioxide (CO_2), The reaction can be written as follow:

$$CaCO_3 \xrightarrow[T=900^{\circ}C]{} CaO + CO_2 \quad (1)$$

CO_2 is an exhaust gas which can have a negative impact on human health, therefore living soil stabilization with lime (quicklime) or calcium oxide (CaO) which is generally obtained from the above reaction can have a negative impact on the environment.

4 Soil physical properties test

4.1 Water content (w).

Water content (Moisture content) is the ratio between weight of water to weight of the soil grains. The water content is percentage of water a material that can be expressed by wet weight (wet basis) or based on the dry weight (dry basis). The water content of the wet weight has a theoretical maximum limit of 100 percent, while the water content by dry weight can be more than 100 percent [6].

4.2 Specific gravity (G).

Specific gravity is the ratio of the density of a substance to the density of a reference substance; equivalently, it is the ratio of the mass of a substance to the mass of a reference substance for the same given volume. Apparent specific gravity is the ratio of the weight of a volume of the substance to the weight of an equal volume of the reference substance. [7]

4.3 Attegberg limit.

Atterberg limits introduced by Albert Atterberg in 1911 with the aim of fine-grained soil to classify and determine the nature of the soil property index. Atterberg limits comprise the liquid limit, plastic limit, and shrinkage limit. [8]

Fine grained soils usually have plastic properties. The plastic properties of the soil's ability to adjust the shape change of the ground after mixed with water at a fixed volume. The land will be a liquid, plastic, semi-solid or solid depending on the amount of water that is mixed in the soil. Atterberg limits show the shape of the solid ground to become viscous fluids according to their water content. Atterberg limits of test parameters will be obtained liquid limit, plastic limit, sticky limit and limit cohesion which is the state of the soil consistency. Atterberg boundaries can be seen in the following Table 1 below:

Table 1. Plasticity limit of soils

PL(%)	Character	Type of soils	Cohesif
0	Non plastic	Sand	Non cohesif
< 7	Low plasticity	Silt	Cohesion partly
7 - 17	Medium plasticity	Clayed silt	Cohesif
> 17	High plasticity	Clay	Cohesif

Liquid limit (LL) is the water content of soil for the values above, the soil will behave as a viscous liquid (the boundary between the liquid state and a plastic state), which is the upper limit of the plastic area.

Plastic limit (PL) is the water content to values below, the ground is no longer influential as the plastic material. Soil will be a plastic material in a water content ranging between LL and PL. This range is called the plasticity index.

Plasticity Index (PI) is an interval of water content, the soil still is plastic. Therefore, the plastic index indicates the nature of the soil keplastisitas. If the soil has a water content interval small plastic area, then this state is called invitation thin soil. In contrast, jka soil water content interval area has a large plastic called peat. Plasticity index value can be calculated with the following equation:

$$IP = LL - PL \quad (2)$$

Restrictions on plastic index, the nature, kinds of soil and cohesion provided by Atterberg contained in the following Figure 1:

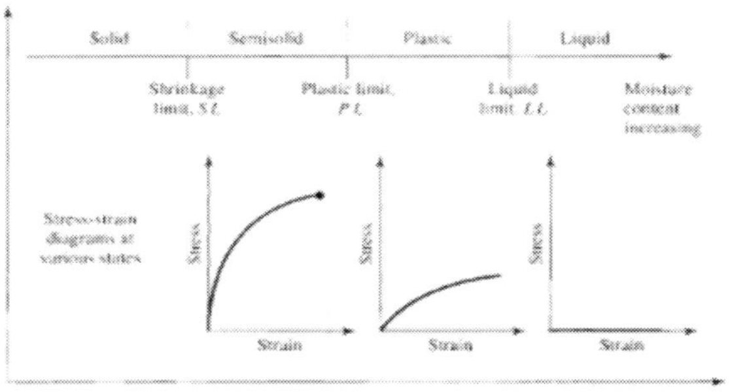

Figure 1. Attegberg limits

Shrinkage Limit (SL). Condition water content at the position between the semi-solid and solid areas, namely the percentage of moisture content where further reduction of water content does not lead to changes in the volume of soil is called Limit Losses.

$$SL=(V_0/(W_0-1/Gs))x100\% \quad (3)$$

where :

SL = Shrinkage limit
V_0 = Volume of specimen dry
W_0 = Weight of specimen dry
Gs = Specific gravity

5 Mechanical analysis of soil

Gradation of soils. Mechanical analysis is the determination of the size range of particles present in a soil, expressed as a percentage of the total dry weight. Two methods are generally used to find the particle size distribution of soil:

Sieve analysis – for particle sizes larger than 0,075 mm in diameter.

Hydrometer analysis – for particle sizes smaller than 0,075 mm in diameter.

The basic principle of sieve analysis and hydrometer analysis are briefly described in the following two section.

Modified proctor test. Modified Proctor compaction test is a laboratory method to determine the optimal experimental moisture content where in a particular type of soil will be the most solid and achieve maximum dry density. The term proctor is to honor RR Proctor, who in 1933 showed that the dry density for a given compaction effort depends on the amount of groundwater during soil compaction and the original tests often referred to as a standard proctor compaction test [9]. Laboratory tests generally consist of soil compaction on water levels known to the cylinder mold dimensions using the standard amount of compaction effort is controlled. The soil is usually compacted into molds with a certain amount of the same layer, each received a blow from a hammer-weighted standard at a certain height. The process is then repeated for a range of water content and dry density determined for each. Graphic relationship of the dry density of the water content is then plotted to form a curve compaction. Finally maximum dry density obtained from the cusp of the curve compaction and moisture content are appropriate, also known as the optimum moisture content.

The formula used in the calculation modified proctor compaction as follows:

$$ZAV=Gs/(1+(Gs+ w) / 80) \quad (4)$$

$$\gamma =(\text{weight of wet soil} / (\text{volume mold}) \quad (5)$$

$$\gamma_d=\gamma/(1+w) \quad (6)$$

where ;

Z.A.V = zero void water
w = moisture content
Gs = specific gravity
γ_d = Unit weight dry (g/cm^3)
γ = Unit weight (g/cm^3)

5.1 Soil sampling test

Soils sampling sites taken in district Gowa, Sunggu- minasa, Subdistrict Pallangga, Poros Limbung.street in South Sulawesi Province, country Indonesia.

Testing was conducted in laboratory test soils mechanics Universitas Atmajaya Makassar.

5.2 Flow diagram

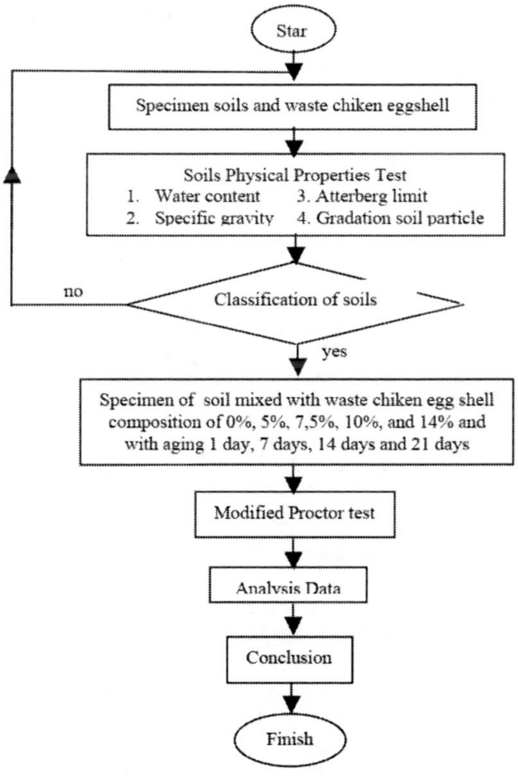

Figure 2. Flow diagram

Above diagram is shown in the following Figure 2:

6 Methodology

Types of research. Research was an experimental study, to make trial samples and then test in soil mechanics laboratory. Samples for testing are made as many as 85 samples, consisting of soil samples no mixture is 5 sample, sample soils with the addition of 5%, 7.5%, 10% and 14% respectively of 20 samples consisting of 5 samples for soaked 1 day, 5 samples for soaked 7 days, 5 samples for soaked of 14 days and 5 samples for soaked of 21 days.

Standard of Reference. Table 2 shows the reference laboratory test

Table 2. Standard of reference

Test	Reference
Water Content	ASTM D-2216-98
Unit Weight	ASTM D-854-02
Sieve Analysis	ASTM D 1140
Liquit Limit	ASTM D-4318-00
Plastis Limit	ASTM D-4318-00
Modified Proctor	ASTM D-698

6.1 Analysis data

Soil physical mechanical properties. Soil Physical Properties Test results are shown in Table 3:

According to the Unified Soil Clasification System (USCS), including fine-grained soil, this is expressed by percent through sieve 200 more than 50%. Liquid limit values obtained less than 50% causes the soil has a low plasticity. From the results of soil classification by USCS then district Gowa, Sunggu-minasa, Subdistrict Pallangga, Poros Limbung.street in South Sulawesi Province belonging to the types of silt organic and clay organic with low plasticity (OL).

Table 3. Testing physical and mechanical properties of soils

No	Soils test	Value
1	Water content (w)	27,61%
2	Unit weight (γ)	2,475
3	Liquid Limit (LL)	31,92%
4	Plastic Limit (PL)	26,48%
5	Index Plasticity (IP)	5,44%
6	Shrinkage Limit (SL)	19,44%
7	Percent sieve 200	60,24%
8	Unit weight dry (γd)	1,58 gr/cm^3
9	Water content optimum (w_{opt})	20,33%

Testing chemical properties of waste chiken eggs shell. Testing chemical properties of waste chiken eggs shell results are shown in Table 4:

Chemical test results show that waste chiken eggs shell contains Calcium (Ca) is 37, 30% and Carbonate (CO_3) is 58,0% which is large around 95,30%.

Modified proctor test. Modified Proctor test with the original soil, water content optimum (w_{opt}) is 20,33% and unit weigth of dry is 1,58 gr/cm^3. Result modified proctor test in the addition of a waste chicken egg shell with aging 1 day, 7 days, 14 days and 21 days are shown in Table 5 to 8 below:

The test results of modified proctor test showed the addition of waste chicken eggs as much as 10% obtained value unit weight dry maximum where the aging 1 day obtained is 1,612 gr/cm^3, aging 7 days obtained is 1,618 gr/cm^3, aging 14 days was obtained 1,630 gr/cm^3 and 21 days of aging was obtained 1,625 gr/cm^3. Unit weight dry value in the addition of 10% waste chicken egg shell has increased a maximum of 1.58 gr/cm^3 into 1,630 gr/cm^3 at 14 days of aging (increase occurred at 3,164%).

Graph addition of waste chicken egg shell to the value of the unit weight dry shown in Figure 3 below.

Figure 3 shows the maximum value of unit weight dry of 10% occurred on the addition of waste chicken egg shell on curing 14 days and the addition of up to 14% showed a decline of 1,630 gr/cm^3 be 1,628 gr/cm^3.

Table 4. The absolute and relative weight of the mineral constituent of egg shell

Mineral	% of the total weight	g / total weight
Calcium (Ca)	37,30	2,30
Magnesium (Mg)	0,38	0,02
Phosphorus (P)	0,35	0,02
Carbonate (CO_3)	58,00	3,50
Manganese (Mn)	7,00	ppm

Source [3]

Table 5. Result Modified Proctor test with aging 1 day

No	Waste chiken eggs shell	Unit weight dry (gr/cm^3)	Optimun Moistuire Content (%)
1.	5 %	1,599	16,46
2.	7,5%	1,605	17,29
3.	10 %	1,612	17,36
4.	14 %	1,610	17,31

Table 6. Result Modified Proctor test with aging 7 day

No	Waste chiken eggs shell	Unit weight dry (gr/cm^3)	Optimun Moistuire Content (%)
1.	5 %	1,602	17,54
2.	7,5%	1,609	18,96
3.	10 %	1,618	21,46
4.	14 %	1,615	19,61

Table 7. Result Modified Proctor test with aging 14 day

No	Waste chiken eggs shell	Unit weight dry (gr/cm^3)	Optimun Moistuire Content (%)
1.	5 %	1,622	19,55
2.	7,5%	1,625	20,48
3.	10 %	1,630	22,08
4.	14 %	1,628	20,13

Table 8. Result Modified Proctor test with aging 21 day

No	Waste	Unit weight	Optimun

	chiken eggs shell	dry (gr/cm^3)	Moistuire Content (%)
1.	5 %	1,618	19,10
2.	7,5%	1,620	20,20
3.	10 %	1,625	22,18
4.	14 %	1,619	20,10

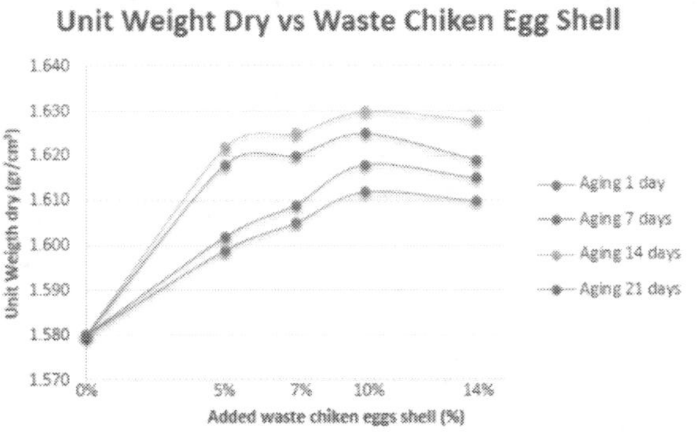

Figure 3. Unit weight dry vs waste chiken egg shell

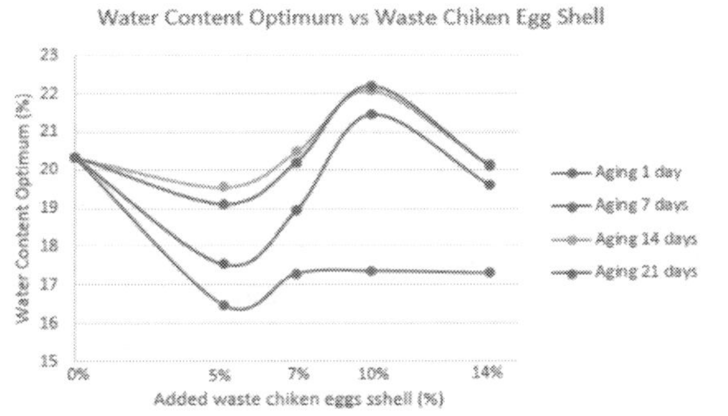

Figure 4. Water content optimun vs waste chiken egg shell

Graph addition of waste chicken egg shell to the value of the water content optimun shown in Figure 4 below

Figure 4 shows the maximum value of water content optimun of 10% occurred on the addition of waste chicken egg shell on curing 21 days and the addition of up to 14% showed a decline of 22,18% be 20,10%.

7 Conclusion

Results of research on soils is added to the waste chiken eggs shell as follows:

Soil sample from district Gowa, Sungguminasa, Subdistrict Pallangga, Poros Limbung street in South Sulawesi Province is classification by USCS is the types of silt organic and clay organic with low plasticity (OL).

Unit weight dry value in the addition of 10% waste chicken egg shell has increased a maximum of 1.58 gr/cm^3 into 1,630 gr/cm^3 at 14 days of aging (increase occurred at 3,164%)

Water Content Optimun in the addition of 10% waste chicken egg shell has increased a maximum of 22,18% into 20,33% at 21 days of aging (decrease occurred at 1,85%)

Acknowledgment

The author wish to express their gratitude and deep appreciation to Gladys Gabriella Tecoalu Faculty of Engineering Study Program Civil Engineering for their assistance and help.

This research was supported by Universitas Atmajaya Makassar, South Sulawesi. The author are grateful to the authorities for their support.

References

1. Hardiyatmo, Hary Christady. Treatment of Landslide and Erosi.Gadjah Mada University Press. Yogyakarta, 2006
2. P. Beukering, M. Schaafsma, A. Usup, and M. Embang, 2009, The Economic Value Peatlands in Central Kalimantan: Public Perception Of Rehabilitation and Revitalization Ex PLG Central Kalimantan, 2009
3. T. Yuwanta, Eggs and quality egg. Gadjah Mada University
   Press, Yogyakarta, 2010
4. T. O. Soine, & Wilson, C. O. Roger's Inorganic Pharmaceutical Chemistry.Seventh Edition. Philadelphia: Lea & Febiger. Pages. 390, 391.,1961
5. Ditjen POM, Kalsium Dalam Cangkang Telur, Ditjen POM Press, Jakarta, 1995
6. Nurul Gofar, Khairul Annar Kassim," Introduction to Geotechnical Engineering" Prentice Hall, Pearson Education South Asia Pte, Ltd, Singapore, 2007
7. E. Bowles, Joseph, Physical and Geotechnical Properties of Soils, Second Edition, McGraw-Hill Book Company, 1985
8. A. Cassagrande, "Research on the Atteberg Limitsof Soils," ASCE J. Geotech. Eng. Div. Vol 90, Sept 1964
9. F. Craig, S. Robert, Budi Susilo, Soil Mechanics, Fourth Edition, Erland., 1987

Determination of Model Kinetics for Forced Unsteady State Operation of Catalytic CH_4 Oxidation

Mohammad Effendy [1] and Endarto Y. Wardhono [2]

[1] *Surabaya State University, in Surabaya, Indonesia*
[2] *University of Sultan Ageng Tirtayasa, Indonesia*

Abstract. The catalytic oxidation of methane for abating the emission vented from coal mine or natural gas transportation has been known as most reliable method. A reverse flow reactor operation has been widely used to oxidize this methane emission due to its capability for autothermal operation and heat production. The design of the reverse flow reactor requires a proper kinetic rate expression, which should be developed based on the operating condition. The kinetic rate obtained in the steady state condition cannot be applied for designing the reactor operated under unsteady state condition. Therefore, new approach to develop the dynamic kinetic rate expression becomes indispensable, particularly for periodic operation such as reverse flow reactor. This paper presents a novel method to develop the kinetic rate expression applied for unsteady state operation. The model reaction of the catalytic methane oxidation over Pt/ -Al_2O_3 catalyst was used with kinetic parameter determined from laboratory experiments. The reactor used was a fixed bed, once-through operation, with a composition modulation in the feed gas. The switching time was set at 3 min by varying the feed concentration, feed flow rate, and reaction temperature. The concentrations of methane in the feed and product were measured and analysed using gas chromatography. The steady state condition for obtaining the kinetic rate expression was taken as a base case and as a way to judge its appropriateness to be applied for dynamic system. A Langmuir-Hinshelwood reaction rate model was developed. The time period during one cycle was divided into some segments, depending on the ratio of CH_4/O_2. The experimental result shows that there were kinetic regimes occur during one cycle: kinetic regime controlled by intrinsic surface reaction and kinetic regime controlled by external diffusion. The kinetic rate obtained in the steady state operation was not appropriate when applied for unsteady state operation. On the other hand, the kinetic rate expression obtained in the unsteady state operation fitted quite well. It was proven that in one cycle period the kinetic rate would always shift according to the ratio of CH_4/O_2.

1 Introduction

The increase of methane emission rate and high global warming potential, if not treated, can induce a negative impact on the environment. The catalytic treatment of methane emissions is one of the most reliable alternatives to be applied. However, since the concentration of emitted methane is low, the flow rate and concentration changes over time, and the temperature is ambient, the conventional treatment using once-through fixed bed reactor operating under steady state conditions is less appropriate. For the methane content of 0.5%-v, the adiabatic temperature rise is approximately 115°C. As a result, extra external energy is required to increase the temperature of the feed gas. Reverse flow reactor is a new solution that has been well proven to abate the exhaust gas emission auto thermally with characteristics as aforementioned. Reverse flow reactor may be defined as a fixed bed reactor in which the flow direction is altered periodically. Moreover, the reverse flow reactor could also improve the average methane conversion compared to the one obtained from steady-state operation.

Increased conversion of methane may be induced by changes of the reaction rate of methane over catalyst surface due to dynamic system operated at proper time scale. Therefore, there are two important aspects to be considered, (1) reactor design, (2) operation method. In the dynamic reactor, the mechanism and the rate-controlling step are always changing at each operating cycle. This change indicates that the reactor which is designed based on the steady state will give a different result compared with reactor which is designed based on the dynamic state. To the best of our knowledge, operating the dynamic reactor is usually designed on the basis of steady state condition due to no approach that has been developed, implementing the dynamic system in designing a reactor.

Development of dynamic operation is divided into two important aspects studied by many researchers, aspect of the design and method of operation. On aspects of the operation method, the goal is to get the proper operating procedures in order to obtain a better objective function of the steady state [1], [2]. However, on the research aspects of operating procedures was not studied.

In the designing aspect, to the lean methane oxidation on dynamic conditions, research led to the determination of catalyst for improved performance of the dynamic reactor [3]-[5]. Research related with the reaction rate in dynamic conditions is still limited to the suitability of the use of models of the reaction mechanism. Lopez et al. (2001) studied the effect of kinetic model (power law and LH) for methane combustion reaction against auto thermal operation on the catalytic reactor [6].

It is shown in different steady-state conditions, research on the reaction rate has been done, such as: determining step reaction mechanism for methane oxidation using Pt catalyst [7] developed a kinetic model for the methane oxidation using a catalyst Pd/Al_2O_3 in dry and wet conditions [8], studied and proposed a Langmuir-Hinshelwood reaction mechanism for the methane oxidation [9], study the reaction mechanism of the catalytic oxidation of a mixture of methane - water vapor using a fixed bed reactor [10], and study the reaction mechanism of the catalytic oxidation of a mixture of methane - hydrogen using a fixed bed reactor [11].

Based on previous research overview (state of the art), the reaction mechanism of methane oxidation on the dynamic conditions in one cycle has not been fully explained. Reaction mechanisms and rate controlling step reaction that occurs during dynamic conditions become a fundamental requirement for the design of the reactor is operated under dynamic conditions. Therefore, the methods to get the reaction rate on dynamic conditions need to be developed further.

2 Materials and Methods

The target research is to get a reaction mechanism that occurs in a dynamic period. The technique used by the approach steady state, i.e. the segment the dynamic condition in a steady state conditions. Study includes;

1) to obtain the equation lean methane oxidation kinetics under conditions of excess oxygen using a catalyst $Pt/\gamma\text{-}Al_2O_3$,
2) to test the suitability of kinetic equations for a single cycle of dynamic conditions,
3) to identify the effect of disturbance (feed composition modulation) on the response of the reactor (conversion of methane) and define the points disturbance which allegedly has a different reaction mechanism,
4) to establish equality for lean methane oxidation kinetics under the same conditions with the results of disturbance points identified in step 3,
5) to develop a model of lean methane oxidation kinetics in dynamic conditions and validate the dynamic model of kinetic equations is obtained.

Kinetic equations for lean methane oxidation in excess oxygen conditions obtained tested for compliance to dynamic conditions. Conformance testing is done by modeling and simulation. The model used is the pseudo-homogeneous continuum model 1D with the completion of the model using FlexPDE software version 6. These results provide direction, whether further steps need to be done or not.

Identify the effects of disturbance on the response of the reactor is expressed in feed concentration ratio CH_4/O_2 toward the methane conversion. Identification is done by analyzing and setting points of the feed concentration ratio CH_4/O_2 who allegedly had a different reaction mechanism.

Determination of reaction mechanisms and kinetics parameters on the conditions of the points that have been assigned disturbance conducted at steady state condition. Model of the reaction mechanism used model-based Langmuir-Hinshelwood (LH). The method used is the technique of fitting parameters on LH models with experimental data is carried out with the help of the software matlab.

Reaction mechanism in dynamic conditions is obtained by identifying changes in the mechanisms and rate controlling step reaction, did segmentation in the range of change reaction mechanisms, and validate the model of mechanism reaction in one cycle of dynamic operations.

3 Results and Discussions

3.1 Catalytic oxidation of lean methane in excess oxygen conditions

Model of the reaction mechanism used is the model of Langmuir-Hinshelwood (LH). The experiments were performed at steady state with a variable range of the experiment are as follows: minimum feed linear velocity of 7.5 e^{-4} m/sec, the catalyst mass 0.5 grams with a diameter of 2 mm catalyst, and reaction temperature 500°C, 520°C and 540°C. Range is based on the results of experiments testing the effect of external and internal mass transfer.

There are three models were tested based on the assumption LH reaction mechanism, which is derived based on the assumption of the model LH1 controlling phase adsorption of methane, LH2 models based on oxygen adsorption stage controller, and LH3 models based on surface reaction between methane with one atom of oxygen adsorbed. Results obtained by using the technique of fitting parameters LH models with experimental data is carried out with the help of matlab software are shown in Table 1.

The kinetics parameter of lean methane oxidation under conditions of excess oxygen has been obtained (see Table 1) tested for compliance to one cycle of dynamic conditions in the next step.

3.2 Suitability test for one kinetics equation on dynamic conditions

Disturbance created to make dynamic condition is obtained by modulating the composition of the feed. This resulted in the concentration of the feed gas changes each time as shown in Figure 1.

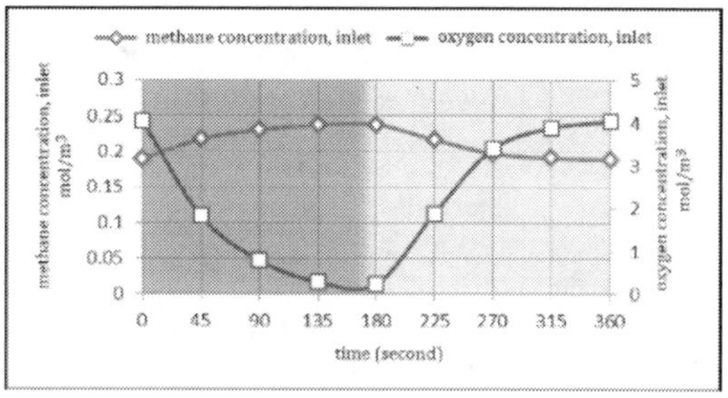

Figure 1. Behaviour of the oxygen concentration and the concentration of methane in the input due to interference open / close the flow of oxygen at the periodic switching time of 180 seconds. Initial conditions: a linear flow rate of CH_4 0.00030 m/sec at a concentration of 1% methane in air-mole.

The model used to test the suitability of kinetic parameters in dynamic operating one-cycle is pseudo-homogeneous 1D continuum model. Pseudo-homogeneous 1D continuum model equation and its boundary conditions are shown by Equation 1-3.

$$\varepsilon_{bed}\frac{\partial C_{CH4}}{\partial t} = D_{eff}\frac{\partial^2 C_{CH4}}{\partial z^2} - u\frac{\partial C_{CH4}}{\partial z} - \eta\,\rho_{kat}\,r_{CH4} \quad (1)$$

$$\varepsilon_{bed}\frac{\partial C_{O2}}{\partial t} = D_{eff}\frac{\partial^2 C_{O2}}{\partial z^2} - u\frac{\partial C_{O2}}{\partial z} - \eta\,\rho_{kat}\,r_{O2} \quad (2)$$

with boundary conditions:

z = 0, at t = 0 – 180 seconds,

$$C_{O2}^{0} = a.t^4 + b.t^3 + c.t^2 + d.t + e$$
$$C_{CH4}^{0} = f.t^4 + g.t^3 + h.t^2 + i.t + j$$
$$u^0 = k.t^4 + l.t^3 + m.t^2 + n.t + t \quad (3)$$

at t = 180 – 360 seconds,

$$C_{O2}^{0} = a_1.t^4 + b_1.t^3 + c_1.t^2 + d_1.t + e_1$$
$$C_{CH4}^{0} = f_1.t^4 + g_1.t^3 + h_1.t^2 + i_1.t + j$$
$$u^0 = k_1.t^4 + l_1.t^3 + m_1.t^2 + n_1.t + t$$

If z = L, $\frac{\partial C_{H_4}}{\partial z} = \frac{\partial O_2}{\partial z} = 0$

where the polynomial equation for the change in concentration and flow rate in the boundary condition refers to a pattern of behavior shown in Figure 1 and the rate equation for the reaction rate in Equation 1 and 2 refer to Table 1. Simulation results are shown in Figure 2.

The controlling stage is methane adsorption (A), the controlling stage is oxygen adsorption (B) and the controlling stage is reaction at the surface (C). Initial conditions: a linear flow rate of methane 0.00045 m / sec at a concentration of 1% methane in air-v.

of feed gas. This is shown by the comparison of experimental data and simulation on dynamic conditions, at the

From Figure 2 can be expressed, obtained methane conversion in dynamic conditions is not sufficiently influenced by changes in the concentration and flow rate

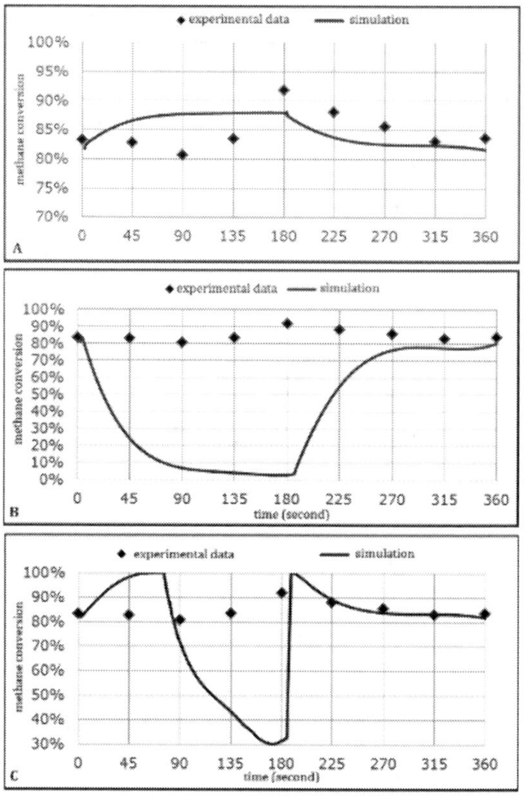

Figure 2. Comparison of methane conversion on dynamic conditions between the experimental results with simulations

condition of small CH_4/O_2 feed concentration ratio, ie the excess oxygen conditions, the reaction rate is controlled by surface reaction between CH_4 with 1 atom O_2. However, at the feed concentration ratio greater CH_4/O_2, ie the lack of oxygen conditions, the conversion of methane simulation results with experimental data is different on all three kinetic parameters were tested. So that the alleged change in the reaction mechanism at dynamic conditions will be identified at a later stage.

Table 1. Methane oxidation reaction kinetics parameters for three models

controlling phase	reaction rate equation	parameter	error
adsorption of CH_4	$\frac{k_1 C_{CH4}}{1+\sqrt{K_2 C_{O2}}}$	$k_1 = 1488 exp\left(\frac{-82245.41}{RT}\right)$ $K_2 = exp\left(-\frac{8373.61}{R}+\frac{44640.9}{RT}\right)$	3.7%
adsorption of O_2	$\frac{k_2 C_{O2}}{\left(1+\sqrt{K_1 C_{CH4}}\right)^2}$	$k_2 = 1.26 exp\left(\frac{-48201.58}{RT}\right)$ $K_2 = exp\left(-\frac{231.94}{R}+\frac{3048.74}{RT}\right)$	7.0%
Surface reaction between 1 atom of O_2 with 1 atom of CH_4	$\frac{k_3 K_1 C_{CH4}\sqrt{K_2 C_{O2}}}{\left(1+K_1 C_{CH4}+\sqrt{K_2 C_{O2}}\right)^2}$	$k_3 = 112726 exp\left(\frac{-9204.71}{RT}\right)$ $K_1 = exp\left(-\frac{2.30}{R}+\frac{39192.47}{RT}\right)$ $K_2 = exp\left(-\frac{159.22}{R}+\frac{235774.78}{RT}\right)$	3.6%

Table 2. Kinetics parameters in three conditions CH_4/O_2 feed gas concentration ratio (φ)

φ	reaction mechanism	controlling phase	reaction rate equation	Parameter
0.05	Total oxidation reaction $r_{CH4\text{-}t}$ $CH_4 + O_2 \rightarrow CO_2 + H_2O$	Surface reaction between 1 atom of O_2 with 1 atom of CH_4	$\frac{k_3 K_1 C_{CH4}\sqrt{K_2 C_{O2}}}{\left(1+K_1 C_{CH4}+\sqrt{K_2 C_{O2}}\right)^2}$	$k_3 = 369743 exp\left(\frac{-9204.71}{RT}\right)$ $K_1 = exp\left(-\frac{2.30}{R}+\frac{39192.47}{RT}\right)$ $K_2 = exp\left(-\frac{159.22}{R}+\frac{235774.78}{RT}\right)$
1	Partial oxidation reaction $r_{CH4\text{-}pl}$ $CH_4 + O_2 \rightarrow CO_2 + 2H_2$	Surface reaction between 1 atom of O_2 with 1 atom of CH_4	$\frac{k_3 K_1 C_{CH4}\sqrt{K_2 C_{O2}}}{\left(1+K_1 C_{CH4}+\sqrt{K_2 C_{O2}}\right)^2}$	$k_3 = 4850 exp\left(\frac{-9204.71}{RT}\right)$ $K_1 = exp\left(-\frac{21.79}{R}+\frac{53526.41}{RT}\right)$ $K_2 = exp\left(-\frac{31.86}{R}+\frac{91828.98}{RT}\right)$
2.5	Decomposition reaction of methane, $r_{CH4\text{-}p}$ $CH_4 + O_2 \rightarrow C + 2H_2O$	External diffusion	$r = A_o exp\left(\frac{-E_a}{RT}\right) C_{CH4}$	$A_o = -0.0821\varphi^2 + 3172\varphi - 0.2112$ For $1 \leq \varphi \leq 1.92$ $A_o = 0.0064 ln(\varphi) + 0.0918$ For $1.92 \leq \varphi \leq 3.88$ $E_a \equiv 6300$

3.3 Identify the influence of disturbances (feed composition modulation) on the response of the reactor (methane conversion)

The feed concentration is changing due to the feed composition modulation expressed in concentration ratio feed CH_4/O_2. The relationship concentration ratio feed CH_4/O_2 with methane conversion is shown in Figure 3. The changes in feed concentration ratio CH_4/O_2 (φ) possess different trend pattern with methane conversion. This difference is thought to be caused by changes in the reaction mechanism. In order to identify the reaction mechanisms involved in the dynamic operating cycle, the three point determined CH_4/O_2 feed concentration ratio for the reaction mechanism analyzed by steady-state conditions. The third point CH_4/O_2 feed concentration ratio determined by extreme points in Figure 4, which is the smallest φ (φ = 0.05), the culmination of φ at a flow rate of methane 3.0e-4 m/sec (φ = 1), point average and peak flow rate of methane φ at 10.5e-4 m/sec and 4.5e-4 m/sec (φ = 2.5).

DOI: 10.1051/matecconf/20166003001

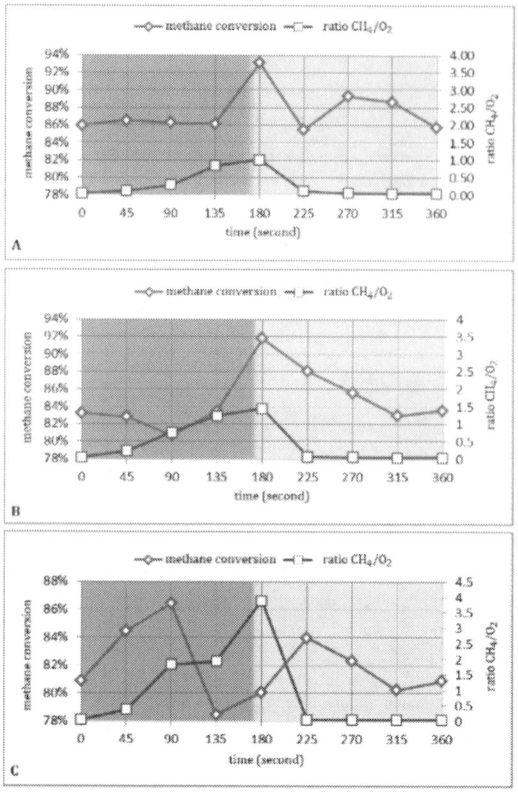

Figure 3. Behavioral response (conversion of methane) due to disturbances (CH_4/O_2 ratio) for one full cycle at a feed flow rate of methane with methane concentration of 1%-v in air. Methane flow rate of 3.0 e^{-4} m / sec (A); methane flow rate 4,5 e^{-4} m/detik (B); methane flow rate 10,5 e-4 m/second (C).

3.4 Determine the kinetics equation of lean methane oxidation at the conditions dots feed concentration ratio (ϕ) results identified in step 3

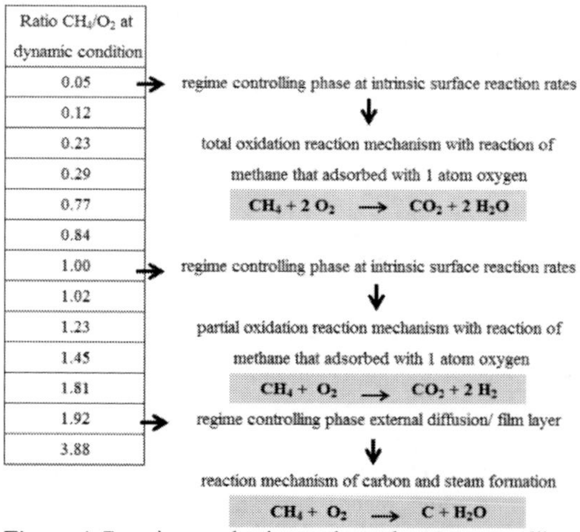

Figure 4. Reaction mechanism and reaction rate controlling step in the three-point CH_4/O_2 feed gas concentration ratio (φ = 0.05, 1, 2.5) are included in the dynamic range.

The identification of the reaction mechanism obtained three reaction mechanisms involved, the total oxidation of methane to CO_2 and H_2O, the partial oxidation of methane to CO_2 and H_2, and the reaction of methane decomposition into C and H_2O. Table 2 shows the kinetic parameters obtained from the identification.

The patterns of transition from one mechanism to another reaction mechanism in order to get a reaction mechanism in a single cycle of dynamic conditions described at a later stage

3.5 The develop a model of lean methane oxidation kinetics in dynamic conditions

Determination of the transition pattern of the reaction mechanism in dynamic conditions by analyzing the reactions that occur in feed concentration ratio ranges between 0.05 and 1 CH_4/O_2, ie in the range of total oxidation reaction mechanisms leading to the partial oxidation reaction mechanism. Set ϕ = 0.2 for analyzing the mechanism of the reaction that occurs in these conditions. The results obtained by using the same technique as the determination of kinetic parameters in other conditions, parameter kinetikanya not obtained. Therefore, is suspected in these conditions, there are two reaction mechanisms, that is the feed gas partial oxidation mechanism following total and some others follow the mechanism of partial oxidation.

The identification was done by using the following optimization percentage of the feed gas, partially from feed gas following oxidation reaction mechanism total and partly following the partial oxidation mechanism. Optimization techniques using FlexPDE software. The results obtained following the 51% feed gas total oxidation reaction mechanism and the rest follow the partial oxidation reaction mechanism. Figure 5 shows the equation that connects the feed concentration ratio CH_4/O_2 with the percentage number of moles of methane oxidation reaction mechanism following total and partial oxidation reaction mechanism.

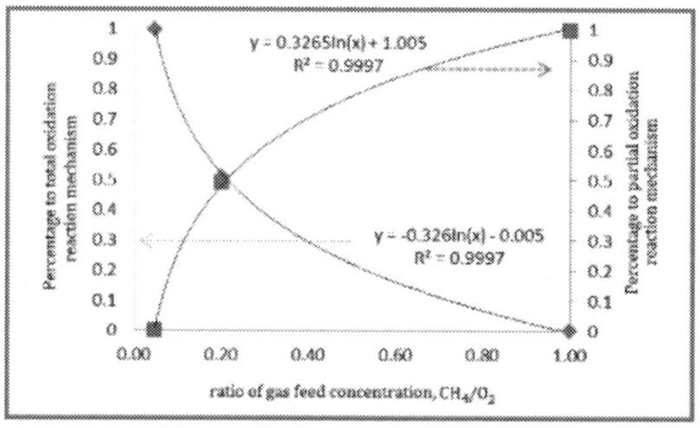

Figure 5. Equation connecting the feed concentration ratio CH_4/O_2 with the percentage number of moles of methane oxidation reaction mechanism following total and partial oxidation reaction mechanism.

From Figure 5 it can be stated, in the region 0.05 <φ<1, the percentage change in the number of moles of methane partial oxidation reaction mechanism followed a logarithmic function of with the value of the function y = 0.326 ln (φ) + 1.005. At φ> 1, significant changes, where the reaction rates are no longer being in the kinetic regime but was on the external diffusion regime as controlling the rate of reaction with the the reaction occurring at the surface of the catalyst is the reaction of methane decomposition into C and H_2O. At this range, ϕ changes that affect the acquisition methane conversion was approached by changing the value of impact factor on kinetic equations for external diffusion control stage shown in Table 2.

Overall, the transition mechanism of the reaction in one cycle of dynamic operations, which are based on changes in the concentration ratio CH_4/O_2 is caused by modulation of feedback given, as shown in Figure 6.

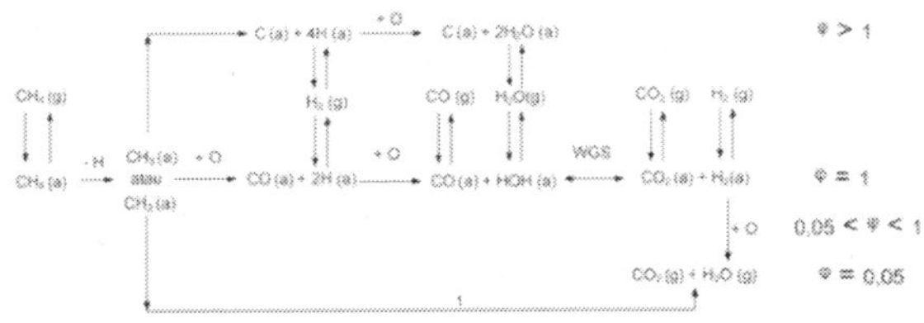

Figure 6. The transition mechanism of the reaction in one cycle of dynamic operations.

The reaction mechanism has been obtained for the dynamic conditions will be validated with the experimental results. The method is performed with the modeling and simulation. The model used to illustrate transition mechanism of the reaction in one cycle of dynamic operation is model of pseodo-homogeneous continuum 1D. Model of pseoudo-homogeneous continuum 1D with boundary conditions indicated by Equation 1-3, with the changes kinetikanya mechanisms and parameters which is a function of the concentration ratio changes CH_4/O_2 (ϕ) follow the restrictions ϕ as follows:

$\varphi \leq 0.05$; reaction mechanism following total oxidation of methane to CO_2 and H_2O with the stage control surface reaction between methane with the one atom of oxygen.

$0.05 < \varphi < 0.8$; reaction mechanisms is partially methane feed gas follow oxidation total methane into CO_2 and H_2O and others following the partial oxidation of methane to CO_2 and H_2.

$0.8 < \varphi < 1.1$; reaction mechanism is the following partial oxidation of methane to CO_2 and H_2 with the stage control surface reaction between methane with the one atom of oxygen.

$\varphi \geq 1.1$; reaction mechanism is followed methane decomposition reaction to CO_2 and H_2O with the external diffusion control stage.

Comparison of simulation results with the experimental data for various methane flow rate on methane concentration of 1%-v in air is shown in Figure 7.

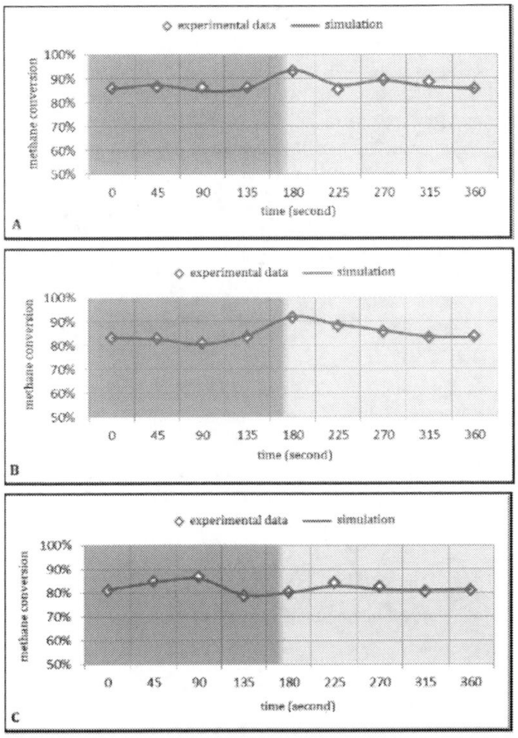

Figure 7. Comparison profile of the methane concentration at dynamic conditions between the experimental data with the the model for a linear flow rate of methane (a) 0.00030 m / sec, (b) 0.00045 m / sec, and (c) 0.00105 m / sec at methane concentration 1%-v in the air.

Figure 7 shows a model built with the attention to changes in reaction mechanisms that occur during dynamic operation in compliance. Methane conversion results obtained from simulations on dynamic conditions have values coincide with the values of methane conversion of experimental data that has noticed changes in the reaction mechanism during dynamic operation. This shows the importance of building a model of the reaction mechanism in dynamic conditions. Reaction mechanism in dynamic conditions can be approximated by the steady state reaction mechanism, with respect to kinetic regime of gas-phase reactions that might occur in porous catalysts in dynamic operation.

4 Conclusions

Equations should be centred and should be numbered with the number on the right-hand side.

The new method was developed to determine the mechanisms and rate controlling stage reaction that occurs in dynamic operation showed excellent results. This is indicated by the model and stage change mechanisms controlling the rate of reaction in a dynamic cycle that has been obtained to describe the characteristics of methane conversion obtained quite well, as shown in Figure 7.

Developed a new method can approximated by do segmentation the disturbance and the response to be seen reaction mechanism at steady state condition. Disorders that cause dynamic conditions in the reactor with the the response given by the reactor is expressed with the the ratio of the concentration of the feed gas with the methane conversion CH_4/O_2 obtained.

Reaction mechanism obtained from the segmentation of the dynamic conditions that obtain three reaction mechanisms, reaction of total oxidation of methane to CO_2 and H_2O, the partial oxidation of methane to CO_2 and H2, and the reaction of methane decomposition into C and H_2O, the reaction mechanisms of transition in one operating cycle of the dynamic based on changes in feed concentration due to interference ratio modulation feedback given.

Nomenclature

ε_{bed} = porosity of packed bed
C = Concentration
t = time
D_{eff} = effective dispersion
z = axial coordinate
u = linier velocity
ρ_{kat} = rho catalyst
φ = feed concentration ratio CH_4/O_2

References

1. M. Cittadini, M. Vanni, A.A. Barresi, A.A. Chem. Eng. Proc. **41**, 5 (2002)
2. S. Salomons, R.E. Hayes, M. Poirier, H. Sapoundjiev, Comp. Chem. Eng. **28**, 9 (2004)
3. M.A.G. Hevia, S. Ordóñez, F.V. Díez. Chem. Eng. Journal, **129**, 1-3 (2007)
4. P.A. Carlsson, E. Fridell, M. Skoglundh. Cat. Lett. **115**, 1 (2007)
5. E. Becker, P.A. Carlsson, L. Kylhammar, M.A. Newton, M. Skoglundh, J. Phy. Chem. C, **115**, 4 (2010)
6. A. Lopez, D.P. Harrison. I&E Chem. Res. **40**, 23 (2001)
7. P. Aghalayam, Y.K. Park, N. Fernandes, V. Papavassiliou, A.B. Mhadeshwar, D.G. Vlachos, J. Cat. **213**, 1 (2003)
8. J. Cortes, E. Valencia, P. Araya, Cat. Lett. **112**, 1 (2006)
9. P. Hurtado, S. Ordonez, H. Sastre, F.V. Diez, App. Cat. B. Environ. **51**, 4 (2004)
10. S.A. Shahamiri, I. Wierzba. Chem. Eng. J, **149**, 1-3 (2009)
11. S.A. Shahamiri , I. Wierzba. Iran. J. Chem. Eng. **7**, 2 (2010)

Study on the Chemical Constituents and anti-HBV Activity of *Hypericum japonica*

Gao-yun Chen[1], Min Liu [1], Meng-bin Yu[1], Chao Zheng[2] and Si-chen Wei[3]

[1] *Institute of Chemical Defense, Beijing, PR China*
[2] *Department of Biology,Shantou University, Shantou ,PR China*
[3] *Faculty of Science of Beijing University of Chemical Technonlogy, Beijing ,PR China*

Abstract. The most common cause of chronic viral hepatitis worldwide is from the Hepatitis B virus (HBV). Screening the safe and effective anti-HBV compounds in traditional Chinese medicine provides a large opportunity for the development of anti-HBV drugs. We found the ethanol extract of *H. japonica* had the besr inhibitory effects on HBeAg secretion from 20 species of medicinal plants In this study, silica gelSephadex LH-20. and a thin layer of chromatography were used to screen and isolate active anti-hepatitis B virus compounds from *Hypericum japonica* that was extracted with 60% ethanol. Four compounds were separated from the ethyl acetate extract of *H•japonica* and their structures were identified by spectral chemical and structural analyses. In addition, using HepG2 2.2.15 cell line with the cytotoxicity and inhibitory effecting on the secretion of HBeAg and HBsAg of the 4 isolated compounds were examined by MTT and enzyme-linked immunosorbent assays. The four isolated compounds were determined and 1,3,6,7-tetrahydroxy-9H-xanthen-9-one (Compound 4) exhibited the highest activity in inhibiting secretion of HBeAg in HepG2 2.2.15cells. The results indicated a 67.66% of inhibition rate, 40.25 μg /mL of IC50 value and TI＞2. We conducted a qualitatively evaluated research using duck hepatitis B virus model in vivo, the results indicated that ethanol extracts of *H. japonica* had anti-HBV effects and almost no sign of disease rebound after drug withdrawing.But the compound 4 has no direct inhibitory effect on HBV-DNA in duck experiment.

1 Introduction

The Hepatitis B virus (HBV) was discovered in 1966, but even after 50 years, it is still a major global public health problem[1], especially in Asian and African countries. Approximately 350-400 million people are affected with HBV in the world, and up to 1 million suffer from are HBV-related complications every year, including viral hepatitis and hepatocellular carcinoma(HCC)[2]. According to a national serological survey in 2006, China has one of the highest carrier rates of the Hepatitis B surface antigen (HBsAg), reaching 7.18% of the general population [2-7]. Currently, a great progress is being made in the treatment of the HBV with the development of effective and well-tolerated nucleotide analogues (NAs). However, the treatments of HBV, relying on some limited available antiviral drugs, are not successful in most of the cases. There are still some issues with the progress. First, there are still no highly effective anti-HBV drugs to remove the HBV-DNA from chronic infections completely. Second, the clinical anti-HBV drugs have some side effects.

However, anti-HBV drugs of constituents from traditional Chinese herbs have some unique advantages such as high efficiency, low toxicity, and multi-target points. Therefore, screening safe and effective anti-HBV constituents in herbs is an important and necessary aspect within the research of anti-HBV drugs.

We screened out four herbs within this study: (*Hypericum japonica*, *Fagopyrum dibotrys* (D.Don) *Hara*, *Areca catechu* L and *Parmelia saxatile* (L.) Ach) with anti-HBV activity from twenty kinds herbs. The result indicated that *Hypericum japonica* has the potential to suppress the secretion of HBeAg effectively in a dose-dependent method. In addition, its inhibition rate was the highest out of the four herbs that were screened.

2 Materials and Methods

In this Duck-Model, Beijing ducks were infected with duck hepatitis B virus (DHBV) used to evaluate anti-HBV activity in this study. HepG2 2.2.15 cell line was used in this study to screen the ethanol extracts from these plants using in vitro assays. The cytotoxicity was examined by MTT assay, and the effects on the secretion of HBeAg and HBsAg was examined by enzyme-linked immunosorbent assay.

DOI: 10.1051/matecconf/20166003002

3 Results

3.1 Screening of the herbs with anti-HBV activity

Within this study, the "Records of Li Folk Medicine," was used to identify 20 species of medicinal plants (table 1) and were selected to screen the herds with anti-HBV activity in our study. HepG2 2.2.15 cell line was used to screen the ethanol extracts from these plants using in vitro assays[8-11]. The cytotoxicity was examined by MTT assay, and the effects on the secretion of HBeAg and HBsAg were examined by enzyme-linked immunosorbent assay（ELISA）.

The data indicated that *H.japonica* could suppress the secretion of HBeAg effectively in the dose-dependent method and its inhibition rate was 67.66%, the IC50 value was 40.25 μg/mL.

Table 1. Inhibitory effects of 20 species of herbs on HBsAg and HBeAg secretion in highest non-toxic concentration in HepG2 2.2.15 cells ($\bar{x} \pm s$, n=3)

Ethanol extraction	Concentration (μg/mL)	HBeAg OD450	HBeAg Inhibition rate (%)	HBsAg OD450	HBsAg Inhibition rate (%)
Acanthus ilicifolius L.	80	0.759 ±0.055***	34.90	2.746 ±0.136	NA
Curcuma longa L.	20	1.109 ±0.027	4.89	2.595 ±0.167	NA
Zornia gibbosa Spanog	40	0.784±0.051***	32.78	1.815±0.264**	30.0
DMSO	0	1.166±0.046		2.597±0.153	
Hypericumjaponica Thunb.ex Murray	40	0.538±0.018***	69.06	2.443 ±0.197	NA
Clausena Lansium(Lour.)Skeels	20	1.347±0.077***	22.55	2.461 ±0.096	1.0
Bidens pilosa L.	160	1.438 ±0.073**	17.29	2.623±0.102	NA
DMSO	0	1.739±0.109		2.487±0.193	
Ciausena lansium(Lour.)Skeels(seed)	ND	NT	NT	NT	NT
Plumbago zeylanica L.	ND	NT	NT	NT	NT
Lygodium japonica(Thunb.)Sw.	160	1.574±0.192	NA	2.009±0.146**	22.93
DMSO	0	1.456±0.080		2.607±0.057	
Fallopia multiflora(Thunb.)Harald.	160	1.234 ±0.060	17.80	2.070 ±0.101	10.5
Uncaria macrophylla Wall	40	0.973±0.106***	35.21	2.321±0.051	NA
Geum aleppicum Jacq.	10	1.574±0.0.192	NA	2.411±0.304	NA
DMSO	0	1.502±0.044		2.314±0.217	
Phyllanthus urinaria L.	40	0.835±0.053***	44.4	2.965±0.116	NA
Mallotus apelta(Lour.)Muell.-Arg.	40	0.949±0.076***	36.78	2.436 ±0.369	11.16
Dichondra repens Forst.	80	1.150±0.194**	23.4	2.196 ±0.041*	19.8
DMSO	0	1.502±0.044		2.742±0.170	NA
Emilia sonchifolia(L.)DC.	80	1.279±0.035**	20.96	2.769±0.119	NA
Blumea balsamifera(Linn.)DC.	20	1.198±0.023***	25.96	2.751±0.049	NA
Ciausena lansium(Lour.)Skeels(leaves)	10	1.368±0.037**	15.45	2.892±0.192	
DMSO	0	1.618±0.142		2.771±0.123	
Atropa belladonna L.	160	1.593 ±0.046	1.50	2.622 ±0.068	4.35
Tylophora ovata(Lindl.)Hook.ex Stend.	160	1.613 ±0.156	0.29	2.577 ±0.220	6.01
DMSO	0	1.618±0.142		2.838±0.220	
3TC(Posive control)	80	1.548±0.055***	30.16	1.234±0.04***	43.31
DMSO	0	2.217±0.122		2.174±0.10	

Note: NA, Inhibitory effects of 20 species of herbs on HBsAg and HBeAg secretion not tested. NT, toxic to cells and not tested. ND, the highest non-toxic concentraion can not be determined. ***P<0.001;**P<0.01,*P<0.05, as compared with the control group (DMSO).

3.2 The inhibitory effects of Ethanol extract of H. japonica on HBsAg and HBeAg

In addition, the inhibitory effects of *H.japonica* on HBeAg and HBsAg had better results compared with 3-TC and Kushenin.. The results indicated that the ethanol extract of *H. japonica* did not have any inhibition on HBsAg; whereas, the ethanol extract of *H. japonica* had obvious inhibitory effects on HBeAg secretion, and the highest non-toxic concentration of inhibition rate was 69.06%, which was higher than that of Kushenin. The IC50 value was 14.27μg/mL and the therapeutic index (TI) > 2 (table 6); therefore, the ethanol extract of *H. japonica* had the further research value.

Table 2. The inhibitory effects of H. japonica, 3-TC and Kushenin on HBsAg and HBeAg Secretion in HepG2 2.2.15 cells ($\bar{x} \pm s$, n=3)

Drugs	Concentration (ug/mL)	HBeAg OD450	HBeAg Inhibition rate (%)	HBsAg OD450	HBsAg Inhibition rate (%)	Cell viability (%)
Ethanol extract of *H. japonica*	80	----	----	----	----	51.33±0.113
	40	0.538±0.018**	69.06	2.443±0.197	1.17	104.70±0.253
	20	0.847±0.123**	51.64	1.937±0.211	0.73	111.91±0.189
	10	0.917±0.004**	47.28	1.941±0.420	0.52	97..35±0.218
	5	1.166±0.241**	32.92	1.944±0.006	0.34	101.20±0.412
	2.5	1.421±0.081**	18.27	1.954±0.014	-0.13	95.05±0.319
	1.25	1.532±0.190**	11.89	1.959±0.332	-0.43	98.07±0.362
	0	1.739±0.109	0	1.951±0.042	0	100
3-TC	80	1.330±0.060	0.58	1.234±0.040**	43.33	103.47±0.114
	40	1.278±0.050	4.47	1.320±0.140**	39.30	106.43±0.067
	20	1.375±0.140	-2.82	1.638±0.090**	24.65	105.18±0.312
	0	1.338±0.090	0	2.174±0.100	0	100
Kushenin	3000	1.352±0.090**	37.40	0.883±0.110**	64.99	108.3±0.215
	1500	1.748±0.110**	19.06	1.321±0.330**	47.66	96.7±0.163
	7500	1.878±0.160**	13.04	1.888±0.180**	25.17	101.4±0.068
	0	2.159±0.070	0	2.523±0.070	0	100

Note： HepG2 2.2.15 cells were cultured in the presence of Hypericum japonica,3-TC and Kusheninat at various concentrations for 7 days. The HBsAg and HBeAg in the supernatants were quantified using specific ELISA kits. Data are presented as mean ±S.D. of three experiments. **P< 0.01, *P< 0.05, as compared with the no drug control group.

3.3 Inhibitory effects of different fractions from *H. japonica* on HBeAg

The H.japonica ethanol extracts were acquired by trichloromethane, ethyl acetate, n-butanol, and water, respectively, to obtain their different fractions. The four fractions from H.japonica ethanol extracts were tested for anti-HBV activity by MTT and ELISA methods, respectively. Results indicated it had the highest inhibition rate of ethyl acetate fraction on the secretion of HBeAg in HepG2 2.2.15 cell line(table 3).This fraction showed 63.46% of inhibition rate,111.07 μg/mL of IC50 value and TI>2(table 6). Consequently, it is necessary to isolate further the ethyl acetate fraction of *H. japonica*, which has the highest inhibition rate on the secretion of HBeAg in HepG2 2.2.15 cell line.

Table 3. The inhibitory effects of different fractions from *H. japonica* on HBeAg Secretion in HepG2 2.2.15 cells (n = 3, $\bar{x} \pm s$)

Fractions	Concentration (ug/mL)	OD_{450}	Inhibition rate (HBeAg,%)	Cell viability (%)
	320	----	----	----
	160	----	----	----
	80	----	----	----
	40	----	----	55.30±0.092
Chloroform fraction	20	0.726±0.213**	40.27	95.00±0.120
	10	0.768 ±0.085**	39.66	98.00 ±0.069
	5	0.810±0.087**	36.36	103.90±0.043
	2.5	0.949±0.149**	25.47	103.81±0.055
	0	1.216±0.042	0	100
	320	0.444±0.138**	63.46	109.60±0.145
	160	0.559±0.079**	54.03	116.5±0.080
	80	0.562±0.167**	53.78	123.60±0.095
	40	0.828±0.105**	31.88	122.30±0.124
Ethylacetate fraction	20	0.915±0.07**	24.72	107.60±0.074
	10	0.978±0.095**	19.60	123.30±0.170
	0	1.216±0.042	0	100
	320	----	----	----
	160	----	----	----
	80	----	----	44.11±0.332
	40	0.842±0.068**	52.08	105.80±0.145
n-Butanol fraction	20	0.982±0.011**	44.09	122.90±0.202
	10	1.109±0.051**	36.84	106.000.220
	5	1.270±0.034**	27.67	101.90±0.103
	2.5	1.429±0.096**	18.64	99.20±0.299
	1.25	1.436±0.078**	18.24	98.60±0.182
	0	1.948±0.063	0	100
	320	0.892±0.114	26.64	110.53±0.091
	160	0.934±0.153	22.33	109.21±0.013
Water fration	80	0.993±0.201	18.32	100.76±0.021
	40	1.077±0.116	11.41	101.53±0.011
	20	1.146±0.103	5.75	99.21±0.073
	0	1.216±0.042	0	100

Note: ----, toxic to cells and not tested,**P<0.01,*P<0.05,as compared with the no drug control group.

3.4The inhibitory effects of different parts of ethyl acetate fraction on HBeAg

The next step was to isolate the ethyl acetate fraction, as well as obtain 3 parts (YZ1,YZ2 and YZ3) with the active anti-hepatitis B virus by the way of silica gel chromatography. Looking at Table 4, we see that, YZ2 is of the highest inhibition rate on the secretion of HBeAg in HepG2 2.2.15 cell line. This part indicated that it had 71.57%% of inhibition rate, 85.21 μg/mL of IC50 value and TI>2(table 6).

Table 4. The different inhibition parts of ethyl acetate fraction from *H. japonicaon* on HBeAg secretion in HepG2 2.2.15 cells(n = 3, $\bar{x} \pm s$)

Different parts of ethyl acetate fraction	Concentration (μg/mL)	OD_{450}	Inhibition rate (HBeAg,%)	Cell viability (%)
	160	----	----	60.40±0.151
	80	0.776±0.028**	44.90	103.30±0.208
YZ_1	40	1.029±0.015**	26.90	105.50±0.012
	20	1.031±0.039**	26.70	98.40±0.398
	10	1.134±0.065**	19.40	121.10±0.359
	0	1.407±0.009	0	100
	160	0.380±0.080**	71.57	95.60±0.148
	80	0.818±0.017**	38.85	104.80±0.262
YZ_2	40	0.907±0.045**	32.20	108.90±0.319
	20	1.061±0.058**	20.66	111.50±0.217
	10	1.077±0.069**	19.51	103.10±0.570
	0	1.338±0.085	0	100
	160	----	----	40.00±0.071
	80	----	----	78.70±0.197
YZ_3	40	0.798±0.033**	40.40	107.20±0.117
	20	0.991±0.042**	25.90	107.50±0.352
	10	1.034±0.018**	22.70	97.70±0.059
	0	1.338±0.085	0	100

Note: ----, toxic to cells and not tested,**P<0.01,*P<0.05,as compared with the no drug control group.

3.5 The isolation of the compounds with anti-HBV activities

Four compounds were separated from the ethyl acetate fraction of *H. japonica*. Their structures were identified as Vincetoxin side B, (2R,3R) double hydrogen quercetin-7-O-α-L-rhamnoside, quercitrin, and 1,3,6,7-tetrahydroxy-9H-xanthen-9-one(figure 1). Among the 4 compounds, 1, 3, 6, 7-tetrahydroxy-9H-xanthen-9-one exhibited the highest activity in inhibiting secretion of HBeAg in HepG2 2.2.15 cells (table 5). The results indicated a 67.66% of inhibition rate, 40.25 μg /mL of IC50 value and TI＞2 (table 6).

Table 5. The inhibitory effect of the four compounds of the ethyl acetate fraction from *H. japonica* on HBeAg secretion in HepG2 2.2.15 cells (n = 3, $\bar{x} \pm s$)

Different parts of ethyl acetate fraction	Concentration (μg/mL)	OD_{50}	Inhibition rate (HBeAg,%)	Cell viability (%)
	40	----	----	28.10±0.131
	20	----	----	35.80±0.151
	10	1.103±0.055	-0.094	100.10±0.145
Compound 1	5	1.055±0.063	-0.046	109.30±0.020
	2.5	1.027±0.046	-0.018	101.90±0.395
	0	1.009±0.034	0	100
	40	0.798±0.042**	22.60	119.40±0.076
	20	0.857±0.070**	17.00	100.40±0.111
	10	0.851±0.025 **	17.50	114.90±0.242
Compound 2	5	0.852±0.065**	17.40	122.70±0.019
	2.5	0.971±0.074**	5.80	108.60±0.542
	0	1.032±0.182	0	100
	80	0.803±0.249**	22.10	95.80±0.086
	40	0.900±0.243**	12.80	103.40±0.137
	20	0.974±0.257**	5.60	101.50±0.035
Compound 3	10	1.032±0.132	-0.012	105.60±0.149
	5	1.035±0.223	-0.321	105.50±0.157
	2.5	1.033±0.246	-0.145	111.80±0.140
	0	1.032±0.182	0	100
	80	0.334±0.057**	67.66	101.80±0.577
	40	0.571±0.101**	44.65	117.30±0.081
	20	0.682±0.171**	33.81	103.10±0.234
Compound 4	10	0.853±0.098**	17.31	112.50±0.341
	5	0.960±0.161**	6.94	105.40±0.264
	2.5	1.019±0.042**	1.23	99.10±0.077
	0	1.032±0.182	0	100

Note: ----, toxic to cells and not tested,**P<0.01,*P<0.05, as compared with the no drug control group.

DOI: 10.1051/matecconf/20166003002

Table 6. Evaluation of anti-HBV efficacy of the active site and isolated compound of H. japonica

Extract	IC_{50} (μg/mL)	TC_{50} (μg/mL)	TI
H. japonica **ethanol extract**	14.27	>40	>2
Ethylacetate extract	111.07	>320	>2
YZ_2	85.21	>180	>2
Compound 4	40.25	>100	>2

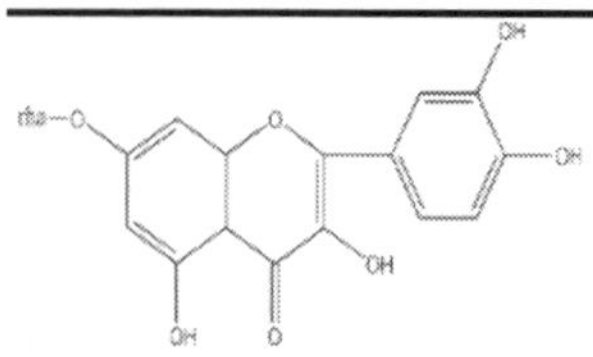
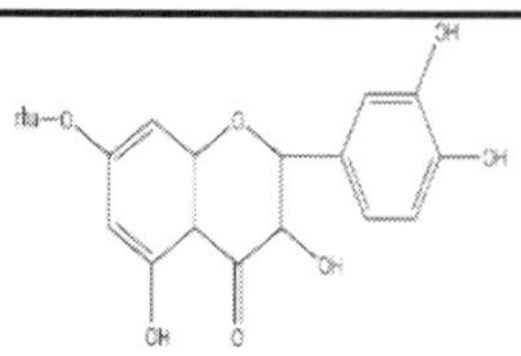

Compound 1(vincetoxicoside B) Compound 2((2R,3R)-taxifolin-7-O- α -L-rhamnoside)

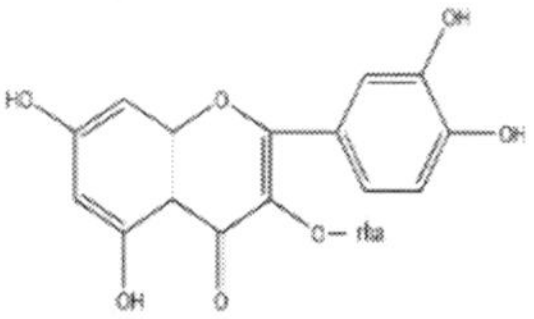
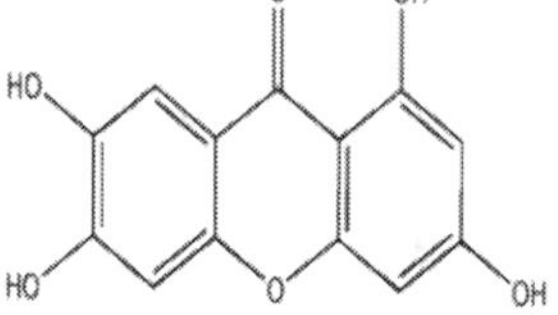

Compound 3 (quercitrin) Compound 4 (1,3,6,7-tetrahydroxy-9H-xanthen-9-one)

Fig.1. Chemical structures of the compounds.

vincetoxicoside B

mp 174 ~ 180 ℃ , ESI-MS m/s:447[M-H]-, 1H-NMR(400MHz,DMSO-d6) δ:1.14 (3H,d,J=6.0Hz, rha-Me), 5.55(1H,J=1.7Hz, rhaH-1), 6.41(1H,m,H-6), 6.79(1H,d,J=2.1Hz, H-8),6.90(1H,d,J=8.1Hz, H-5), 7.57(1H,dd,J=8.6Hz, H-6′), 7.72(1H,d,J=2.1Hz, H-2′). 13C-NMR(100MHz, DMSO-d6)δ:18.0(C-6″), 69.9(C-5″), 70.0(C-2″), 70.2(C-3″), 71.6(C-4″), 94.2(C-8), 98.4(C-6), 98.8(C-1″), 104.6(C-10), 115.2(C-6′), 115.6(C-5′), 120.1(C-2′), 121.8(C-1′), 136.1(C-3), 145.1(C-3′),147.5(C-4′), 147.9(C-2), 155.7(C-9), 160.4(C-7), 161.4(C-5), 176.0(C-4).

（2R,3R）-taxifolin-7-O-α-L-rhamnoside

mp 198~200℃.1H-NMR(400MHz,DMSO-d6)δ:6.76 (1H, s, J=11.60Hz, H-2′), 6.75(1H, d, J=8.1Hz, H-5′), 6.75(1H, dd,J=1.0,8.0 Hz,H-6′), 6.17 (1H, d, J=1.9Hz, H-8),6.12(1H, d, J=1.8Hz, H-6), 5.45(1H, d, J=10.3Hz, H-2), 3.92(1H, d, J=11.6Hz, H-3),3.38(1H,dd,J=1.7,3.4 Hz, RhaH-2),3.36(1H, dd, J=3.1,9.4 Hz, RhaH-3),3.28(1H,m,RhaH-5),3.26(1H,t,J=10.0 Hz, RhaH-4),1.10(3H, d, J=6.0 Hz, Rha-Me); 13C-NMR(DMSO-d6, 100MHz)δ:17.9(C-6″), 68.9(C-5″), 69.8(C-2″), 70.2(C-3″),71.5(C-4″), 75.5(C-3), 83.1(C-2), 95.4(C-8),96.5(C-6), 98.2(C-1″),102.0(C-10), 115.1(C-5′), 115.4(C-2′), 119.4(C-6′), 127.8(C-1′),145.0(C-3′), 145.8(C-4′),162.3(C-9),162.8(C-5), 164.1(C-7), 198.6(C-4).

Quercitrin

mp 180~181℃. ESI-MS m/z: 447[M-H]-. 1H-NMR (400 MHz, DMSO-d6)δ:7.30(1Hd, J=2.0HzH-2′)7.29(1HddJ=8.42.0 HzH-6′) 6.86 (1H, d, J=8.4 Hz, H-5′), 6.39(1H, d, J=2.0Hz, H-8), 6.20(1H, d, J=2.0Hz, H-6), 5.24(1H, d, J=1.6Hz, H-l″), 3.97(1H, brsH-2″), 3.51(1H, dd, J=9.53.2HzH-3″), 3.15(1H, t, J=10Hz, H-4″), 3.22(1H, dp, J=9.56.1HzH-5″), 0.88(3H, d, J=6.1Hz, H-6″); 13C-NMR (DMSO-d6, 100MHz) δ:17.7 (C-6″), 70.0 (C-5″), 70.3(C-2″), 70.6(C-3″), 71.5(C-4″),101.1(C-1″), 120(C-6′), 115.5(C-5′), 148.4(C-4′),145.2(C-3′), 115.6(C-2′),1 21.1(C-1′), 104.1(C-10), 157.3(C-9), 93.6(C-8),164.2(C-7), 98.7(C-6),161.3(C-5),177.8(C-4), 134.2(C-3), 156.4(C-2).

1,3,6,7-tetrahydroxy-9H-xanthen-9-one

mp>300 ℃ , ESI-MS m/z: 260[M-H]-. 1H-NMR (400MHz, DMSO-d6) δ:7.36(1H, d, J=8Hz, H-7), 6.54(1H, d, J=8Hz, H-5), 6.32(1H,s, H-4), 6.14 (1H, s, H-2); 13C-NMR(DMSO-d6, 100MHz) δ:93.6(C-4), 97.6(C-2), 101.5(C-9a), 102.5(C-5), 107.8(C-8), 111.5(C-8a), 143.9(C-7), 151.2(C-4b), 154.7(C-6), 157.3(C-4a), 162.5(C-1), 164.6(C-3), 178.8(C-9).

3.6 The inhibitionry effects on serum DHBV-DNA of compound 4

we conducted a qualitatively evaluated research using duck hepatitis B virus (DHBV) model in vivo [12-13]. The results showed that the compound 4 has no direct inhibitory effect on HBV-DNA in duck experiment. This indicated that compound 4 had the effect of anti-HBV activity through inhibiting the secretion of HBeAg .

4 Discussion

In this study, we conducted a qualitatively evaluated research using DHBV model in vivo, Ethanol extracts from *F. dibotrys*, *P. saxatile* and *H. japonica*, as well as the compound 4. Beijing ducks were infected with duck hepatitis B virus (DHBV) and they were used to evaluate anti-HBV activity. Treatments began on the 7th-day post infection. The ducks were infected with 26 mg/kg (high dose groups), 6.5 mg/kg (low dose groups) of ethanol extracts from *F. dibotrys*, *P. saxatile*, respectively. Drugs were given twice a day for ten consecutive days. The results indicated that the high dose groups, which were treated with ethanol extracts of *P. saxatile* or *H. japonica*, had anti-HBV effects and almost no sign of disease rebound after drug withdrawing. Treatment with ethanol extract from *F. dibotrys* also had certain levels of inhibitory effect on HBV-DNA, but the disease rebound after the withdrawing of treatment. The compound 4 has no direct inhibitory effect on HBV-DNA in duck experiment.

To our knowledge, this is the first report to show the 1, 3, 6, 7-tetrahydroxy-9H-xanthen-9-one has anti-HBV activity and low cytotoxicity in the in vitro assay, and our *P. saxatile* has anti-HBV activity in vivo.

References

1. María L. Cuestas & Verónica L. Mathet. Drug delivery systems and liver targeting for improved pharmacotherapy of the Hepatitis B virus (HBV) infection. Pharm Res, **27**, 1184–1202 (2010)
2. Su-zhen Jiang, Jia-jia Zheng. Hepatitis B Virus S Promoter Deletionin Hepatocellular Carcinoma. Infection Inte national, **1**, 14-23 (2012)

3. *Hepatitis B. World Health Organization.* World Health Org. **204** (2012)
4. Park NH, Song IH. Chronic hepatitis B in hepatocarcinogenesis.Pos. Med J, **82**(970), 507–515.(2006)
5. Westland C, Delandy W. Hepatitis B virus genotypes and virologic response in 694 patients in phase III studies of adefovir dipivoxil.Gas, **125**(1), 107-116 (2003)
6. Wright TL.Introduction to chronic hepatitis B infection.Am J Gastl, **101**, 1-6 (2006)
7. Ceen-Ming Tang, Tung On Yau. Management of chronic hepatitis B infection: Currenttreatment guidelines, challenges, and new developmentsWorld J Gastroenterol, **20**(20), 6262-6278 (2014)
8. Han YX, Xue R,Zhao W. Antiviral therapertic efficacy of foscamet inhepatitis B virus infection. Antiv Res, **68**(3), 147-153 (2005)
9. Shin MS, Kang EH. A flavonoid from medicinal plants blocks hepatitis Bvirus-e antigen secretion in HBV-infected hepatocytes. Antiv Res, **67**(3), 163–168 (2005)
10. Zembower DE, Lin YM. Robustaflavone, a potential non-nucleoside anti-hepatitis B agent. Antiv Res, **39**(2),81-88 (1998)
11. Zhou Z,Zhang Y,Ding XR. Protocatechuic aldehyde inhibits hepatitis Bvirus replication both in vitro and in vivo. Antiv Res, **74**(1), 59–64 (2007)
12. Addison WR, Wong WW. A quantitative competitive PCR assay for thecovalently closed circular form of the duck hepatitis B virus. Antiv Res, **48**(l), 27-37 (2000)
13. Le Guerhier F, Thermet A. Antiviral effect of adefovir in combination with aDNA vaccine in the duck hepatitis B virus infection model.J. Hepa, **38**(3), 328-334 (2003)

Effects of Puerarin Derivative P on Learning, Memory and MPO Activity in Vascular Dementia Model Mice

Pei Jiang[1,2] and Yubin Ji[1,2]

[1]Research Center on Life Sciences and Environmental Sciences,Harbin University of Commerce,Harbin 150076,China;
[2]Engineering Research Center of Natural Anticancer Drugs,Ministry of Education,Harbin 150076,China

Abstract. Pharmacological experiments confirmed that Puerarin could not only expand the arterial vessels, increase the local blood flow, but also could protect the myocardium, improve the blood supply of ischemic tissue, which was used to treat the cerebrovascular dementia disease in clinical practice.Its own structure particularity caused the lower bioavailability and poorer solubility. In order to improve the deficiency, the related pharmacological experiments of the newly synthesized puerarin derivative (P), so as to expect the better curative effect than puerarin in the process of treating vascular dementia. The vascular dementia model was established by permanently ligating the common carotid artery in mice. The effects of puerarin derivative(P) on learning and memory in mice by water maze, Ymaze and new object discrimination methods;The myeloperoxidase activity in the ischemic cerebral cortex was evaluated in mice by biochemical method. The experimental results showed that the mice spontaneous alternation responseaccuracy in Y maze could be obviously improved in 100mg/kg puerarin derivative group; In water maze, the swimming time to the safe platform would be significantly decreased in puerarin derivative group.Meanwhile, the MPO activity in the cerebral cortex of dementia mice was significantly decreased in 100mg/kg puerarin derivative group by permanently ligating the unilateral common carotid artery.

Vascular dementia (VD) is a kind of intellectual disabilitydisease caused by a variety of cerebrovascular diseases (ischemic and hemorrhagic cerebrovascular disease, chronic and acute hypoxic cerebrovascular disease, etc.). It seriously damages the brain parenchyma,hinders the patients' learning ability and memory function to a certain extent, and then affects the life and health of patients[1]. At present, VD is the second common dementia disease worldwide except Alzheimer's Disease(AD) and showing a rising trendin some countries and regionsyear by year. With the continuous improvement of people's living standards, the morbidityof VD will rise perpendicularly with the increase of the age[2],[3]. Ischemic and hypoxic hypoperfusion was considered to be one of the VD pathogenesis[4].

In general, the white blood cells do not exist in the brain tissue. Theinflammatory cellsinfiltrationonly occurs at time of cerebral ischemia or hypoxia[5], [6]. Among them, the activity ofmyeloperoxidase (MPO)representing the marker of neutrophil infiltration is also increasinggradually at time of cerebral ischemia, which also proves"inflammatory response theory" in cerebral injurydue to cerebral ischemia.

Puerarinis white powdery substance extracted from the root of Radix Puerariae. Itschemical name is 7, 4'-dihydroxy-β-C-D-pyranoidisoflavone. Puerarin has wide pharmacological actions, which is mainly used to treat the cardiovascular and cerebrovascular diseases [7]. Studies showed that [8], [9], puerarincould alleviatethe brain edema, remove lipid peroxidation, enhance antioxidant capacity, improve the antioxidant activity of brain tissue andalleviate focal cerebral ischemic injurydegree in cerebral ischemia reperfusion injury rats. However, the development of new drugs to treat central nervous system diseases was restricted by many factors. One of the main factors was how to effectively pass through the blood brain barrier [10]. Puerarinapplication was limited due to its lower water solubility and lipid solubility. The Puerarin structure was modified aiming to its defects in the paper and the novel puerarin derivative P was designed and synthesized.Its lipophilicity (logp value was increased) was increased.The effects of the new compound on learning memory functions and MPO activity were investigated preliminarily in vascular dementia mice, so as to improve the permeability of blood brain barrier, enhance its pharmacological activity, which could provide the reference for further developing and utilizingpuerarin and its derivatives in the treatment of vascular dementia.

1 Experimental

1.1 Reagents and Instruments

MS spectra were obtained with an Agilent G6300 Series (HPLC-MS) LC-MS analyzer. ^{1}H NMR spectra were obtainedwith aBruker DMX600MHz / 300MHz

DOI: 10.1051/matecconf/20166003003

NMR spectrometer(DMSO-d6 as solvent-, tetramethylsilaneas internal standard-). AutoDockVina molecular docking software was used. Puerarin pharmaceutical raw materials (≥98%) were purchased from Xi'an Frierson Biotechnology Co., Ltd.All reagents and solvents were of analytical grade. The silica gel used for column chromatography (reagent grade, 200 to 300 mesh)was purchased from Qingdao HaiyangChemical Co., Ltd.

1.2 Experimental animals

Male Kunming mice (SCXK (Army) 2013-006, weighing 25 ~ 30g) were used in this study.

1.3 Synthesis of P

Puerarin (15g, 36mmol) was dissolved in anhydrous DMF (200ml) to which K_2CO_3 (29.8g, 216.3mmol) was then added at room temperature. After stirring for 90min, ethyl bromoacetate (35.8g, 216mmol) was added and the mixture was stirred at room temperature for an additional 4h.The mixture was filtered to remove insoluble solids and then solvents were removed using a rotovapor until the mixturewas in anoily state; the oily solute was then solubilized with anhydrous ether.After again rotovappingtheether solution,the remains werespin-dried then dissolved in anhydrous methanol and isolated by silica gel column chromatography.

P(ethyl2-((3-(4-hydroxyphenyl)-4-oxo-8-((2S,3R,4R,5S,6R)-3,4,5-trihydroxy-6-(hydroxymethyl)tetrahydro-2H-pyran-2-yl)-4H-chromen-7-yl)oxy)acetate) ： logP-0.23; ESI-MS: m/z503.1 $[M+H]^+$, ^{1}H-NMR (300MHz, DMSO-d_6)δ_H 8.43 (1H, s,H-2), 8.07 (1H, d, J =9.0 Hz, H-5), 7.18 (1H, d, J =9.0 Hz, H-6), 7.43 (1H, d, J =8.7 Hz, H-2', 6'), 6.81 (1H, d, J =8.7 Hz, H-3', 5'), 4.88 (2H, s, H-1''), 4.22 (2H, m, H-3''), 1.25 (3H, m, H-4''),4.91 (1H, d, J = 9.6 Hz, H-1'''), 3.95 (1H, m, H-2'''), 3.69 (1H, m, H-3'''), 3.47 (1H, m, H-4'''), 3.30 (1H, m, H-5'''), 3.43 (1H, m, H-6a'''), 3.65 (1H, m, H-6b'''), 9.55 (1H, s, -OH).

1.4 Preparation,grouping and drug administration scheme of animal model

The healthy male Kunming mice were intraperitoneally injected with 2mg/kg atropine.3.5% chloralhy drate was intraperitoneally injected 5min later to anesthetize the mice. After the fixation and disinfection, the aseptic surgery was performed. The cervical median incision wasmade, the unilateral common carotid arterywas separated (not including the vagus nerve), permanently ligated and sutured with the operative thread. The mice occurred to convulsion, slow deep breathing andaccelerating heart beat.0.3 ml blood was discharged 1 cm way from the tail tip andstopped by compression. The drug was infused in the stomach 48h later.The distilled water was infused in the model group and sham operation group.

The mice were randomly divided into 5 groups, 12/group, including the sham operation group, model group, 100mg/kg puerarinpositive group; 100mg/kg puerarin derivative P group and25mg/kg puerarinderivative P group. All the mice were treated in the postoperative 48h. Thedistilled water was given in the sham operation group and model group.

1.5 Measurement of MPO activity in the cerebral cortex in mice

The mice were rapidly decapitated. The brain was removed, placed on the watch glass with ice blocks. The surface blood was sucked up. The bilateral hippocampiand cortexes wereseparated, weighed and frozen at -80℃ for standby. The biochemical indexes were detected.4 ℃ saline was added according to 1:9weight/volume ratio.10% homogenate of brain tissue was made and centrifuged by 3000 r/min at low temperature for 15 min. The supernatant was removed and MPO activity was detected by UV-2102Cultravioletandvisiblespectrophotometeraccording to the kit.

1.6 Statistics processing

The experimental data were expressed usingx±SD.The one-way[11] analysis of variance was performed using SPSS18.0 statistical software. $P<0.05$ indicated that the difference hadthe statistical significance.

2 Results

2.1 Water maze test

The experimentalresults showed that compared with the sham operation group, the swimming time to the safe platform was significantly prolonged in the mice of the experimental groupfrom the first day to the fifth day; Compared with the model group, the swimming time was shortened significantly in varying degrees in allPadministration groups and positive administration group. And the swimming time to the safe platform was alsodecreased significantly in the fifth day compared with that of the first day.

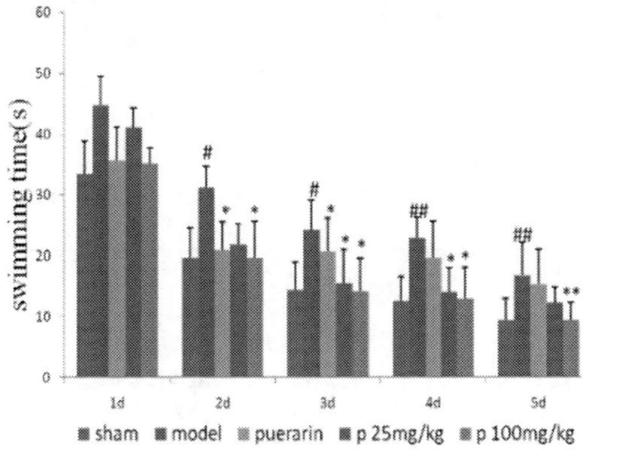

Figure 2-1. Effect of P on swimming time in water maze test in mice with Permanent ligation of the left common carotid artery （n=12, X±SD）
#P <0.05, ##P < 0.01 vs. sham group; *P < 0.05, **P < 0.01 vs. model group

2.2 Y maze test

The experimental results showed that compared with the sham operation group,the spontaneous alternating responseaccuracywas decreased significantly in the model group; Compared with the model group, themice spontaneous alternate response accuracy was increased in varying degrees in high dose P group, lowdose P group and positive drug administrationgroup.

Table 2-2. Effect of P on the spontaneous alternation behavior in the Y-maze test in the mice with Permanent ligation of the left common carotid artery（n=12, X̄±SD）

Group	Dose(mg/kg)	Alternation Behavior(%)
Sham	--	71.76±10.31
Model	--	41.27±8.77##
Puerarin	100	65.21±11.44*
P	25	61.37±13.87
	100	70.36±10.11**

##P <0.01 vs. sham group; *P <0.05,**P <0.01vs. model group.

2.3 New object discrimination experiment

The experimental results showed that compared with the sham operation group, the priority index and discrimination coefficient of the mice in the model group were decreased significantlyin two training periods (1h and 24h); Compared with the model group, the priority index and discrimination coefficient of the mice inhigh dose P group and positive drug administrationgroup were significantly enhanced in two periods (1h and 24h).

Table 2-3. Effect of P on discrimination index for the new object of 1h and 24h in the novel object recognition in the mice with Permanent ligation of the left common carotid artery （n=12, X̄±SD）

Group	Dose(mg/kg)	Discriminati on index(1h)	Preferential index(24h)
Sham	--	0.47±0.08	0.86±0.08
Model	--	0.36±0.09##	0.46±0.06##
Puerari	100	0.46±0.07*	0.76±0.09*
n	25	0.38±0.05	0.66±0.06
P	100	0.45±0.09*	0.81±0.11* *

##P<0.01 vs. sham group; *P <0.05,**P <0.01 vs. model group.

2.4 Effects of puerarin derivative P on MPO activity in VD mice cerebral cortex

The results were shown in Table 2-4. Compared with the sham operation, the MPO activity in the mice cortex was significantly increased in themodel group and the inflammatory reaction wasobvious; Compared with the model group, the MPO activity was decreased in all P administration groups and positive drug administration group by ligating the unilateral common carotid artery.Among them, the MPO activity was significantly decreased in the high dose P administration group and there was significantly different.

Table 2-4. Effect of P on the activity of MPO in the mice with Permanent ligation of the left common carotid artery （n=12, X̄±SD）

Group	Dose(mg/kg)	MPO activity(U/g)
Sham	--	0.26±0.16
Model	--	0.66±0.21##
Puerarin	100	0.38±0.08*
P	25	0.42±0.10
	100	0.36±0.15*

##P<0.01 vs. sham group; *P <0.05vs. model group.

3 Discussion

In this paper ,learning and memory obstacle model was established by permanently ligating the unilateral common carotid artery. The principle of this model was to ligate the common carotid artery to causecerebral ischemia, insufficient cerebral blood supply and oxygen supply and ultimatecerebral injury. The experiments showed that the cerebral injury symptomsof the animal model were clear and definite, could cause the mice memory obstacle,could simulate the pathogenesis of VD and then could targetedly investigatethe anti-vascular dementia activity of puerarinderivative.

Y maze and water maze were the common behavioralapproachto study the neurodegenerative diseases. The working memory and spatial memory were investigated respectively. New object discrimination experiment was a learning and memory test method based on the principle of exploratory tendency [12], [13]. The learning and memory tests could be performed in mice under free state, and could more approximately simulate the behaviors of human learning and memory. Meanwhile, the experiment could test thelong-term or short-term memory mechanism and evaluate the memory in particular stages by flexible conversion of the new object (shape and size), which was widely used in the learning, memory, new drug evaluation and other fields. In this paper, the behavioral experiment results showed that 100mg/kg puerarin derivative P could improve the learning and memory abilities of model mice in varying degrees, suggesting that it could improve the memory impairment caused by VD.

Studies [14], [15] showed that myeloperoxidase was a very important and special peroxidase secreted in neutrophil (PMN),most of which was located in PMN. The brain tissue hypoxia and ischemia would lead tothe inflammatory response of PMN infiltration. The myeloperoxidase (MPO) activity could indirectly reflect the PMN infiltration in cerebral ischemic areas. The cerebral cortex was the most sensitive to ischemia and was the important part of the human brain system. So the MPO activity in the cerebral cortex could indirectly reflect the degree of inflammatory damage. The experimental results showed that the MPO activity of the ischemic brain cortex was increased significantly in the model group, proved that PMN could indeed be related toa series of processes of the ischemic brain injury once again; The myeloperoxidase activity in the mice cerebral cortex could significantly decrease in high dose puerarin derivative group, so as to alleviate the inflammatory response and to improve the learning and memory abilities.

4 Conclusions

The structural modification of puerarin was conducted aiming to the solubility defects of puerarin, so as to obtain better lipophilic Puerarinderivatives P. 100mg/kg puerarin derivative P showed better improving the learning, memory and all eviating the inflammatory response, which could provide the experimental basis to study the vascular dementia pathogenesis and therapeutic effects for natural drug structural modification.

References

1. Roman GC. Vascular dementia revisited: diagnosis, pathogenesis, treatment, and prevention. *Med Clin North Am*, **86**(3): 477- 499(2002).
2. TorreillesF, et al[J]. *Brain Res*, **30**: 153-163(1999)
3. LongaZ, WeinsteinPR, CarsonSeta.l Revesible middle cerebra artery without craniotomy in rats. *Stroke*, **20**: 84- 91(1989)
4. Dohmen C,Kumura E,Rosner G,et al. Extracellular correlates of glutamate toxicity in short-term cerebral ischemia and reperfusion:a direct in vivo comparison between white and gray matter.*Brain Res*, **1037**(1-2): 43-51(2005).
5. Poornima Venkat,Michael Chopp,Jieli Chen.*Experimental Neurology*.**272**:97-108(2005)
6. Xu Chen,Rong Li,Tao Liang,Kefeng Zhang,Ya Gao,Lingyuan Xu.*Phytomedicine*.**20**:849-852(2013)
7. Mohammad Abul Kaisar,Shika Prasad,Luca Cucullo.*Brain Research*.**19**(1627):90-100(2015)
8. Fatma Emel Kocak,Aysegul Kucuk,Filiz Ozyigit,Murat Tosun,Cengiz Kocak, et al.*Journal of Sugical Research*.**199**:393-401(2015)
9. Uzakov S,Frey JU ,KorzV. Reinforcement of rat hippocampal LTP by hole board training. *Learn -Mem*.**12**:165-171(2005).
10. Kano T,Shimizu-Sasamata M,Huang PL,et a1.Effects of nitric oxide synthase gene knockout on neuro transmitter release in vivo.*Neuro seience*.**86**(3):695-699(2011).
11. Rosenberg GA. Ischemic brain edema. *Prog Cardiovasc Dis*, **42**(3): 209-216(1999).
12. Luo J, Wang YH, YasudaRP, etal. Thema jority of N2methyl2D-2aspartaterece ptorcom plexesina dultrat cerebral cortex contain at least three different subunits (NR1 /NR2A /NR2B)[J] *MolPharma-col*,**51** (1) :79-86(1997).
13. Nathan C. Nitric oxide as a secretory product of mammalian cells.*Faseb J.* **6**(12): 3051(1992)
14. Roman Gustavo C.Cholinergic dysfunction in vascular dementia. *Curt Psychiat, Rep,* 7:18-26(2005).
15. HuT,Fu Q,et al.Increased acetylcholinesterase and caspase-3 expression in the brain and peripheral immune system of focal cerebral hehemicmt8. *Neuroimmunol*, **211** (1-2):84-91(2009).

Effects of Fatty Acid Salts against Trichophyton Violaceum

Mariko ERA [1], Takayoshi KAWAHARA [2], Takahide KANYAMA [2] and Hiroshi MORITA [3]

[1] *Graduate school of Environmental Engineering, The University of Kitakyusyu*
[2] *Shabondama Soap Co., Ltd*
[3] *Faculty of Environmental Engineering, The University of Kitakyusyu*

Abstract. *Trichophyton violaceum* is an anthropophilic fungus. Dermatophytosis (Tinea) is fungal infection that can infect the scalp, glabrous skin, and nails. In general, Tinea can be spread by skin-to-skin contact or bathroom or floor materials. The treatments of Tinea need antifungal medication and good hygiene environment. The effective antifungal medication and infection prevention, and the creation of antifungal medication with high safety are required. In this study was focused on the antifungal effect of fatty acids potassium salts. The antifungal activity of nine fatty acid salts (butyrate, caproate, caprylate, caprate, laurate, myristate, oleate, linoleate, and linolenate) was tested on the spores of *Trichophyton violaceum* NBRC 31064. The results show that C6K, C8K, C10K, C12K, C18:2K, C18:3K was the most inhibit 4-log unit (99.99 %) of the fatty acids potassium incubated time for 10 min. It was observed that C12K and C18:3K was most high antifungal activity MIC. Commercially soap was lowest antifungal activity. This is because of the oleic acid is a major component of soap. Although further investigation is necessary to make clear antifungal mechanisms, our results suggest that fatty acid potassium will use to the development of a coating agent such as furniture.

1 Introduction

Dermatophytes are parasitic fungi that infect skin, hair and nails of both humans and animals [1]. David *et al.* suggest that dermatophytes infections are caused by 40 species of fungi which are grouped into three genera;*Trichophyton*, *Microsporum* and *Epidermophyton* [2]. *Tricophyton violaceum* is an anthropophilic dermatophyte. Dermatophyte infections are common worldwide, and infections of skin, hair, and nails caused by *Trichophyton* .spp [3], [4]. Antifungal agents such as itraconazole, ketoconazole, miconazole, clotrimazole, voriconazole, terbinafine, fluconazole is used as a therapeutic agent. However, although these antifungal therapy is usually well tolerated, the drug may cause gastrointestinal side effects and must be administered for a period of several weeks, making successful treatment of young children potentially difficult. We were searching for an alternative to the more secure and antifungal agents [3], [5], [6].

In this study, we focused on fatty acid salts, which are the main component of soap. Fatty acids, the raw material for the production of fatty acid salts, have been reported to show some antibacterial and antifungal activity. Fatty acids vary in length and degree of saturation, and naturally occurring fatty acids have a chain length of 4 to 28 carbons, which may be saturated or unsaturated [7]. Saturated fatty acids are straight chains and consist of a carbon chain with single bonds, while unsaturated fatty acids contain one or more carbon-carbon double bonds (C=C), which introduce fixed bends into the carbon chain [8]. Some report on the antimicrobial and antifungal effect of the fatty acid. For example, *Neisseria gonarrhoese* and *Helicobacter pylori*, *Aspergillus* spp. and *Penicillium* spp [9]-[11]. A previous study (Era *et. al.*, 2015) indicated that the antifungal activity of nine fatty acid salts was tested on the spores of *Penicillium pinophilum* NBRC 6345 and *Penicillium digitatum* NBRC 9651. Potassium caprate showed the strongest antifungal activity at 4 log-units. At incubation times of 180 min, potassium caprylate and potassium laurate showed antifungal activities of 2 log-units against *P. pinophilum* NBRC 6345 [12]. The results confirm the effectiveness of the ability of C10K to inhibit fungal growth on orange rind. C10K effectively inhibited *P. pinophilum* NBRC 6345 growth on orange rind [12]. Thus, C10K shows promise as an antifungal agent. This result shows that fatty acid salts prevented fungal growth, and fatty acid salts can be forecasted from its effective antifungal agent for dermatophytes. Dermatophytes can be infected by towels and clothing, while Foot ringworm is infected through shoes, socks and the floor or bathroom floor mats. We believe that the fatty acid salt serve the prevention of dermatophytes infection.

2 Materials and Methods

2.1 Fungus stains

The spore suspension of *T. violaceum* NBRC 31064 was obtained from the NBRC (Biological Resource Center, NITE, Tokyo, Japan). The fungus was initially grown on Sabouraud dextrose agar (Nissui Pharmaceutical Co., Ltd., Tokyo, Japan) in the dark and the culture stocks were stored at 4°C. Plant-pathogenic fungi were routinely sub-cultured on Sabouraud dextrose agar in slants in the dark at 27°C so that the fresh cultures were available for later use.

2.2 Preparation of spore suspensions

Fungus was grown on Sabouraud dextrose agar slants at 27°C until well sporulated, approximately 10-14 d. Spores were harvested in sterile distilled water using a sterile inoculation loop and gentle agitation; 0.9% NaCl were added to aid wetting of the spores. The spore concentration was determined by counting using a hemacytometer (Thoma, Sunlead Glass Corp., Saitama, Japan). The initial spore concentration was adjusted to 3.0×10^4 spores/mL.

2.3 Source of fatty acids

Nine fatty acids, butyric acid (C4:0), caproic acid (C6:0), and linoleic acid (C18:2) were obtained from Wako Pure Chemical Industries, Ltd., Osaka, Japan. Caprylic acis (C8:0), capric acid (C10:0), lauric acid (C12:0), myristic acid (C14:0), oleic acid (C18:1), and Linolenic acid (C18:3) were obtained from Tokyo Chemical Industry Co., Ltd. Potassium salts of fatty acids were also prepared by mixing fatty acid salt with the KOH: potassium hydroxide pellets (Wako Pure Chemical Industries, Ltd., Osaka, Japan) and water. Samples were prepared at 350 mM concentrations. The samples were then stirred for 2h at 75°C. KOH aq was added to yield theoretical neutralization at pH 10.5 of fatty acid salts. Potassium butyrate (C4K), potassium caproate (C6K), potassium caprylate (C8K), potassium caprate (C10K), potassium laurate (C12K), potassium myristate (C14K), potassium oleate (C18:1K), potassium linoleate (C18:2K), potassium linolenate (C18:3K), and the blank were all adjusted using a KOH pH-adjusted solution (pH 10.5) [12]. All fatty acid salts and the KOH pH- adjusted solution were filter-sterilized at low temperature (4-6°C) using a 0.20-μm Millipore filter (Toyo Roshi Kaisya, Ltd., Tokyo, Japan).

2.4 Effect of fatty acid salts on fungal spores

Solutions of 400 μL of fatty acid salts (final concentration of 175 mM in the tubes) and 400 μL of the spore suspension (3.0×10^4 spores/mL) were prepared in 1.5 mL plastic tubes. Spores mixed with the KOH pH-adjusted solution were used as controls [12]. Final pH of all samples were a range of pH 9.2-10.8. The mixtures were incubated at 25°C. Samples were counted at 0, 10, 60, and 180 min by plating (100 μL) on Sabouraud dextrose agar. Fungal colonies were counted after incubation for 3 d or 7 d at 27°C. Viable counts (log10 CFU) of spore was subtracted from the viable count of the control (log10 CFU), and the difference was used as a measure of the antifungal activity. All experiments were performed at least thrice.

2.5 Determination of minimum inhibitory concentrations (MICs)

The MIC is defined as the lowest concentration of drug sufficient for inhibiting visible growth of spores after 10 min of incubation. MICs against fungi were determined using the two-fold dilution method [13], [14]. Each fatty acid salt was separately inoculated with 400 μL of *T. violaceum* NBRC 31064 at 3.0×10^4 spores/mL. 1.5 mL plastic tubes containing 400μL of each of fatty acid salts were inoculated separately with 400 μL of the fungi. The tubes, each containing a total volume of 800 μL, were incubated at 25°C for each organism for 10 min. After incubation, samples were plating on Sabouraud dextrose agar, incubated at 27°C for 7 d, and then examined for the growth of spores. Following incubation, the end point was visually assessed and expressed in mM. The lowest concentration of the antifungal treatment that inhibited visible growth of the fingi after incubation was taken as the MIC of the treatment [14].

2.6 Effect of C12K combined with other fatty acid salts

C12K was mixed with short-chain fatty acid salts (C4K, C6K), medium-chain fatty acid salts (C8K, C10K) or long-chain fatty acid salts (C14K, C18:1K, C18:2K, C18:3K); final concentrations of each fatty acid salt of 0.175, 1.75, 17.5, 35, 87.5, 130 mM. Final concentration of C12K was 5.5 mM. These samples were used to measure the antifungal activity of C12K mixed with other fatty acid salts [12].

3 Results and discussion

3.1 Antifungal activity of saturated fatty acid salts

Antifungal effects of saturated fatty acid potassium salts against *T. violaceum* NBRC 31064 are shown for Figure 1. The average initial population of fungi at 0 min in all samples was approximately 3.0×10^4 spores/mL. Fungus was incubated for 3 d. Final concentration of fatty acid salts were 175 mM. In the saturated fatty acid C6K, C8K, C10K, C12K produced a 4 log-units reduction in the growth of *T. violaceum* NBRC 31064 after incubation for 10 min. Thus, C6K, C8K, C10K, C12K suppressed 99.99% of fungal growth. C4K produced a 1 log-units reduction in the growth of *T. violaceum* NBRC 31064 after incubation for 180 min.These results show that the compound with the 6 to 12-C chain produced combination the highest antifungal effect. Saturated fatty acid salts exerted an antifungal effect, and no effect was produced by the pH-adjusted solution alone.

No growth samples were cultured up to 10 d. Than this, the high antifungal effect samples was confirmed that there is a persistent.

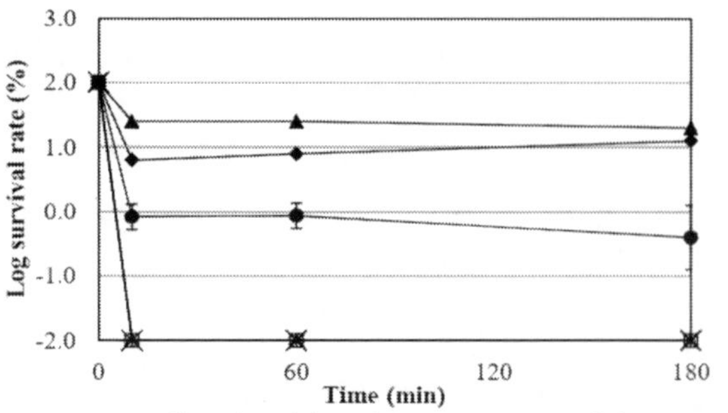

Figure 1. Antifungal activity of 175 mM saturated fatty acid salts against *T. violaceum* NBRC 31064. Spores were counted at the time of inoculation (0 min) and after 10, 60, 180 min of incubation by means of plating 100 μL portions of the same samples. Log survival rate (expressed log CFU/mL) of the samples stored at 27°C for 3 d were enumerated at the specified time points on Sabouraud dextrose agar. Symbols: ◆, C4K; □, C6K; Δ, C8K; ×, C10K; ∗, C12K; •, C14K; ▲, Control (KOH pH-adjusted). Experiments were performed in triplicate, and the error bars represent the standard deviation.

In previously reported, Against *P. pinophilum* NBRC 6345, C8K and C12K produced an antifungal effect of 2 log-units (suppressing 99% of growth) following incubation for 180 min. Further, C6K was ineffective after 180 min [12]. Desbois *et al.* has been reported for bacteria, it remains unclear exactly how free fatty acids exert their antibacterial activities but the prime target seems to be the bacterial cell membrane [15]. Further the various essential processes that occur within and at the membrane and this allows them to interact with the cell membrane to create transient or permanent pores of variable size [15]. Believes that the effect is also similar in *T. violaceum*.

3.2 Antifungal activity of unsaturated fatty acid salts

Antifungal effects of unsaturated fatty acid potassium salts against *T. violaceum* NBRC 31064 are shown for Figure 2. The average initial population of fungi at 0 min in all samples was approximately 3.0×10^4 spores/mL. Fungus was incubated for 3 d. Final concentration of fatty acid salts were 175 mM. In the unsaturated fatty acid C18 :2K, C18 :3K produced a 4 log-units reduction in the growth of *T. violaceum* NBRC 31064 after incubation for 10 min. Thus, C18:2K, C18:3K suppressed 99.99% of fungal growth. C18 :1K produced a 1 log-units reduction in the growth of *T. violaceum* NBRC 31064 after incubation for 180 min.These results show that the compound with the 18-C cahin has more than 2 double bonds produced combination the highest antifungal effect. Unsaturated fatty acid salts exerted an antifungal effect, and no effect was produced by the pH-adjusted solution alone.

No growth samples were cultured up to 10 d. Than this, the high antifungal effect samples was confirmed that there is a persistent.

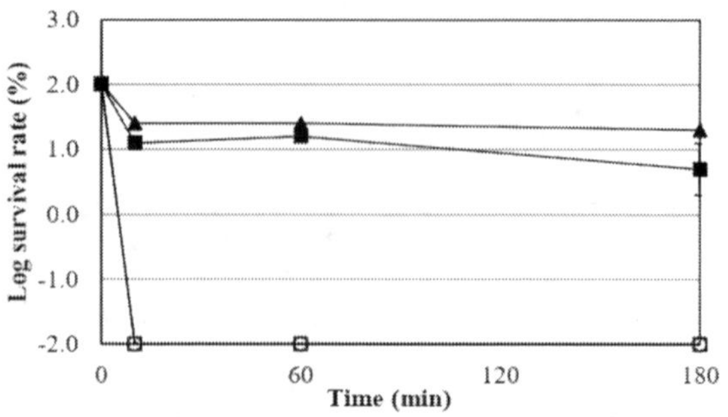

Figure 2. Antifungal activity of 175 mM unsaturated fatty acid salts against *T. violaceum* NBRC 31064. Spores were counted at the time of inoculation (0 min) and after 10, 60, 180 min of incubation by means of plating 100 μL portions of the same samples. Log survival rate (expressed log CFU/mL) of the samples stored at 27°C for 3 d were enumerated at the specified time points on Sabouraud dextrose agar. Symbols:■, C18:1K; ○, C18:2K; ◇ , C18:3K;▲, Control (KOH pH-adjusted). Experiments were performed in triplicate, and the error bars represent the standard deviation.

In previously reported, Against *P. pinophilum* NBRC 6345 and *P. digitatum* NBRC 9651, C18:2K and C18:3K were ineffective after 180 min [12].

3.3 Antifungal effect of short contact time in the case of saturated fatty acid salts

Figure 3. showed the antifungal effect against *T. violaceum* NBRC 31064 of fatty acid salts to shorten the incubation. The results of C4K was similar to the results of Figure 1. However, C14K produced an antifungal effect of 1log-units (suppressing 90% of growth) following incubation for 60 sec. These results suggest that compound with the 6 to 12-C chain produced and the highest antifungal effect. These results raise the possibility that fatty acid salts was affecting the surface of the cell.

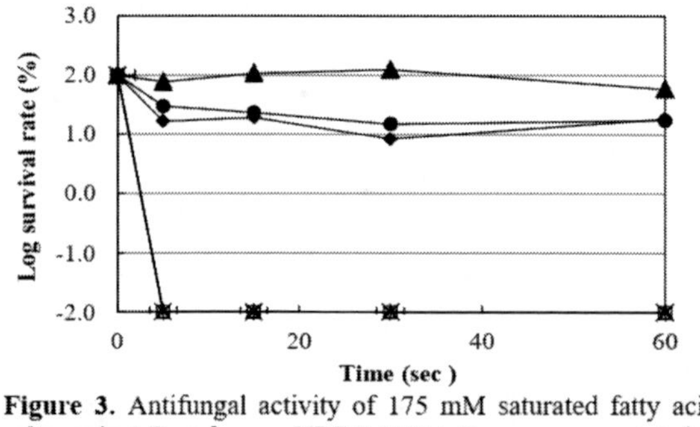

Figure 3. Antifungal activity of 175 mM saturated fatty acid salts against *T. violaceum* NBRC 31064. Spores were counted at the time of inoculation (0 min) and after 5, 15, 30, 60 sec of incubation by means of plating 100 μL portions of the same samples. Log survival rate (expressed log CFU/mL) of the samples stored at 27°C for 3 d were enumerated at the specified time points on Sabouraud dextrose agar. Symbols: ◆, C4K; □, C6K; Δ, C8K; ×, C10K; ∗, C12K; •, C14K; ▲, Control (KOH pH-adjusted).

3.4 Antifungal effect of short contact time in the case of unsaturated fatty acid salts

Figure 4. showed the antifungal effect against *T. violaceum* NBRC 31064 of unsaturated safatty acid salts to shorten the incubation. The results of C18 :1K were similar to the results of Figure 1.

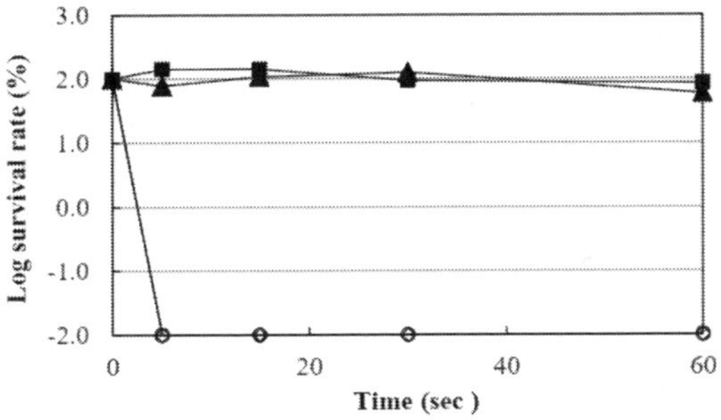

Figure 4. Antifungal activity of 175 mM unsaturated fatty acid salts against *T. violaceum* NBRC 31064. Spores were counted at the time of inoculation (0 min) and after 5, 15, 30, 60 sec of incubation by means of plating 100 μL portions of the same samples. Log survival rate (expressed log CFU/mL) of the samples stored at 27°C for 3 d were enumerated at the specified time points on Sabouraud dextrose agar. Symbols:■, C18:1K; ○, C18:2K; ◇, C18:3K; ▲, Control (KOH pH-adjusted).

3.5 MICs of fatty acid salts and other reagents tested

Two-fold dilution samples of the 175 mM solution inoculated with fungi were incubated for 10 min, and then applied to the agar medium, and MIC were determined after 7 d of culture. The experimental results showed that, form the nine fatty acid potassium salts tested, C12K and C18 :3K had the best antifungal effect against *T. violaceum* NBRC 31064 (Table 1). The MIC of C12K and C18 :3K were 5.5 mM (Table 1). The peak of antifungal effect of saturated fatty acid salts was 12 carbon chain and one more thing the peak of antifungal effect of unsaturated fatty acid salts was 3 double bonds in 18 carbon chain. Similarly, Isaacs *et al.* reported that fatty acids and monoglycerides containing 8-12 carbons showed stronger antiviral and antibacterial activities than their long-chain counterparts [16]. Desbois *et al.* investigated the C12 has the best balance between hydrophobic and hydrophilic groups among saturated fatty acids [17]. Moreover, Zheng *et al.* suggest that the anti bacterial effect of long-chain unsaturated fatty acids was due to their inhibition of fatty acid biosynthesis [10].

Table 1. MICs of fatty acid salts and other reagents tested against *T. violaceum* NBRC 31064.

T. violaceum NBRC 31064 MIC (mM)	
Potassium butyrate (C4K)	>175
Potassium caprote (C6K)	175
Potassium caprylate (C8K)	43.8
Potassium caprate (C10K)	21.8
Potassium laurate (C12K)	5.5
Potassium myristate (C14K)	>175
Potassium oleate (C18:1K)	>175
Potassium linoleate (C18:2K)	21.8
Potassium linonate (C18 :3K)	5.5

The MIC is the lowest concentration of a drug inhibiting visible growth of spores after 10 min of incubation. MIC, minimum inhibitory concentration.

Suggest that cis double bond increased the activity of a straight-chain fatty acid and, the addition of a second double bond further increased the toxicity of the compounds to grampositive bacteria [18]. Unsaturated fatty acid was potassium linonate (C18 :3K) was effective. Kodicek reported consistent with us, the toxicity in increasing order: oleic acid < linoleic acid < linolenic acid [19]. However, this does not agree with Kabara *et al.* Bayliss and Fuller *et al.* reported that the MIC of linoleic acid was somewhat lower than that of linolenic acid[18], [20], [21]. Thus, these results were suggested that similar to previous results. In the results *P. pinophilum* NBRC 6345 MIC was 175 mM [12]. MIC was better of *T. violaceum* NBRC 31064 that comparison of *T. violaceum* NBRC 31064 with *P. pinophilum* NBRC 6345. The difference of the effect of the fungi strain has not been clarified. Antifungal effect was not know the different reasons by the fungi strain.

3.6 Effect of C12K mixed with other fatty acid salts

Soap can be produced using coconut, palm, or olive oils, various fatty acids. Carbon number of fatty acids are generally used the range of 8 to 22 [10]. The main component of soap is fatty acid salts bearing the carbon number and the salts formed from them. The fatty acid salts are characterized by micelle formation at high concentration. The micelle formation depends on the concentration of them, the critical micelle concentration (CMC) of long-chain fatty acid salt is low. Thus, we investigated the effect on the antifungal activity of C12K further by studying the antifungal effect of C12K, which showed the highest antifungal activity, against *T. violaceum* NBRC 31064 when mixed with other salts. Figure. 5 (A)~(H) shows the effect of the antifungal activity with mixing C12K and other fatty acid salts (C4K, C6K, C8K, C10K, C14K, C18:1K, C18:2K, or C18:3K).

C12K mixed with C4K showed the same antifungal activity (4 log-units) as C12K alone (Figure. 5A). Thus, addition of C4K did not affect the activity of C12K. In addition, the effect of mixing C12K with C4K did not change, regardless of concentration. Similar results were obtained when C12K was mixed with short-chain fatty (C6K) and medium-chain fatty acid salts (C8K or C10K) and log-chain fatty acid saits (C14K, C18 :2K or C18 :3K). However addition of long-chain fatty acid salts (C18:1K) inhibited the antifungal activity of C12K (Figure. 5F). The antifungal activity of C12K decreased when mixed with long-chain fatty acid salts, and decreased more strongly as the concentration of long-chain fatty acids increased. However, addition of 3.5 mM of long-chain fatty acid salts did not affect the antifungal activity of C12K. This may have been due to the low concentration; addition of 3.5 mM of any compound did not inhibit C12K.

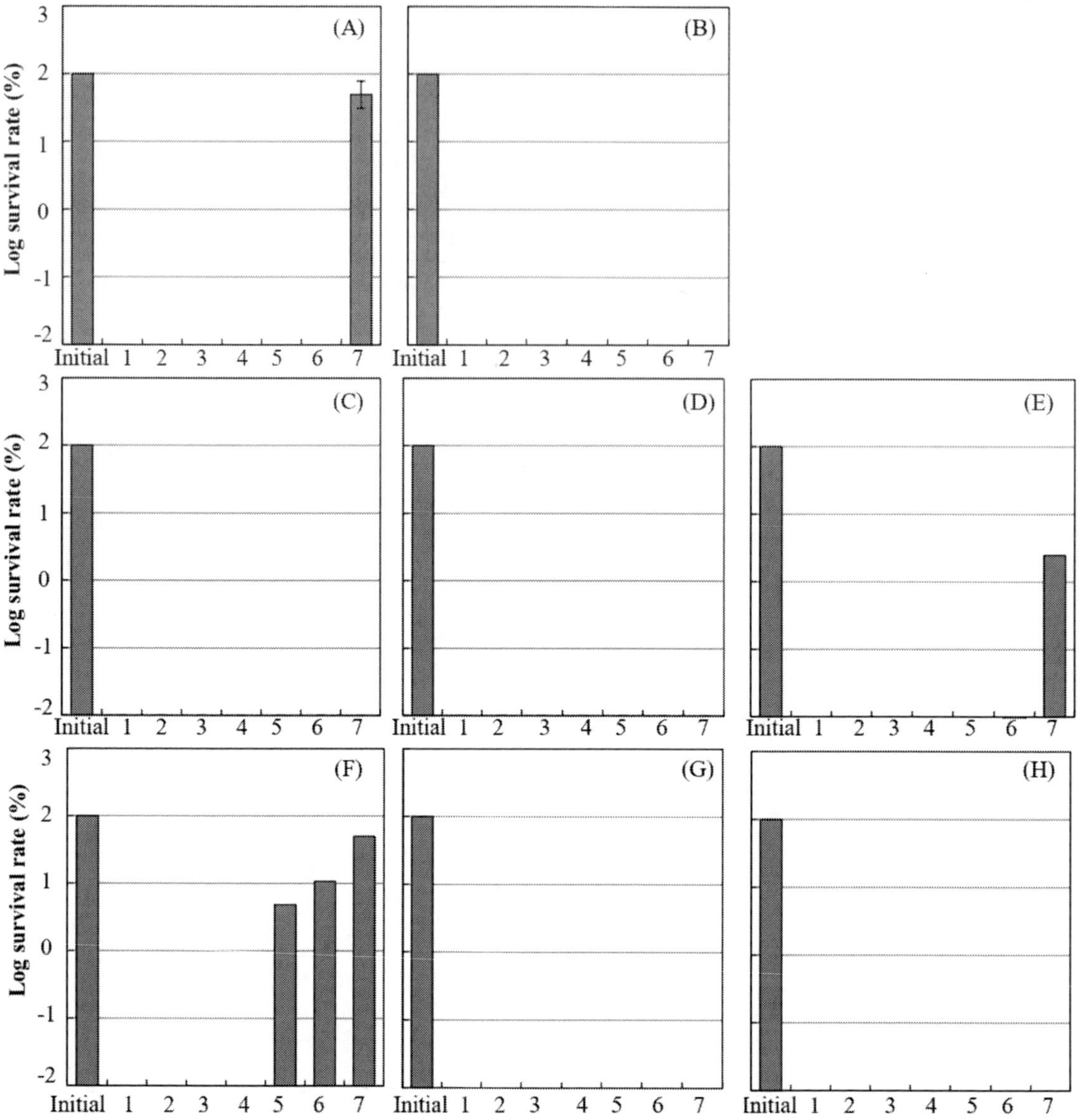

Figures 5. The effect of mixing with other fatty acid salts (C4K, C6K, C8K, C10K, C14K, C18:1K, C18:2K or C18:3K) on the antifungal activity of C12K. A C12K with C4K, B C12K with C6K, C C12K with C8K, D C12K with C10K, E C12K with 14K, F C12K with 18:1K, G C12K with 18:2K, H C12K with 18:3K. Samples: initial, fungal suspension; sample 1, 5.5 mM C12K and 0.175 mM other fatty acid salts; sample 2, 5.5 mM C12K and 1.75 mM other fatty acid salts; sample 3, 5.5 mM C12K and 17.5 mM other fatty acid salts; sample 4, 5.5 mM C12K and 35.0 mM other fatty acid salts; sample 5, 5.5 mM C12K and 87.5 mM other fatty acid salts; sample 6, 5.5 mM C12K and 130 mM other fatty acid salts ; sample 7, 175 mM other fatty acid salts alone. Spores were counted at the time of inoculation (0 min) and after 10 min of incubation by means of plating 100 µL portions of the same samples. Log survival rate (expressed log CFU/mL) of the samples stored at 27°C for 3 d were enumerated at the specified time points on Sabouraud dextrose agar.

4. Conclusions

Of the nine fatty acid salts, saturated fatty acid salts tested, potassium lauric (C12K) had the most antifungal effect on *T. violaceum* NBRC 31064, and MIC was 5.5 mM. This is in agreement with the results of several other investigators [18], [20], [22]. C6K, C8K, C10K, C12K and C18:2K, C18:3K can inhibit the ability of *T. violaceum* to take up nutrients, such as amino acids, thereby effectively starving the *T. violaceum* of the nutrients it requires to remain viable [23], [24]. A control solution at the same pH as the fatty acid salt solutions did not affect fungal growth, and we concluded that the antifungal activity was due to the fatty acid salts themselves, not pH.

The type of fatty acid salts that act is different it has been confirmed by the species of fungi [12]. Also, it has antifungal effect was revealed that the carbon is from 10 to 12 saturated fatty acid salts.

C4K mixed with C12K, was shown to increase the antifungal effect of C4K (Figure 5A). However, C18 :1K showed a reduction in antifungal effect in the mixture of 35 mM (Figure 5F). The kind of oil used for soap making are beef fas, palm oil, coconut oil, olive oil. It is known that the soap contained many of medium-chain fatty acids and long chain fatty acids [10]. From this study, the soap comprising oleic acid has been shown that low antifungal effect. The experimental results showed the effectiveness of lauric acid salts was prevention for dermatophytes infections. We believe that the fatty acid salt serve the prevention of dermatophytes infection. However, further experiments are required to determine their precise antifungal mechanism.

References

1. R. Chermette, L. J .G .Ferreiro ; Dermatophytoses in animals. *Mycopathologia*. **166**., 385-405(2008)
2. D. Greenwood, R.C. B. Slack, M. R. Barer, W. L Irving ; Medical microbiology. Aguide to microbial infections palliogenesis, immunity, laboratory diagnosis and control. *Ed. 15th ELST publishers*. ,558-564 (1997)
3. S.S. Magill, L. Manfredi, A. Swiderski, B. Cohen, W.G. Merz ; Isolation of Trichophyton violaceum and Trichophyton soudanense in Baltimore, Maryland. *Journal Clin Microbiol*., **45**, 461-465 (2007)
4. K. Giddey, M. Monod, J. Barblan, A. Potts, P. Waridel, C. Zaugg, M. Quadroni, Comprehensive Analysis of Proteins Secreted by Trichophyton rubrum and Trichophyton violaceum under in Vitro Conditions. *Journal of Proteome Research*., **6**, 3081-3092 (2007)
5. B. Fernández-Torres, A. J. Carrillo, E. Martín, A. Del Palacio, M. K. Moore, A. Valverde, M. Serrano, J. Guarro1 ; In Vitro Activities of 10 Antifungal Drugs against 508 Dermatophyte Strains. *ANTIMICROBIAL AGENTS AND CHEMOTHERAPY*., **45**, 2524-2528 (2001)
6. A. Panáčeka, M. Kolářb, R. Večeřováb, R. Pruceka, J. Soukupováa, V. Kryštofc, P. Hamalb, R. Zbořila, L. Kvíteka ; Antifungal activity of silver nanoparticles against Candida spp. *Biomaterials*., **30**, 6333-6340 (2009)
7. S. L. Sado-Kamdem, L. Vannini, M. E. Guerzoni ; Effect of α-linolenic, capric and lauric acid on the fatty acid biosynthesis in *Staphylococcus aureus*. *J. Food Microbiol*., **129**, 288-294 (2009)
8. C. H. Pohl, J. L. F. Kock, V. S. Thibane ; Science against microbial pathogens: Commmunicating current Research and technological advances. *Formatex Reseach Center*., **1**, 61-71 (2011)
9. S. Khulusi, H. A. Ahmed, P. Patel, M. A. Mendall, T. C. Northfield, The effects of unsaturated fatty acids on *Helicobacter pylori* in vitro. *J. Med. Microbiol*., **42**, 276-282 (1995)
10. C. J. Zheng, J. S. Yoo, T. G. Lee, H. Y. Cho, Y. H. Kim, W. G. Kim ; Fatty acid synthesis is a target for antibacterial activity of unsaturated fatty acids. *FEBS Lett*., **579**, 5157-5162 (2005)
11. C. Altieri, D. Cardillo, A. Bevilacqua, M. Sinigaglia, Inhibition of *Aspergillus* spp. and *Penicillium* spp. by fatty acids and their monoglycerides. *Journal of Food Protection*., **70**, 1206-1212 (2007)
12. M. Era, S. Sakai, A. Tanaka, T. Kawahara, T. Kanyama and H. Morita ; Antifungal Activity of Fatty Acid Salts Against *Penicillium pinophilum*. *Japan Journal of Food Engineering*., **16**, 99-108 (2015)
13. A. Kollanoor, P. Vasudevan, M. K. M. Nair, T. Hoagland, K. Venkitanarayanan ; Inactivation of bacterial fish pathogens by medium chain lipid molecules (caprylic acid, monocaprylin and sodium caprylate). *Aquaculture Res*., **38**, 1293-1300 (2007)
14. K. Takashima ; Mold inspection manual color atlas (Kabi Kensa Manuaru Kara Zufu). *Techno System Co., Ltd*., 214-215 (2009)
15. A. P. Desbois and V. J. Smith, Antibacterial free fatty acids: Activities, mechanisms of action and biotechnological potential. *Applied Microbiology and Biotechnology*. **85**, 1629-1649 (2009)
16. C. E. Isaacs, R. E. Litov, H. Thormar, Antimicrobial activity of lipids added to human milk, infant formula, and bovine milk. *J. Nut. Biochem*., **6**, 362-366 (1995)
17. A. P. Desbois, V. J. Smith ; Antibacterial free fatty acids: activities, mechanisms of action and biotechnological potential. *Appl Microbiol Biotechnol*., **85**, 1629-1642 (2010)
18. J. J. Kabara, D. M. Swieczkwowski, A. J. Conley, and J. Truant, Fatty acids and derivatives as antimicrobial agents. *Antimicrob Agents Chemother*., **2**, 23-28 (1972)
19. E. Kodicek, The effect of unsaturated fatty acids on gram-positive bacteria. *Soc. Exp. Biol. Symp*., **3**, 217-232 (1949)
20. M. Bayliss, Effect of the chemical constitution of soaps upon their germicidal properties. *J. Bacteriol*. **31**, 489-504(1936)
21. R. Fuller, J. H. Moore, The inhibition of the growth of Clostridium welchii by lipids isolated from the contentsof the small intestine of the pig. *J. Gen. Microbiol*. **46**, 23-41 (1967)
22. O. Wyss, B. J. Ludwig, and R. R. Joiner, The fungistaticand fungicidal action of fatty acids and related compounds. *Arch. Biochem*., **7**, 415-425 (1945)
23. H. Galbraith, T. B. Miller, Effect of long chain fatty acids on bacterial respiration and amino acid uptake. *Journal of Applied Bacteriology*., **36**, 659-675 (1973b)
24. I. Shibasaki, N. Kato, Combined effects on antibacterial activity of fatty acids and their esters against Gram-negative bacteria. In: Kabara JJ (ed) Symposium on the pharmacological effect of lipids. *The American Oil Chemists' Society, Champaign*,**19**, 15-24 (1978)

Antibacterial Activity of Cetyltrimethylammonium Bromide Modified Silver-Bentonite

Nik Ahmad Nizam Nik Malek , Wan Nur Azalisa, Clara Chong Yieh Lin

Faculty of Biosciences and Medical Engineering, Universiti Teknologi Malaysia, 81310 UTM, Skudai, Johor, Malaysia

Abstract. Organo-Ag-bentonite was prepared by the attachment of cationic surfactant cetyltrimethylammonium bromide on silver(A)g-exchanged bentonite. The prepared samples were characterized by X-ray diffraction (XRD), Fourier transform infrared (FTIR) spectroscopy and energy dispersive X-ray (EDX) analysis and antibacterial assay was performed against Escherichia coli in different percentage of saline solutions through minimum inhibition concentration (MIC) method. Organo-Ag-bentonite showed higher antibacterial activity than organo-bentonite and Ag-bentonite especially in saline solution suggesting that the precipitation of AgCl in the presence of Ag-bentonite in saline solution could be avoided by the attachment of cationic surfactant on Ag-bentonite surfaces, hence increased their antibacterial activity.

1 Introduction

For centuries, silver (Ag) has been well known for its antibacterial [1] and medicinal properties [2]. Because of this, Ag based products have been widely developed nowadays as antibacterial agent. The examples included Ag nanoparticles [3], Ag nanocomposite [4] and Ag sulfadiazine [5]. On the other hand, there are also various innovative Ag based products which have been produced and studied recently such as sutures modified by Ag-loaded montmorillonite [6], Ag nanoparticle–ionic silsesquioxane [7], nano-Ag fluoride in treating dental decay in children [8] and incorporation of silver nanoparticles into acrylic bone cement [9].

Exploring an improved Ag based products that can control the release of Ag ions into the environment is crucial since there were some reports on the toxicity of silver nanoparticles on a common grass, *Lollium multiflorum* [10], freshwater alga *Ochromonas danica* [11] and aquatic custaceans (*Daphnia magna* and *Thamnocephalus platyurus*) [12]. The use of inorganic carrier to support antibacterial compounds could increase their antibacterial efficiency and could control the release of Ag ions. In previous studies, some of the ideal carrier system for antibacterial agents were montmorrilonite (a type of clay) [13-15], kaolinite [16], zeolite [17-19] and bentonite [20-21].

Besides that, the antimicrobial activity of Ag based products has shown to decrease in solution containing electrolyte [17]. The present of chloride ions (Cl^-) in saline solution tend to reduce the number of Ag ions due to the formation of AgCl precipitate. As Ag ions are the main agent in killing the bacteria cells, reduction in the number of Ag ions will eventually lower its antibacterial activity. In order to solve this issue, combination of Ag ions with cationic surfactant QAC might be useful [17]. Among all the surfactant, quaternary ammonium compounds (QAC) is the choice as it is known to be the most useful antiseptics and disinfectants and it is claimed to have a great affinity towards chloride or bromide ions. Therefore, combination of cationic surfactant and Ag could be developed on the support system such as bentonite. Cationic surfactant and Ag supported on the bentonite could firstly be prepared by loading bentonite with Ag ions [13] from $AgNO_3$ solution through ions exchange technique [15] so that the Ag ions are located inside the bentonite framework structure followed by the attachment of the cationic surfactant on the prepared Ag-bentonite [17, 20].

The aim of this study was to prepare and characterize organo-Ag-bentonite; cationic surfactant QAC loaded on Ag exchanged bentonite. The antibacterial activity of the organo-Ag-bentonite was studied against *Escherichia coli* in different percentage of saline solutions through minimum inhibition concentration (MIC) method in order to prove their antibacterial efficiency in different concentration of electrolyte (chloride ions).

2 Materials and Methods

2.1 Materials

Bentonite was purchased from Sigma-Aldrich. Source of Ag ions and surfactant were taken from $AgNO_3$ (Merck) and cetyltrimethylammonium bromide (CTAB) (Merck), respectively. For the antibacterial assay, saline solution was prepared by dissolving an appropriate amount of

NaCl (Merck) in distilled water. Luria-Bertani (LB) broth and nutrient agar (NA) as growth medium for bacteria were purchased from Sigma-Aldrich.

2.2 Preparation of Samples

Silver modified bentonite (AgBent-S1) was firstly prepared by mixing bentonite (Bent-S) (10 g) with 1000 ppm $AgNO_3$ solution (1000 mL). The suspension was stirred using a magnetic stirrer for 16-17 hours and the solid fraction was collected by filtration system using Whatman filter paper. The solid residue was dried in an oven at 90°C overnight. The dried AgBent-S1 was further treated with $AgNO_3$ solution again to produce AgBent-S2 for the higher amount of Ag ions in the bentonite than AgBent-S1. CTAB-bentonite (CTAB-Bent-S) and CTAB modified Ag-bentonite (CTAB-AgBent-S-1 and CTAB-AgBent-S-2) were prepared by mixing 2.0 mM CTAB solution (200 mL) with bentonite (5.0 g), AgBent-S1 (5.0 g) and AgBent-S2 (5.0 g), respectively. The suspensions were stirred, filtered and dried similar to that of preparation of AgBent-S1.

2.3 Characterization

Bent-S and AgBent-S1 were characterized by X-ray diffraction (XRD) and energy dispersive X-ray (EDX) analyzer. XRD patterns were recorded using a Bruker D8 Advance diffractometer with these parameters: $CuK\alpha_1$ radiation with λ = 1.5418 Å, 40 kV, 20 mA, 2θ = 5° to 50°, scanning speed of 0.05° per second. The major elements especially the presence of Ag was detected by EDX (CARL ZEISS 35 VP Supra) with JED-2300 analysis station as a detector. All samples were characterized by Fourier-transform infrared (FTIR) spectroscopy (Nicolet™ iS™5 FT-IR Spectrometer) equipped with attenuated total reflectance (ATR) technique.

2.4 Antibacterial Assay

In the antibacterial assay, *E. coli* was used as a model bacteria through the determination of minimum inhibitory concentration (MIC) in distilled water and different concentrations of saline solution. *E. coli* was taken from the collection of bacteria culture in Microbiology Laboratory, Faculty of Bioscience and Medical Engineering, UTM and the bacteria was originally supplied by Institute Medical Research (IMR), Kuala Lumpur, Malaysia. Method for determination of MIC values for each sample was carried out based on procedure described elsewhere [17] but with different solutions. In the last stage, the *E. coli* cells were dispersed in distilled water and three different concentrations of saline solution: 0.1, 1.0 and 5.0 %. The *E. coli* suspension was added with the samples at different concentration: 0.1 g/L, 0.5 g/L, 1.0 g/L, 2.0 g/L, 5.0 g/L, 8.0 g/L and 10.0 g/L. The suspension was incubated in incubator shaker with agitation rate of 100 rpm for 30 minutes. After that, 10 µL of solution from the suspension was dropped on the NA plate of nutrient agar using the dropped plate technique [22]. The plate was then incubated in incubator at 37°C for 8 hours and finally, the plate was observed for the growth of bacteria. MIC value was determined by the lowest concentration that can kill the whole colony of bacteria.

3 Results and Discussion

3.1 Characterization of Studied Materials

Fig. 1 shows XRD patterns of raw bentonite (Bent-S) and Ag-bentonite (AgBent-S1). The XRD patterns shows that the major phase is montmorillonite (M) and other minor phases present in the sample are illite (I), quartz (Q) and kaolinite (K) [23]. There were no changes in the XRD patterns of both samples indicating that the structure of bentonite was stable after loaded with Ag cations [24]. Characterizations of both samples were further performed by elemental analysis qualitatively using EDX (Fig. 2). The EDX spectrum of Ag-Bent-S indicates the presence of Ag but absence in raw bentonite (Bent-S).

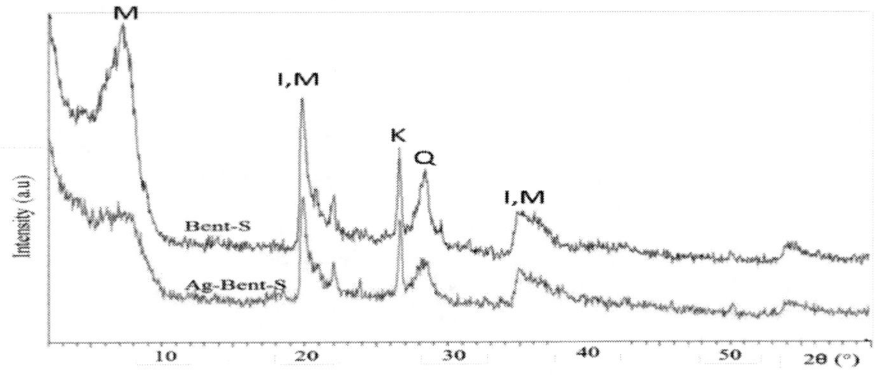

Figure 1. XRD of Bent-S and Ag-Bent-S (M: Montmorillonite, I: Illite, K: Kaolinite, Q: Quartz).

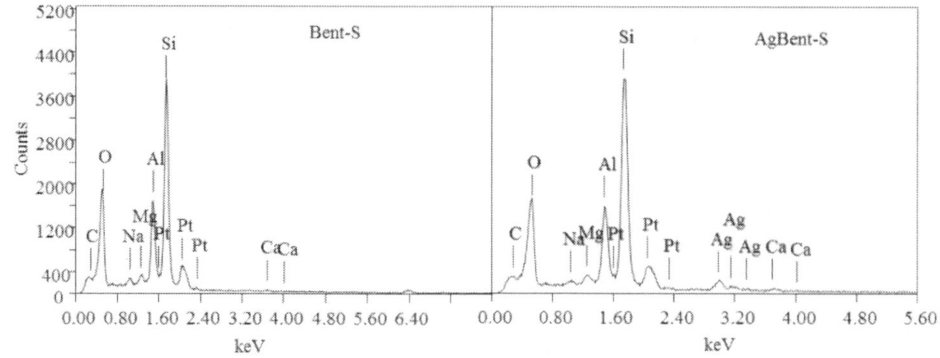

Figure 2. EDX spectra of Bent-S and Ag-Bent-S.

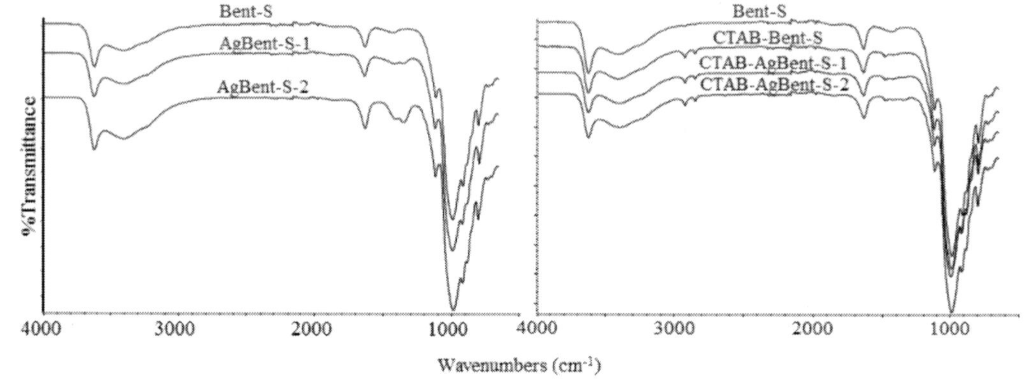

Figure 3. FTIR spectra of the samples.

FTIR spectra of Ag-bentonite (AgBent-S), CTAB modified bentonite (CTAB-Bent-S) and CTAB modified Ag-bentonite (CTAB-AgBent-S) are given in Fig. 3. Each FTIR spectra was compared with that of raw bentonite (Bent-S). Two important findings can be obtained from the FTIR spectra of the samples: (1) the structural stability after modification, and (2) the presence of CTAB molecules on the bentonite. All FTIR spectra showed the presence of similar peaks in the range of 1100–600 cm^{-1}. In this range, there are peaks related to stretching vibration of Si-O groups, Al-O-Si and Si-O-Si bending vibrations and Al-O and Si-O out of plane vibrations of bentonite framework structure [24]. Thus, these FTIR spectra show the structural stability of bentonite after modification with Ag and CTAB [17]. On the other hand, the presence of CTAB in CTAB modified bentonite and Ag-bentonite can be proven by the emergence of two additional peaks at 2850 and 2920 cm^{-1}. These two peaks are assigned to the asymmetric (2850 cm^{-1}) and symmetric (2920 cm^{-1}) stretching vibrations of C-H in the alkyl chain of CTAB [25]. Fig. 3 clearly indicates that the FTIR spectra of CTAB-Bent-S and CTAB-Ag-Bent-S possess these two additional peaks confirming the attachment of CTAB molecules on bentonite and Ag-bentonite surfaces.

Findings from XRD, EDX and FTIR validated the modification of bentonite with Ag (AgBent-S) and CTAB (CTAB-Bent-S) and also, the attachment of CTAB on Ag-bentonite creating organo-Ag-bentonite (CTAB-AgBent-S). Additionally, results from XRD and FTIR show that the bentonite structure remains stable after modification since the modification process of Ag and CTAB on bentonite occurred via ion exchanged process without affecting the structure of bentonite [17].

3.2 Antibacterial Activity

Table 1. MIC values of studied samples against *E. coli.*

Sample	MIC [g/L]			
	Distilled water	Saline solution [%]		
		0.1	1.0	5.0
Bent-S	>10	>10	>10	>10
AgBent-S1	0.5	0.8	3	>10
AgBent-S2	0.1	0.5	2	>10
C-Bent-S	>10	>10	>10	>10
CTAB-AgBent-S-1	0.5	2	2	5
CTAB-AgBent-S-2	0.1	0.5	1.0	1.0

The antibacterial activity of each sample was performed against *E. coli* (Gram negative bacteria) through the determination of minimum inhibition concentration (MIC) values in distilled water and saline solution with different concentrations (0.1, 1.0 and 5.0 %). The lower MIC value indicates better antibacterial activity since only small amount of sample is needed to kill the whole bacteria or to inhibit the bacterial growth. Table 1 gives MIC values for each sample.

Raw bentonite (Bent-S) and CTAB modified bentonite (C-Bent-S) have no antibacterial activity against *E. coli* in distilled water and saline solutions (MIC values >10 g/L). Raw bentonite did not show any antibacterial activity because there were no antibacterial elements e.g. antibacterial metal (Cu, Zn and Ag) present in the sample according to the elemental analysis of raw bentonite by EDX (Fig. 2). CTAB on its own (free CTAB) has antibacterial activity [13] however, the antibacterial activity of CTAB was reduced after loaded on bentonite because of the attachment of CTAB molecules on bentonite prevents the movement of CTAB molecules in the solution. This result is in agreement with the work done by Malachova *et al.* [13] where CTAB modified montmorillonite showed lower antibacterial effects by 18 orders of magnitude as compared to CTAB itself.

Generally, the antibacterial activity of the Ag-bentonite depends on the amount of Ag in the sample where the antibacterial activity followed these trends: AgBent-S2>AgBent-S1 and CTAB-AgBent-S-2>CTAB-Agbent-S-1. This result showed that the antibacterial activity of bentonite was enhanced by the Ag ions inside the bentonite. The antibacterial assay was carried out in solution where the samples in powder form were added in bacteria solution either in distilled water or saline solution. Thus, the Ag ions could release into the solution from bentonite framework and eventually kill the bacteria cells [13,15]. Additionally, the antibacterial activity of the Ag-bentonite was also affected by different concentrations of saline solution (0.1 % > 1.0 % > 5.0 %). At 5.0 % of saline solution, Ag-bentonite showed no antibacterial activity. This is mainly due to the reaction between Ag ions that released from bentonite with the chloride ions in the saline solution forming precipitate AgCl [13,26]. Precipitation of the AgCl in the bacteria solution reduced the antibacterial activity since the free Ag ions are no longer exist in the bacteria solution to kill the bacteria.

From MIC values of the samples, it is observed that the modification of Ag-bentonite by CTAB (CTAB-AgBent-S) could increase the antibacterial activity of Ag-bentonite especially in the highest concentration of saline solution (5.0 %). The antibacterial activity of CTAB modified Ag-bentonite did not affected by the saline solution and this is certainly due to the presence of CTAB on the bentonite surfaces. This can be understood by the mechanism of surfactant attachment on clay framework. It is well-known for a long time that surfactant modified clay or organoclay has the ability to adsorb and retain anionic compound [27-29]. Due to the molecular structure of cationic surfactant like CTAB that consists of two essential parts: (1) head positive group of quaternary ammonium and (2) hydrophobic tail of hydrocarbon chain (C16) and the capability of this surfactant to form micelles in the solution, the attachment of cationic surfactant on clay creates organic rich layer on the clay surfaces [16]. The configuration of cationic surfactant molecules on clay surfaces is affected by the concentration or the amount of cationic surfactant in the initial solution (before attachment on clay surfaces). The excess amount of cationic surfactant molecules creating double layer configuration of the molecules on the clay surfaces generated by two main bonding in two surfactant layers: (1) electrostatic bonding between positive head group of cationic surfactant with negative charge framework of clay at the first layer and (2) hydrophobic bonding among cationic surfactant molecules through tail-tail interaction of hydrocarbon tail of surfactant at the second layer [16]. Because of the attachment of cationic surfactant on clay surface, the surfactant modified clay is capable of adsorbing and retaining anionic compounds. In relation to this study, possible reason for the enhanced antibacterial activity for CTAB modified Ag-bentonite is due to the uptake of chloride ions by surfactant molecules on bentonite layer and thus prevents the formation of silver chloride in the solution. This increased the Ag ions in the high concentration of saline solution and eventually kill the bacteria by the action of Ag ions.

4 Conclusion

Organo-Ag-bentonite was prepared without affecting the original structure of bentonite and showed higher antibacterial activity than both CTAB-bentonite and Ag-bentonite especially in saline solution. Therefore, organo-Ag-bentonite can be an alternative antibacterial agent that can replace other Ag based products due to its stability and high antibacterial activity for wide spectrum of bacteria. This modification is also useful in preventing the reduction of antibacterial activity of Ag based products in solution that contain chloride ions.

Acknowledgements

Authors would like to thanks Ministry of Education (MoE), Malaysia and Universiti Teknologi Malaysia (UTM) for financial support under Fundamental Research Grant Scheme (FRGS) (Vot No: 4F514) and Faculty of Biosciences and Medical Engineering, UTM.

References

1. P. Praus, M. Turicová and M. Valášková: J. Brazil. Chem. Soc. Vol. 19 (2008), p. 549.
2. B.S. Atiyeh, M. Costagliola, S.N. Hayek and S.A. Dibo: Burns. Vol. 33 (2007), p.139.
3. S. Rajakannu, S. Shankar, S. Perumal, S. Subramanian and G.P. Dhakshinamoorthy: Int. J. Curr. Microbiol. Appl. Sci. Vol. 4 (2015), p. 944.
4. G. Yang, C. Wang, F. Hong, X. Yang and Z. Cao: Carbohyd. Polym. Vol. 115 (2015), p. 636.
5. C. Aguzzi, G. Sandri, C. Bonferoni, P. Cerezo, S. Rossi, F. Ferrari, C. Caramella and C. Viseras: Colloid Surface B. Vol. 113 (2014), p. 152.
6. G.F. Cao, Y. Sun, J.G. Chen, L.P. Song, J.Q. Jiang, Z.T. Liu and Z.W. Liu: Appl. Clay Sci. Vol. 93 (2014), p. 102.
7. A.C. Schneid, E.W. Roesch, F. Sperb, U. Matte, N.P. da Silveira, T.M.H. Costa, E.V. Benvenutti and E.W. de Menezes: J. Mater. Chem. B. Vol. 2(8) (2014), p.1079.

8. A.G.R. Targino, M.A.P. Flores, V.E. dos Santos Junior, F.D.G.B. Bezerra, H. de Luna Freire, A. Galembeck and A. Rosenblatt: J. Mater. Sci.-Mater. Med. Vol. 25(8) (2014), p. 2041.
9. J. Slane, J. Vivanco, W. Rose, H.L. Ploeg and M. Squire: Mater. Sci. Eng. C. Vol. 48 (2015) p. 188.
10. L. Yin, Y. Cheng, B. Espinasse, B.P. Colman, M. Auffan, M. Wiesner, J. Rose, J. Liu and E.S. Bernhardt: Environ. Sci. Technol. Vol. 45(6) (2011), p. 2360.
11. A. J. Miao, Z. Luo, C. S. Chen, W.C. Chin, P.H. Santschi and A. Quigg: PLoS One, Vol. 5(12) (2010), p. e15196.
12. I. Blinova, J. Niskanen, P. Kajankari, L. Kanarbik, A. Käkinen, H. Tenhu, O.P. Penttinen and A. Kahru: Environ. Sci. Pollut. Res. Vol. 20(5) (2013), p. 3456.
13. K. Malachova, P. Praus, Z. Pavlíčková and M. Turicová: Appl. Clay Sci. Vol. 43 (2009), p. 364
14. K. Malachová, P. Praus, Z. Rybková and O. Kozák: Appl. Clay Sci. Vol. 53(4) (2011), p. 642.
15. S.M. Magana, P. Quintana, D.H. Aguilar, J.A. Toledo, C. Ángeles-Chávez, M.A. Cortés, L. León, Y. Freile-Pelegrín, T. López and R.M.T. Sánchez: J. Mol. Catal. A.-Chem. Vol. 281(1) (2008), p. 192.
16. N.A.N.N. Malek, N.I. Ramli: Appl. Clay Sci. Vol. 119-110 (2015), p. 8.
17. N.A.N.N. Malek, C.D. Williams, S. Dhanabal, H.S. Bhall and N. Ibrahim: Appl. Mech. Mater. Vol. 606 (2014), p.29.
18. Y. Matsumura, K. Yoshikata, S.I. Kunisaki and T. Tsuchido: Appl. Environ. Microbiol. Vol. 69(7) (2003), p. 4278.
19. B. Kwakye-Awuah, C. Williams. M.A. Kenward and I. Radecka: J. Appl. Microbiol. Vol. 104(5) (2008), p. 1516.
20. N.A.N.N. Malek, S.A. Ishak and M.R.A. Kadir: Adv. Mater. Res. Vol. 626 (2013), P. 178.
21. M.F. Santos, C.M. Oliveira, C.T. Tachinski, M.P. Fernandes, C.T. Pich, E. Angioletto, H.G. Riella and M.A. Fiori: Int. J. Miner. Process. Vol. 100(1) (2011), p. 51.
22. B. Herigstad, M. Hamilton and J. Heersink: J. Microbiol. Method, Vol. 44(2) (2001), p.121.
23. M. Sarkar, K. Dana, S. Ghatak and A. Banerjee: Bull. Mater. Sci. Vol. 31(1) (2008), p. 23.
24. E. Eren and B. Afsin: J. Hazard. Mater. Vol. 151(2) (2008), p. 682.
25. Z. Li and L. Gallus: Appl. Clay. Sci. Vol. 35(3) (2007), p. 250.
26. O. Choi, K.K. Deng, N.J. Kim, L. Ross Jr, R.Y. Surampalli and Z. Hu: Water Res. Vol. 42(12) (2008), p. 3066.
27. Z. Li and R.S. Bowman: Environ. Eng. Sci. Vol. 15(3) (1998), p. 237.
28. S.A. Boyd, M.M. Mortland and C.T. Chiou: Soil Sci. Soc. Am. J. Vol. 52(3) (1988), p. 652.
29. Z. Li; J. Environ. Qual. Vol. 28(5) (1999), p. 1457.

Estimation of Odds and Probability Risk Factors in Experiencing Heart Attack

Jackie D. Urrutia , and John Lean B. Diaz
Office of the Academic Head, Polytechnic University of the Philippines – Parañaque Campus
Col. E. L. De Leon St. Sto. Niño Parañaque City, Philippines

Abstract. The Heart Attack has produced an alarming rate of deaths and considered as one of the most deadly events for persons with hypertension, diabetes, with family history, a smoker and alcoholic. The study aims to produce a model using the Forward Stepwise Binary Logistic Regression Method for estimating the odds and probabilities when conceded with the given situations, and also the comparisons of the odds in which the factors are arranged by combinations of the significant variables that are included in the equation of every circumstances. Combination of covariates, gender aside, where the outcome of the study indicates that men are two times more likely to have heart attack than women. Hypertension and Diabetes, two covariates giving two of the highest odds among the factors, is 35 times more likely to trigger a heart attack than a person with none. Smoking plus the two mentioned covariates gives 114 times more. Alcohol added, has the ratio of 210 to 1 which is harshly a huge number. Summing up all the risks of the factors considered, including the family history as the last covariate, it shows the ratio 352 times that is, than a person with no complication nor any harmful vices.

1 Introduction

The number of cases of Heart Attack in the Philippines has a consistent escalation that makes the country 29th on the world ranking to have the most number of deaths where 5 out of 10 Filipinos die of heart disease according to the NSO. Furthermore, World Health Organization (WHO) has said that an estimated 17 million people die of cardiovascular diseases every year. Most of these were heart attacks and strokes. Most common causes of these were other related ailments such as diabetes and hypertension that are associated with heart diseases, and harmful vices like smoking and others were recorded alcoholic.

The record data, consisting 447 number of cases from 2009-2014 in Quirino Memorial Medical Hospital, will be processed in a Logistic Regression analysis in which the intervening factors considered and their combinations are regressed against a dichotomous response variable for producing a model for the patients' chances of experiencing heart attack. Each variable and combinations of two, three, and a combination where all of the significant factors are included, will be used in the estimating odds and probabilities. Furthermore, the odds of every combination listed will be compared with the odds of a normal individual with no complications.

1.1 Objective of the study

The main objective of the study is to generate the odds and probabilities estimation to trigger a heart attack on patients from Quirino Memorial Medical Hospital that are having any or a combination of the intervening factors from the equations given using Binary Logistic Regression. Furthermore, it is to identify whether gender, hypertension, smoking, alcohol, diabetes or having a family history are significant and are a potent risks of having a heart attack and the odds of these factors compared with the odds of an individual with no complications.

1.2 Scope and limitations

The medical record from Quirino Memorial Medical Hospital of the patients, those who are at risk of experiencing heart attack, were provided by the medical management of the hospital for our study. The data gathered was limited from a total population 447 patients observed within the span of 66 months from January 2009 until June of 2014, where patients were being recorded by whether he/she(x1) has experienced(y) heart attack, if the patient is hypertensive(x2), prone to smoking(x3) or alcoholic drinks(x4), has a diabetes(x5), has a family history(x6), or no complications at all.

1.3 Statement of the Problem

1) What is the risk profile of the patients from the recorded data in Quirino Memorial Medical Hospital?
2) What is the binary response model that estimates the probabilities and odds for the patients who are at risk of heart attack?
3) What are the intervening factors that are significant and will be included in the model?
4) What are the odds and the probabilities of an individual, when classified with the considered significant intervening factors in the model, to trigger a heart attack?

5) What are the odds ratios with one intervening covariate? Two or more covariates? With combinations of the covariates ?

1.4 Research paradigm

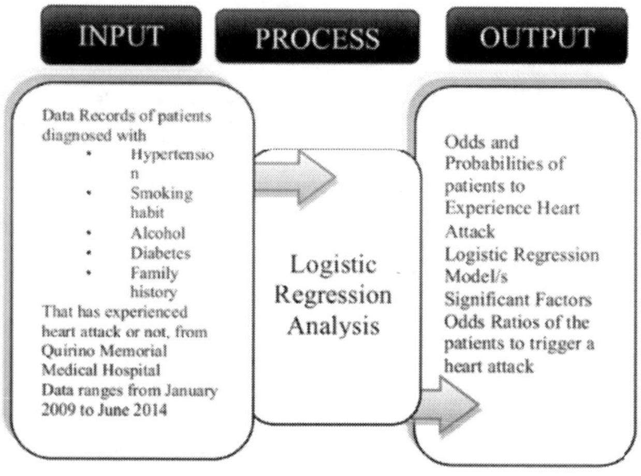

1.5 Significance of the study

The study may be of help in determining and estimating the odds risk of a patient to rouse a heart attack. Outcomes of the study would be a benefit to unaware individual possessing any or some of the factors, so they can make an early stage management of how to deal with their health and promote knowledgeable decisions regarding the chances they have to trigger the heart attack. These might be more possible by being more considerate of the threats of modified intervening risk factors, and might be of help to reduce the probability and the odds of the patients to experience a heart attack.

2 Methodology

2.1 Population and Covariates

The study produces a model for estimating odds and probabilities to trigger a heart attack for when a patient is either male or a female which either has been diagnosed hypertensive, a smoker, alcoholic, diabetic, and has a family history or a combination of these considered factors, by which these patients are from Quirino Memorial Medical Center.

2.2 Logistic Regression

The Binary-Response Logistic Regression Method is mostly used in clinical studies, epidemiology, data mining, social sciences, marketing and engineering, because it models the dependence of a binary response variable from one or more explanatory variables. The model were used in medical and social sciences because it produces and output probability that ranges between 1 and 0 which can be revised as yes or no, positive or negative, and any other dichotomous responses. This model is less restrictive than any other models when it comes to assumptions aside from the dependent variable being dichotomous or categorical, and the predictors not correlating with each other which can be detected through test of multicollinearity. Fitting it to a probability that ranges between 1 and zero, it would take the equation:

$$Probability\ P(y|x_1, x_2, x_3, x_4, x_5, x_6 = \frac{odds}{1+odds} = \frac{\exp[\beta + \beta_1(x_1) + \beta_2(x_2) + \ldots + \beta_n(x_n)]}{1+\exp[\beta + \beta_1(x_1) + \beta_2(x_2) + \ldots + \beta_n(x_n)]}$$

Odds are the likelihood of something happening. The chance that an event will occur divided by the probability that it will not occur. If the probability of an event occurring is p, the probability of the event not occurring is (1-p).

$$Odds = exp\left[Log\left(\frac{p}{1-p}\right)\right] = exp(\beta_0 + \beta_1 x_1 + \beta_2 x_2 + \beta_3 x_3 + \cdots + \beta_n x_n)$$

Where p is the probability of interested outcome, β_0 is the intercept parameter, β is a regression coefficient, and χ is a predictor. Then, it would become e^{β_1} for the Odds estimated for a one unit change in x_1.

$$Logit = \ln(odds) = \ln\left(\frac{p}{1-p}\right) = \beta_0 + \beta_1 x_1 + \beta_2 x_2 + \beta_3 x_3 + \cdots + \beta_n x_n)$$

The chance that one thing will happen rather than another, the ratio of the probability of one event to that of an alternative event. Then, the corresponding odds is a value given by odds of an event ($odds_e$) over odds alternate ($odds_a$), which is simply can be considered as:

$$Odds\ ratio = \frac{odds_e}{odds_a}$$

Also, Forward Stepwise Method is used by researchers and is a method where variables are entered in the equation one at a time to fit a model, but it differ from any other logit methods. If a variable is not significant at a specified level of significance chosen by researchers, the variable is excluded from the next model creation.

3. Results and Discussions

3.1. Profile Risk

Table 1.

For a total of 447 patients from Quirino Memorial

Category		Heart Attack Experience			
Intervening Factors	Classification	1=yes		0=no	
		number	percentage	number	percentage
Gender(x_1)	1=male	124	27.74%	40	8.95%
	0=female	203	45.41%	80	17.90%
Hypertension(x_2)	1=hypertensive	212	47.43%	28	6.26%
	0=nothypertensive	115	25.73%	92	20.58%
Smoking(x_3)	1=smoker	180	40.27%	33	7.38%
	0=not smoker	147	32.89%	87	19.46%
Alcohol(x_4)	1=alcoholic	189	42.28%	57	12.75%
	0=non-alcoholic	138	30.87%	63	14.09%
Diabetes(x_5)	1=diabetic	156	34.90%	17	3.80%
	0=not diabetic	171	38.26%	103	23.04%
Family History(x_6)	1=yes	212	47.43%	58	12.98%
	0=no	115	25.73%	62	13.87%
Total		327	73.15%	120	26.85%
		447(100%)			

Hospital, number of 327 or more than 73% of the total population has been recorded to have experienced heart attack while the rest has not experienced a heart attack. Most of them were female patients with a number of 283, and 164 or 37% of the total number of cases were male patients. On the other hand, these patients were also classified by ailments and harmful vices namely: Hypertension, Smoking, Family History, and whether the patient is alcoholic or not. Hypertensive patients were 240 which is 53.7% of the total population of the patients whereas the rest which is 207 otherwise 46.3% were the non-hypertensive patients. Moreover, 213 patients were classified smokers for 47.7% of the entire population from the collected data, the remaining number were non-smoker patients. Alcoholic patients have a total number of 246 which is 55% of the population and the non-alcoholic patients have a total of 201 patients. Non-diabetic patients have a greater in number which is 274 patients or 61.3%, than the patients with diabetes for a total of 173 patients 38.7% of the entire population. Furthermore, 274 patients or 60.4% of the entire number of patients said that a member of their family has experienced having a heart attack and the rest, 39.6% of the entire number of patients or a number of 177 patients said that no member of their family that has experienced a heart attack.

3.2 Model

These are: male or female(x_1), hypertensive(x_2), a smoker or not(x_3), the patient is alcoholic or not(x_4), diabetic or not(x_5), and whether some of their family members have already experienced heart attack or has no family history of heart attack. In fitting the model into a Binary logistic regression,

$$Odds = exp\left[Log\left(\frac{p}{1-p}\right)\right] = exp(\beta_0 + \beta_1 x_1 + \beta_2 x_2 + \beta_3 x_3 + \beta_4 x_4 + \beta_5 x_5 + \beta_6 x_6)$$

$$odds = exp[-1.61044338719005 + 0.602174493679787(x1) + 1.82822458539002(x2) + 1.17294305751614(x3) + 0.609697762156693(x4) + 1.73475041391757(x5) + 0.51787722202522(x6)]$$

then, getting the probability would be…

$$P(y|x1, x2, x3, x4, x5, x6) = \frac{odds}{1+odds}$$
$$= \frac{exp[-1.6104 + 0.6022(x_1) + 1.8282(x_2) + 1.1729(x_3) + 0.6097(x_4) + 1.7348(x_5) + 0.5179(x_6)]}{1 + exp[-1.6104 + 0.6022(x_1) + 1.8282(x_2) + 1.1729(x_3) + 0.6097(x_4) + 1.7348(x_5) + 0.5179(x_6)]}$$

3.3 Significant Factors

Table 2. Significant Factors

Factors	β	Significance
Gender(x_1)	0.602174493679787	0.047
Hypertension(x_2)	1.82822458539002	0.000
Smoking(x_3)	1.17294305751614	0.000
Alcohol(x_4)	0.609697762156693	0.035
Diabetes(x_5)	1.73475041391757	0.000
Family History(x_6)	0.51787722202522	0.044
Constant(β)	-1.61044338719005	0.000

As seen the table, all variables were statistically significant at 0.05 significance level. Gender(x1), the first variable has been considered substantial at 0.047 by which it has a lesser significance value than the 0.05. And same, is applicable to three variables which are Hypertension(x2), Smoking(x3), and Diabetes(x5) these variables have significance probability value which nearly zero. Also, the two remaining variables, Alcohol(x4) and Family History(x6) were considered significant as well with their significance value, 0.035 and 0.044 respectively.

3.4 Odds and Probabilities of Combined Factors

Table 3. Odds and Probabilities of Two Factors

Category	Male		Female	
	Probabilities	Odds	Probabilities	Odds
No complications	0.267318768	0.364850027	0.166527064	0.199799006
Hypertensive	0.694226935	2.270399238	0.554231126	1.243314998
Smoker	0.541075760	1.179008893	0.392336752	0.645648314
Alcoholic	0.401655689	0.671278530	0.268794848	0.367605243
Diabetic	0.674032692	2.067792308	0.531036801	1.132363483
With family history	0.379801304	0.612386493	0.251135361	0.33534813

Using the model, the probabilities and odds are listed by the combinations of all factors that are included in the equations produced in sighting to the considered circumstances, whether the person is a male or female, hypertensive or not, a smoker or non-smoker, alcoholic or not, diabetic or not, and whether has a family history or still has none of the members of the family experienced heart attack.

Visibly in the table, gender and each of the remaining variables are entered in the equation where male has greater probabilities and odds than the female, which means the male patients has higher risks when compared to females which may perhaps be less prone to trigger a heart attack. Then, the factor that has the highest risk is the hypertension than the rest, where it has the utmost odds with 2.27 for male and 1.24 for female, and probability in which it is 69% more likely to happen for men and 55% probability for women. Diabetes has the second highest risk with 2.06 in its odds and 67% probability to trigger a heart attack for men, and 1.13 in the odds and its probability touched 53% chance to trigger a heart attack for women. Smoking also contributes a high risk to trigger a heart attack in which the results above says that smoking is the third most contributor in the odds and probability chances of male and female.

Table 4. Odds and Probabilities of Three Factors

Category	Male		Female	
	Probabilities	Odds	Probabilities	Odds
no complications	0.267318768	0.364850027	0.166527064	0.199799006
hypertension, diabetes	0.92788904	12.86751741	0.875722426	7.046504034
smoking, hypertension	0.880049474	7.336770426	0.800707841	4.017758884
diabetes, smoking	0.869826385	6.682048338	0.785371813	3.659220272
hypertension, family history	0.792133339	3.81077627	0.676045744	2.086855568
hypertension, alcohol	0.806847314	4.177251331	0.695821606	2.287544474
diabetes, family history	0.776321805	3.470708468	0.655247046	1.900627793
diabetes, alcohol	0.791860924	3.804479859	0.67568348	2.08340753
smoking, family history	0.664307902	1.978920286	0.520083412	1.083695427
smoking, alcohol	0.684465869	2.169229263	0.542943298	1.187912343
Alcohol, family history	0.529791194	1.126714742	0.381575023	0.617011015

The table above has some of the odds and the probabilities from the combinations with three factors including the first variable gender, in which compared to the male or female with no complications, a person that have hypertension and diabetes has the highest odds 12.86 for men and 7.05 for women, and probability chances 93% for men and 88% for women to trigger a heart attack. The other two combinations have highest odds are not less than the 6.68 for men and 3.66 for women, and are does not fall below the percentage of 85% for men and 79% for women to trigger a heart attack. That concludes that having these circumstances, there is already a great risk to develop heart attack. The rest of the combinations did not fall below 50% of their probabilities for both men and women, and odds which fell between 5 and 1.

Table 5. Odds and Probabilities of Four Factors

Category	Male		Female	
	Probabilities	Odds	Probabilities	Odds
no complications	0.267318768	0.364850027	0.166527064	0.199799006
hypertension, diabetes, smoking	0.976515481	41.58124247	0.957931405	22.77070109
hypertension, diabetes, alcohol	0.959472538	23.67462662	0.928390811	12.96468826
hypertension, diabetes, family history	0.955747561	21.59762441	0.922041157	11.82728126
diabetes, smoking, alcohol	0.924778861	12.29413525	0.870675866	6.732508757

diabetes, smoking, family history	0.918137173	11.21555664	0.859980415	6.141858025
hypertension, smoking, alcohol	0.9310285	13.49874221	0.88084138	7.392175071
hypertension, smoking, family history	0.924893811	12.31448204	0.87086195	6.743651055
hypertension, alcohol, family history	0.875177137	7.011352885	0.793369384	3.839553879
diabetes, alcohol, family history	0.86460268	6.385670557	0.777625476	3.496918006
smoking, alcohol, family history	0.784527638	3.640966426	0.665983534	1.993864379

The highest odds among the men has reached 41.58, and are smokers with hypertension and diabetes, which also has the highest percentage reached 98% of the chance to trigger a heart attack. Odds and probabilities of men still show greater risks than that of the women. Most of the probabilities from the men were greater than 90%. In women, have the probabilities that does not fall smaller than 60%. The odds of both sections are having candidly high risks for patients, with the given circumstances, to trigger a heart attack.

Table 6. Odds and Probabilities of Five Factors

Category	Male		Female	
	Probabilities	**Odds**	**Probabilities**	**Odds**
No complications	0.267318768	0.364850027	0.166527064	0.199799006
hypertension, diabetes, smoking, alcohol	0.98709749	76.50429825	0.976687395	41.89524901
hypertension, diabetes, smoking, family history	0.985874207	69.79248822	0.97450263	38.21973066
hypertension, diabetes, alcohol, family history	0.975452252	39.73693429	0.956064662	21.76072188
hypertension, smoking, alcohol, family history	0.957729411	22.65711054	0.925414749	12.40747657
diabetes, smoking, alcohol, family history	0.95377908	20.6352249	0.918700856	11.30025248

Five factors included in the equation, produces high probabilities for both men and women where the chances do not fall lower than 90% chance to trigger a heart attack. This indicates that having these circumstances will surely lead to a higher risks. Hypertensive male patients that are smokers, diabetic and alcoholic at the same time, has the highest risks among the listed combinations above in the table, as well as with its odds which reached 76.50 of the likelihood that a heart attack would happen and the probability that it will trigger the event is approximately 99% the chance. Comparably to the chance of a person with no complications, it shows a huge difference. The odds of the different situations given on the list also demonstrate the higher risk of these intervening factors.

Table 7. Odds and Probabilities of All Factors

Category	Male		Female	
	Probabilities	**Odds**	**Probabilities**	**Odds**
No complications	0.267318768	0.364850027	0.166527064	0.199799006
hypertension, diabetes, smoking, alcohol	0.98709749	76.50429825	0.976687395	41.89524901
hypertension, diabetes, smoking, family history	0.985874207	69.79248822	0.97450263	38.21973066
hypertension, diabetes, alcohol, family history	0.975452252	39.73693429	0.956064662	21.76072188
hypertension, smoking, alcohol, family history	0.957729411	22.65711054	0.925414749	12.40747657
diabetes, smoking, alcohol, family history	0.95377908	20.6352249	0.918700856	11.30025248

The probability when all of the intervening factors are inserted in the equation, has reached approximately or almost a hundred percent, while the odds implies very excessive risk or likelihood for a heart attack to happen in men with 128.41 chance and 70.32 for women. Then compared with a person to be non-hypertensive, non-smoker, non-alcoholic, non-diabetic, and has no family history, the difference is huge and is evidently perceived in the table.

4.3 Odds Ratios of Covariate Combinations

Table 8. Odds Ratio Male/Female

Gender Comparison	Odds Ratio
male/female	1.826085297

Table above shows the comparison of the odds by gender, and the outcome indicates that men, are approximately 2 times the risk to trigger a heart attack than women. Men are usually more prone to these complications may perhaps by their harmful vices and unbalanced diet or conceivably unhealthy habits. The outcome might also imply that men may perhaps trigger a heart attack earlier than most women.

Table 9. Odds Ratio of a Covariate

Complication/No complication	Odds Ratio
Hypertension	6.222828744
Smoking	3.231489116
Alcohol	1.839875235
Diabetes	5.667513095
Family History	1.678460866

As seen in the table, having hypertension or diabetes add up to 6 times greater the risk of having a heart attack, than a person with no hypertension or diabetes. Smoking has an odds ratio of 3 is to 1, means that a smoker is 3 times more likely to trigger a heart attack than that of a non-smoker. Likewise, being an alcoholic increases the odds of a person 2 times than non-alcoholic, the same with having family history, which if a history of an individual related to the person has already experienced a heart attack, then, he or she is also 2 times more likely to trigger a heart attack than a person with no family history.

Table 10. Odds Ratio Two Covariates

Complication/No complication	Odds Ratio
hypertension, diabetes	35.26796339
smoking, hypertension	20.10900336
diabetes, smoking	18.31450688
hypertension, family history	10.44477452
hypertension, alcohol	11.4492285
diabetes, family history	9.512698935
diabetes, alcohol	10.42751699
smoking, family history	5.423928018
smoking, alcohol	5.945536795
alcohol, family history	3.088158579

The table above shows the odds ratio of the following collective groupings of the factorial combinations with two variables where the ratios can be applied to both male and female persons. Hypertensive person which is diabetic at the same time, is 35 times more likely to have a heart attack than a person that is not hypertensive nor diabetic. A hypertensive smoker is more likely having 20 times greater odds than a non-smoker and non-hypertensive individual. A diabetic smoker is 18 times more likely to have a heart attack than a person that is a non-smoker nor diabetic. The situation with the other groupings of the factors included, also had their odds compared with the odds of a person with no complications.

Table 11. Odds Ratio Three Covariates

Complications/No complication	Odds Ratio

hypertension, diabetes, smoking	113.9680398
hypertension, diabetes, alcohol	64.88865243
hypertension, diabetes, family history	59.19589637
diabetes, smoking, alcohol	33.69640764
diabetes, smoking, family history	30.74018307
hypertension, smoking, alcohol	36.99805727
hypertension, smoking, family history	33.75217518
hypertension, alcohol, family history	19.21708197
diabetes, alcohol, family history	17.50217919
smoking, alcohol, family history	9.979350836

The combinations listed were the collective groupings of 3 factors. The ratios are applicable with the male and female individual. One of the comparisons that are included in the table where hypertensive smoker that has diabetes 114 times greater the risk of having a heart attack than an individual with no complications like mentioned. The same with the second highest odds ratio of a hypertensive patient with diabetes and family history, which is approximately 65 times greater the odds than a normal individual to trigger a heart attack. 59 times greater if he or she is hypertensive, diabetic and has a family history. 37 times if the person is hypertensive, smoker and alcoholic. The same applies with the rest of the combinations given.

Table 12. Odds Ratio Four Covariates

Complications/No complications	Odds Ratio
hypertension, diabetes, smoking, alcohol	209.6869741
hypertension, diabetes, smoking, family history	191.2908948
hypertension, diabetes, alcohol, family history	108.9130637
hypertension, smoking, alcohol, family history	62.09979123
diabetes, smoking, alcohol, family history	56.55810154

The listed combinations are having four intervening factors where a hypertensive diabetic person which is a smoker and an alcoholic one has 210 times greater risk than a normal individual with no complications, 191 times if the person is hypertensive diabetic smoker with family history, approximately 109 times greater risk if the person is hypertensive alcoholic with diabetes, 62 times if he or she is a hypertensive smoker and alcoholic with family history, and a diabetic smoker which is an alcoholic with family history has 57 times greater risk.

Table 13. Odds Ratio All Five Covariates

Complications/No complication	Odds Ratio
hypertension, diabetes, smoking, alcohol, family history	351.95138

A person possessing all the intervening factors has the odds 352 times greater than a normal individual's chance of having a heart attack whether the person is male or a female.

4. Conclusion and Recommendation

Results implies that the hypertension, diabetes, smoking, alcohol, and having family history of experiencing heart attack increases the risk of developing a heart attack as indicated in the number of times the patients are compared to an individual with no complications. Additionally, gender comparison of the odds were proven that men is more likely to trigger a heart attack than women, which means men is more prone into heart diseases and may perhaps because men are inclined to having unhealthy lifestyle, unbalanced eating habits, and harmful vices that distress the body.

Investigators as well, recommend a further review of the topics discussed in basis with the study, for generating more researches about heart attack risks. Researchers suggests for patients having any or some of the intervening factors included in the study, such as hypertension and diabetes, to inquire from specialists, so that they could make more ascertained and learnt decisions regarding with these ailments and the situation they are in. As implied by the results, smoking and alcohol, as well as having a member of a family that has experienced heart attack, certainly increases the risks in an individual, then, researchers might as well recommend averting one selves from smoking and alcoholic beverages, or implement self-avoidance from these harmful habits to prevent any more complications and of course, having a healthier lifestyle to reduce the risks of triggering a heart attack whitin one's self.

References

[1] Chen, M. (2014, May 13). Hypertensive Heart Disease. Retrieved September 2015 from https://www.nlm.nih.gov/medlineplus/ency/article/000163.htm
[2] Berstien, S. (2014, June 23). Alcohol and Heart Disease. Retrieved September 2015 from http://www.webmd.com/heart-disease features/alcohol-and-heart-disease /
[3] Goldberg, R. (2013, August 1). Diabetes, Heart Disease and Stroke. Retrieved October 2015, from http://www.niddk.nih.gov/health-information/health-topics/Diabetes/diabetes-heart-disease-stroke/Pages/index.aspx
[4] SoRelle, R. (n.d.). Studying Populations and Heart Disease Risk. Retrieved October 28, 2015, from http://circ.ahajournals.org/content/99/5/598.full
[5]Sindico, R. (2012, July 9). 5 out of 10 Filipinos die of heart disease - NSO. Retrieved August 29, 2015, from http://www.philstar.com/breaking-news/2012/07/09/826043/5-out-10-filipinos-die-heart-disease-nso
[6] Sandhu, R., Jimenez, M., Chiuve, S., Fitzegerald, K., Kenfield, S., Tedrow, U., & Albert, C. (2012). Smoking, Smoking Cessation and Risk of Sudden Cardiac Death in Women. AHA Journals. Retrieved at http://circep.ahajournals.org/content/early/2012/11/16/CIRCEP.112.975219
[7] Klodas, E. (2014, September 10). Alcohol and Heart Disease. Article review. Retrieved at http://www.webmd.com/heart-disease/guide/heart-disease-alcohol-your-heart
[8] Woman and Heart Disease Fact Sheet. (n.d.). Retrieved August 25, 2015, from http://www.cdc.gov/dhdsp/data statistics/fact sheets/docs/fs women heart.pdf
[9] Cardiol, E. (2014, May 21). Diabetes, Heart Disease, and Stroke. Nih Publication No. 13-5094 August 2013. Retrieved August 26, 2015, from http://www.niddk.nih.gov/health-information/health-topics/Diabetes/diabetes-heart-disease-stroke/Pages/index.aspx
[10]Wilkins, J., Gidding, S., Liu, K., Ning, H., Polak, J., & Lloyd-Jones, D. (n.d.). Associations Between a Parental History of Premature Cardiovascular Disease and Coronary Calcium and Carotid Intima-Media Thickness: The Coronary Artery Risk Development In Young Adults (CARDIA) Study. Retrieved August 28, 2015, from http://www.ncbi.nlm.nih.gov/pmc/articles/PMC3779512/
[11]Dev Mukherji, Mythili, T., & Padalia, N. (2013). *A Heart Disease Prediction Model using SVM-Decision Trees-Logistic Regression (SDL)* (Vol. 68). Tamil Nadu: International Journal of Computer Applications. http://research.ijcaonline.org/volume68/number16/pxc3887250.pdf
[12] Fernandes, J., Arts, J., Dimond, E., Hirshberg, S., & Lofgren, I. (n.d.). Dietary factors are associated with coronary heart disease risk factors in college students. Retrieved September 8, 2015, from http://static1.squarespace.com/static/527944f7e4b0a241d38036af/t/529d1c12e4b0981753a2a8a0/1386028050930/Dietary factors are associated with coronary heart disease risk factors in college students.pdf
[13]McGorrian, C., Yusuf, S., Islam, S., Jung, H., Rangarajan, S., Avezum, A., . . . Teo, K. (2011). Estimating modifiable coronary heart disease risk in multiple regions of the world: The INTERHEART Modifiable Risk Score. Retrieved September 10, 2015, from http://www.naos.aesan.msps.es/ca/naos/ficheros/investigacion/estimating modifiable coronary heart.pdf
[14] Fukuoka, Y., Choi, J., Bender, M., & Gonzales, P. (2015, April 20). Family history and body mass index predict perceived risks of diabetes and heart attack among community-dwelling Caucasian, Filipino, Korean, and Latino Americans—DiLH Survey. Retrieved September 10, 2015, from http://www.ncbi.nlm.nih.gov/pmc/articles/PMC4470846/
[15] Ong, W. (2011, August 30). Reducing heart disease and lifestyle diseases in the Philippines. Retrieved October 12, 2015, from http://www.philstar.com/health-and-family/721436/reducing-heart-disease-and-lifestyle-diseases-philippines
[16] Sto. Tomas, M., Baluyot, P., Delos Santos, A., Faustino, J., Licudine, E., Mendoza, E., ... Tiburcio, M. (2015). Interleukin 6 as a Biomarker of Ischemic Heart Disease (Issue 7 ed., Vol. 5, p. 4). Manila, Department of Medical Technology, University of Santo Tomas: International Journal of Scientific and Research Publications. Retrieved at http://www.ijsrp.org/research-paper-0715/ijsrp-p4316.pdf
[17] Goroscope, J., Bismonte, S., Arcilla, R., & Gironella, G. (2014, March 6). Ordinal Logistic Regression Analyses on Anemia for Children Aged 6

Months to 5 Years Old in the Philippines. Retrieved October 4, 2015, from http://www.dlsu.edu.ph/conferences/dlsu_research_congress/2014/_pdf/FNH-II-016.pdf
[18] How Does Smoking Affect the Heart and Blood Vessels? (n.d.). Retrieved September 26, 2015, from http://www.nhlbi.nih.gov/health/healthtopics/topics/smo
[19]Rundel, C. (2013, April 15). Lecture 20 - Logistic Regression Statistics 102. Retrieved August 24, 2015, from https://stat.duke.edu/courses/Spring13/sta102.001/Lec/Lec20.pdf
[20]Noncommunicable Diseases (NCD) Country Profiles. (2014). Retrieved August 28, 2015, from http://www.who.int/nmh/countries/phl_en.pdf

Odds and Probabilities Estimation for the Survival of Breast Cancer Patients with Cancer Stages 2 & 3

Jackie D. Urrutia , Ciandreu Noah M. Balmaceda and Edcon B. Baccay
Office of the Academic Head, Polytechnic University of the Philippines- Parañaque Campus
Col. E. L. De Leon St. Sto. Niño Parañaque City, Philippines

Abstract. Breast cancer is one of the leading causes of death in the Philippines. One out of four who are diagnosed with breast cancer die within the first five years, and no less than 40 percent die within 10 years and it has continous rise as time passes by. Therefore, it is very important to know the factors that can help for the survival rate of the patients. The purpose of this study is to identify the best possible treatment or combination of treatments. The researchers considered four independent variables namely: Completed Surgery, Completed Chemotherapy, Completed Hormonotherapy and Completed Radiotherapy. The researchers limit this study for only 160 patients with stage 2 and 135 with stage 3 for a total of 295 patients considering the data gathered from three hospitals from Metro Manila. The names of the hospitals were not declared due to confidentiality of data. In identifying the best treatment or combination of treatments, odds, probabilities and odds ratios of patients, Logistic Regression Analysis was used.

1. Introduction

According to Philippine Breast CancerNetwork (PBCN), the Philippines has the highest incidence rate of breast cancer in Asia and registered the highest increase of 589% among 187 countries over a 30 year period from 1980 to 2010. Breast cancer is the leading killer of women ages 35 to54 worldwide.[1] Also, according to the Philippine Society of Medical Oncology in 2012, an estimated 3 out of 100 Filipino women will contract the disease before age 75 and 1 out of 100 will die before the age of 75.[2]

Breast cancer is a malignant tumor that starts in the cells of the breast. A malignant tumor is a group of cancer cells that can grow into surrounding tissues or spread to distant areas of the body. The disease occurs almost entirely in women, but men can get it, too.[3]

The first symptom of breast cancer that most women notice is a lump or an area of thickened tissue in their breast. Most breast lumps aren't cancerous but it's always best to have them checked by their doctor. An individual should see her doctor if she notices any of the following: a lump or area of thickened tissue in either breast, a change in the size or shape of one or both breast, discharge from either of the nipples, a lump or swelling in either of their armpits, dimpling on the skin of your breast, a rash on or around the nipple and a change in the appearance of the nipple. Breast pain isn't usually a symptom of breast cancer.[4]

Women with breast cancer may have questions about their prognosis and survival. Prognosis and survival depend on many factors. Only a doctor familiar with a person's medical history, type of cancer, stage, characteristics of the cancer, treatments chosen and response to treatment can put all of this information together with survival statistics to arrive at a prognosis. The researchers considered Completed Surgery, Completed Chemotherapy, Completed Hormonotherapy and Completed Radiotherapy as predictive factors because doctors believe that these factors can affect the survival rate of patients having breast cancer.[5]

Some treatments remove or destroy the disease within the breast and nearby tissues, such as lymph nodes. These include: Surgery, radiation therapy, chemotherapy, hormone therapy and targeted therapy but the researchers only limit the study without the data for targeted therapy due to lack of patients that undergo this treatment.[6]

Surgery aims to remove the breast cancer with a margin (border) of normal tissue to reduce the risk of the cancer coming back in the breast (known as local recurrence) and to try to stop any spread elsewhere in the body. The amount of tissue removed depends on the area of the breast affected and the size of the cancer in your breast. Surgery is usually the first treatment for breast cancer, although sometimes chemotherapy or hormone therapy is offered first. This is to begin treating the whole body or to shrink the cancer so that surgery may be less extensive.[7]

Chemotherapy treatment uses medicine to weaken and destroy cancer cells in the body, including cells at the original cancer site and any cancer cells that may have spread to another part of the body.[8]

Hormone therapy (also called hormonal therapy, hormone treatment, or endocrine therapy) slows or stops the growth of hormone-sensitive tumors by blocking the body's ability to produce hormones or by interfering with hormone action. Tumors that are hormone-insensitive do not respond to hormone therapy. Hormone therapy for breast cancer is not the same as menopausal hormone therapy or female hormone replacement therapy, in which hormones are given to reduce the symptoms of menopause.[9]

Radiotherapy is a treatment for cancer that uses carefully measured and controlled high energy x-rays. In primary breast cancer it aims to destroy any cancer cells that may be left behind in the breast area after surgery.[10]

Breast cancer is still the leading cause of cancer among women, accounting for 28 percent of the total cases. One out of four who are diagnosed with breast cancer die within the first five years, and no less than 40 percent die within 10 years.[11]Therefore, it is very important to know the factors that can help for the survival rate of patients experiencing breast cancer.

1.1. Objective of the Study

The main objective of the study is to make a model that generates probabilities and odds of surviving a breast cancer using Logistic Regression Analysis. This study also aims to identify which among the following treatments or combination of treatments namely surgery, chemotherapy, hormonal therapy and radiotherapy are significant variables that highly affect the survival rate of breast cancer.

1.2. Statement of the Problem

The study aims to estimate the odds of an individual experiencing breast cancer with factors which are considered to be significant contributors to the survival of the disease using logistic regression. Specifically, it seeks answers to the following questions:

1.2.1. What is the profile of the breast cancer patients for stage 2 & 3?
1.2.2. What is the logistic model for stage 2& 3 cancer that can be obtained to determine the probability of surviving breast cancer?
1.2.3. What are the significant variables that can affect the survival rate of breast cancer patients?
1.2.4. What are the odds and probabilities of an individual of surviving a breast cancer for stage 2 and stage 3 patients?
1.2.5. What are the odds ratios of an individual surviving a breast cancer compare to a breast cancer patient with no treatment applied?
  1.2.5.1. Single Factor
  1.2.5.2. Combination of two factors

1.3. Scope and Limitation

The researchers limited the study as there were only 290 health data of female breast cancer patients with 160 of them are experiencing stage 2 of the disease and 135 patients are with 3rd stage. The data were gathered from three hospitals from Metro Manila. The names of the hospitals were not declared due to confidentiality of the data. The researchers applied logistic regression analysis to identify key determinants of breast cancer from the qualifying health survey data and clinic pathological predictors of the patients.

2. Methodology

2.1. Population and Covariates

The model for odds and probabilities of survival rate of patients were obtained using the data gathered from three hospitals in Metro Manila compiled by the specialists and experts to imply female patients with breast cancer disease. The names of the hospitals were not declared because of the confidentiality of the data. About 95% of the 160 breast cancer patients with stage 2 disease survived the disease and 5% didn't. On patients with 3rd stage, there were 87.4% survived the disease and 12.6% didn't. There were four treatments that contribute to the overall model. These are Completed Surgery, Completed Chemotherapy, Completed Hormonotherapy and Completed Radiotherapy.

2.2. Statistical Tool

The researchers used the statistical software Statistical Package for Social Science (SPSS) in identifying the significant predictors that actually affect the survival of patients as well as computing the odds, probabilities and their odds ratios by using the Logistic Regression. SPSS is one of the most popular statistical packages which can perform highly complex data manipulation and analysis with simple instructions.[12]

2.3. Logistic Regression

Logistic regression is a statistical method for analyzing a dataset in which there are one or more independent variables that determine an outcome. The outcome is measured with a dichotomous variable in which there are only two possible outcomes.[13]

Logistic Regression does not make many of the key assumptions of linear regression and generalized linear models that are based on ordinary least squares algorithms.

Firstly, it does not need a linear relationship between the dependent and independent variables. Logistic regression can handle all sorts of relationships, because it applies a non-linear log transformation to the predicted odds ratio. Secondly, the independent variables do not need to be multivariate normal – although multivariate normality yields a more stable solution. Also the error terms (the residuals) do not need to be multivariate normally distributed. Thirdly, homoscedasticity is not needed. Logistic regression does not need variances to be heteroscedastic for each level of the independent variables. Lastly, it can handle ordinal and nominal data as independent variables. The independent variables do not need to be metric (interval or ratio scaled).

However, some assumptions still apply. First, the dependent variable must be binary for binary logistic regression and ordinal for ordinal logistic regression. Second, the outcome variable needs to be coded properly that is, 1 for desired outcome and 0 for the event not occurring. Third, only meaningful variables should be included. Fourthly, independent variables should be independent that is data points should not be from another variable (e.g. before-and-after measurements and matched pairings) and there should be no multicollinearity among the variables. Fifth, logistic regression assumes linearity of independent variables and log odds. Lastly, it requires quite large sample size.[14]

The probability of the outcome is measured by the odds of occurrence of an event. If P is the probability of an event, then (1-P) is the probability of it not occurring.

Odds of success = P / 1-P

The simple logistic model has the form:

$$\text{Logit}(Y) = \text{natural log}(odds) = \ln\left(\frac{\pi}{1-\pi}\right) = \alpha+\beta X \quad (1)$$

Taking the antilog of Equation 1 on both sides, one derives an equation to predict the probability of the occurrence of the outcome of interest as follows:

$$\pi = Probability\ (Y = \text{outcome of interest} \mid X = x, \text{a specific value of } X) = \frac{e^{(\alpha+\beta X)}}{1+e^{(\alpha+\beta X)}} \quad (2)$$

where π is the probability of the outcome of interest or "event", α is the *Y* intercept, β is the regression coefficient and $e = 2.71828$ is the base of the system of natural logarithms. *X* can be continuous or categorical, but *Y* is always categorical.

Extending the logic of the simple logistic regression to multiple predictors, one can construct a complex logistic regression for *Y* as follows:

$$\text{Logit}(Y) = \ln\left(\frac{\pi}{1-\pi}\right) = \alpha+\beta_1 X_1 + \beta_2 X_2 + \beta_3 X_3 + \cdots + \beta_n X_n \quad (3)$$

Therefore ,

$$\pi = Probability\ (Y = \text{outcome of interest} \mid X_1 = x_1, X_2 = x_2, X_3 = x_3, \ldots X_n = x_n) = \frac{e^{(\alpha+\beta_1 X_1+\beta_2 X_2+\beta_3 X_3+\cdots+\beta_n X_n)}}{1+e^{(\alpha+\beta_1 X_1+\beta_2 X_2+\beta_3 X_3+\cdots+\beta_n X_n)}} \quad (4)$$

where π is once again the probability of the event, α is the *Y* intercept, βs are the regression coefficients, and *X*s are a set of predictors.[23]

An odds ratio is a measure of association between an exposure and an outcome. The odds ratio represents the odds that an outcome will occur given a particular exposure, compared to the odds of the outcome occurring in the absence of that exposure. Odds ratios are most commonly used in case-control studies; however they can also be used in cross-sectional and cohort study designs as well.[15]

The odds ratio is given by :[16]

$$OR = \frac{[p/(1-p)]_2}{[p/(1-p)]_1} \quad (5)$$

Where:

$[p/(1-p)]_2$ = odds of success at condition 2

$[p/(1-p)]_1$ = odds of success at condition 1

3. Results and Discussion

3.1. Profile of Patients

Table 1. Profile of Stage 2 & 3 Breast Cancer Patients

		Stage 2	Stage 3
Factors	Classification	Percentage	
Completed Surgery	No	3(1.9%)	27 (20%)
	Yes	157(98.1%)	108(80%)
Completed Chemotherapy	No	32 (20%)	13(9.6%)
	Yes	128 (80%)	122(90.4%)
Completed Hormonotherapy	No	74(46.3%)	46(34.1%)
	Yes	86 (53.8%)	89(65.9%)
Completed Radiotherapy	No	126(78.8%)	43(31.9%)
	Yes	34(21.3%)	92(68.9%)

The table shows the profile of the breast cancer patients with stage 2 or stage 3 diseases. Majority of the stage 2 patients completed a surgery accompanying 98 percent out of 160 while 80% out of 135 patients with stage 3 cancer completed a surgery. Also, majority of stage 2 and stage 3 patients completed a chemotherapy treatment with 80% for stage 2 and 90.4% for stage 3. The hormonotherapy treatment was also completed by 53.8% of stage 2 cancer patients while 65.9% for stage 3 patients. Lastly, only 21.3% of the stage 2 patients completed a radiotherapy treatment and almost 69% for stage patients.

3.2. Logistic Model for Stage 2 Breast Cancer Patients

3.2.1. Significant Parameters

By running the Logistic Regression Analysis, only the completed surgery appears to be significant but the

completed chemotherapy is obviously near the significance level (0.05) which signifies that it is a possible contributor to the model.
The model obtained was:

$$\ln\left(\frac{p}{1-p}\right) = \alpha + \beta_1 X_1 + \beta_2 X_2 + \beta_3 X_3 + \beta_4 X_4$$

$$\ln\left(\frac{p}{1-p}\right) = -2.556 + 4.446\mathbf{X_1} + 1.863\mathbf{X_2} + 0.092\mathbf{X_3} + 17.508\,\mathbf{X_4}$$

Where:

X_1 = Completed Surgery or Not
X_2 = Completed Chemotherapy or Not
X_3 = Completed Hormonal therapy or Not
X_4 = Completed Radiotherapy or Not

3.2.2. Forward Stepwise Method

Running the Logistic Regression again but this time, removing all the insignificant factors and leaving only all significant variables in the final model (Forward Stepwise Method), the results implied that Surgery and Chemotherapy are the significant variables with regression coefficients of 4.812 and 2.173 respectively.

Odds= exp $(\alpha + \beta_1 X_1 + \beta_2 X_2 + \beta_3 X_3 + \beta_4 X_4)$
Odds = exp $(-2.866 + 4.812X_1 + 2.173X_2)$
The final model obtained was:

$$\ln\left(\frac{p}{1-p}\right) = \alpha + \beta_1 X_1 + \beta_2 X_2$$

$$\ln\left(\frac{p}{1-p}\right) = \mathbf{-2.866 + 4.812}\boldsymbol{X_1} + \mathbf{2.173}\boldsymbol{X_2}$$

A person's probability of surviving breast cancer with stage 2 is written by:

$$P(Y = 1|X_1, X_2) = \frac{e^{(-2.866+4.812X_1+2.173X_2)}}{1 + e^{(-2.866+4.812X_1+2.173X_2)}}$$

$$or\ \frac{1}{1 + e^{-(-2.866+4.812X_1+2.173X_2)}}$$

Where:

X_1 = Completed Surgery or Not
X_2 = Completed Chemotherapy or Not

The odds of a stage 2 breast cancer patient who didn't undergo surgery nor chemotherapy is given by $e^{-2.866} = 0.056$ which also means that her probability of surviving breast cancer is Y = (Odds) / (1+Odds) = (0.056) / (1+0.056) = 5.4%.

3.3. Logistic Model for Stage 3 Breast Cancer Patients

3.3.1. Significant Parameters

By running the regression with an alpha of 0.05, the researchers found that among the four initial independent variables, only two are significant: Completed Surgery and Completed Radiotherapy with regression coefficients of 1.479 and 1.292 respectively.
The model obtained was:

$$\ln\left(\frac{p}{1-p}\right) = \alpha + \beta_1 X_1 + \beta_2 X_2 + \beta_3 X_3 + \beta_4 X_4$$

$$\ln\left(\frac{p}{1-p}\right) = -1.466 + 1.479\boldsymbol{X_1} + 1.016\boldsymbol{X_2} + 1.147\boldsymbol{X_3} + 1.292\boldsymbol{X_4}$$

Where:

X_1 = Completed Surgery or Not
X_2 = Completed Chemotherapy or Not
X_3 = Completed Hormonal therapy or Not
X_4 = Completed Radiotherapy or Not

3.3.1. Forward Stepwise Method

Removing all insignificant variables and thus leaving just all the significant factors in the final model (Forward Stepwise Method), the researchers concluded that radio therapy is the only treatment that is significant with a regression coefficient of 1.022.
Odds = $\exp(\alpha + \beta_1 X_1)$
Odds = $\exp(1.329 + 1.022X_1)$

The final model obtained was:

$$\ln\left(\frac{p}{1-p}\right) = \alpha + \beta_1 X_1$$

$$\ln\left(\frac{p}{1-p}\right) = 1.329 + 1.022\boldsymbol{X_1}$$

A person's probability of surviving breast cancer with stage 3 is written by:

$$P(Y = 1|X_1) = \frac{e^{(1.329+1.022X_1)}}{1 + e^{(1.329+1.022X_1)}}$$

$$or\ \frac{1}{1 + e^{(1.329+1.022X_1)}}$$

Where:

X_1 = Completed Radiotherapy or Not

The odds of a stage 3 breast cancer patient who didn't undergo radiotherapy is given by $e^{1.329} = 3.777$ which also means that her probability of surviving breast cancer is $Y = \frac{Odds}{1+Odds} = \frac{3.777}{1+3.777} = 79.05\%$.

3.4. Significant Factors

By using the Logistic Regression Analysis, it was concluded that for stage 2 breast cancer patients, the significant predictors are completed surgery and chemotherapy having p-values of 0.001 and 0.015 respectively while on stage 3, radiotherapy is the only significant factor having a p-value of 0.05.

3.5. Odds and Probabilities

The odds of a stage 2 breast cancer patient that competed a surgery only is $e^{-2.866+4.812(1)} = 7.0006$ resulting to $Y = (7.0006) / (1+7.0006) = 87.50\%$ while the odds of a stage 2 breast cancer patient that completed chemotherapy only is $e^{-2.866+2.173(1)} = 0.5001$ resulting to
$Y = (0.5001) / (1+0.5001) = 33.34\%$. The odds of a stage 2 breast cancer patient that undergo both surgery and chemotherapy is $e^{-2.866+4.812(1)+2.173(1)} = 61.4977$ resulting to
$Y = (61.4977) / (1+61.4977) = 98.40\%$.

For stage 3, the odds of a stage 3 breast cancer patient that completed radiotherapy only is $e^{1.329+1.022(1)} = 10.4961$ to $Y = (10.4961) / (1+10.4961) = 91.31\%$.

3.6. Odds Ratio

3.6.1. Odds Ratio of Single and Combination of Treatments

The model predicts the odds ratio for completed surgery to be $OR = [e^{-2.866+4.812(1)}] / (e^{-2.866}) = 122.978$. Therefore, the survival of a stage 2 breast cancer is likely to happen to a patient that has completed surgery by 123 times compared to a person that didn't undergo of any of these treatments. The model also predicts the odds ratio for completed chemotherapy to be $OR = [e^{-2.866+2.178(1)}] / (e^{-2.866}) = 8.7846$. Out of all single and combination factors, the combination of both treatments is better than applying a single treatment only with an odds ratio of more than a thousand indicating that a breast cancer patient that undergo both surgery and chemotherapy will survive the disease by more likely a thousand times compared to a cancer patient that didn't undergo any of these treatments. It is computed as
$OR = [e^{-2.866+4.812(1)+2.178(1)}] / (e^{-2.866}) = 1080.3064$.

3.6.2. Odds Ratio of Single Treatment

The model predicts the odds ratio for completed radiotherapy to be $OR = [e^{1.329+1.022(1)}] / (e^{1.329}) = 2.7790$.
Therefore, the survival of a stage 3 breast cancer is likely to happen to a patient that has completed radiotherapy only by almost thrice compared to a person that didn't undergo any treatment.

4. Conclusions and Recommendations

Researchers conclude that for stage 2, the combination of surgery and chemotherapy are the best treatments that can be applied to a patient for her to get a better chance of surviving the disease and for stage 3, radiotherapy is preferred. The study found out that the survival of the disease for breast cancer patients with stage 2 is more likely to happen to a patient that undergone the combination of surgery and chemotherapy by more than a thousand times compared to a patient that didn't undergone any of these treatments. Also for stage 3, a patient that undergone radiation is three times more likely to survive the disease compared to a patient that didn't undergone any treatment.

The researchers recommend that adding more independent variables such as age, menopausal status, type of cancer, histological grade, lymphatic and vascular invasion, family history and size of primary tumor will enhance the model because some studies proved that these factors really influence the survival rate of patients with breast cancer. The researchers also recommend adding more participating hospitals in order to increase the sample size for a better model.

References

[1] "*Philippine Breast Cancer Network*"(n.d.) Retrieved October 28, 2015 from http://www.pbcn.org
[2] "*PHL has highest breast cancer rate in Asia*"(2013, October 2). Retrieved October 28, 2015 from http://www.gmanetwork.com/news/story/329126/cbb/phl-has-highest-breast-cancer-rate-in-asia
[3] "*What is breast cancer?*" (2015, June 10). Retrieved October 28, 2015 from http://www.cancer.org/cancer/breastcancer/detailedguide/breast-cancer-what-is-breast-cancer
[4] "*Symptoms of Breast Cancer*" (2014). Retrieved November 10, 2015 from http://www.nhs.uk/Conditions/Cancer-of-the-breast-female/Pages/Symptoms.aspx
[5] "*Breast Cancer*" (n.d.). Retrieved October 28, 2015 from http://www.cancer.ca/en/cancer-information/cancer-type/breast/prognosis-and-survival/?region=on
[6] "*Breast Cancer Treatment*" (n.d.) Retrieved December 11, 2015 from http://www.webmd.com/breast-cancer/guide/breast-cancer-treatment
[7]"*Surgery*" (2014) Retrieved December 11, 2015 from https://www.breastcancercare.org.uk/information-support/facing-breast-cancer/going-through-treatment-breast-cancer/surgery
[8]"*Chemotherapy*" (2015) Retrieved December 11, 2015 fromhttp://www.breastcancer.org/treatment/chemotherapy
[9]"*Hormone Therapy for Breast Cancer*" (2012) Retrieved December 11, 2015 from*http://www.cancer.gov/types/breast/breast-hormone-therapy-fact-sheet*
[10] "*Radiotherapy for Primary Breast Cancer*" (2014) Retrieved December 11, 2015 from https://www.breastcancercare.org.uk/information-support/facing-breast-cancer/going-through-treatment-breast-cancer/radiotherapy-primary-breast-cancer
[11]"*The ABCs of breast cancer*" (n.d.). Retrieved October 28, 2015 from http://psmo.org.ph/the-abcs-of-breast-cancer
[12] "*What is SPSS?*" (n.d.) Retrieved November 11, 2015 from https://www.uwindsor.ca/its/sites/uwindsor.ca.its/files/What%20is%20SPSS.pdf
[13] "*Logistic Regression*" (2015, October 22). Retrieved October 28, 2015 fromhttps://www.medcalc.org/manual/logistic_regression.php
[14]"*Assumptions of Logistic Regression*" (n.d.). Retrieved October 28, 2015 from http://www.statisticssolutions.com/assumptions-of-logistic-regression
[15] " *An Introduction to Logistic Regression Analysis and Reporting*" (n.d.) Retrieved November 12,2015 from https://datajobs.com/data-science-repo/Logistic-Regression-%5BPeng-et-al%5D.pdf
[16] "*Explaining Odds Ratios*" (August, 2010). Retrieved November 12, 2015 from http://www.ncbi.nlm.nih.gov/pmc/articles/PMC2938757/
[17] Youlden, Danny R., Cramb, Susanna M., Cheng Har Yip & Baade, Peter D. (2014). "*Incidence and mortality of female breast cancer in the Asia-Pacific region*" Retrieved from http://www.ncbi.nlm.nih.gov/pmc/articles/PMC4069805
[18] Hassan MS, Ansari J, Spooner D & Hussain SA. (2010). "*Chemotherapy for Breast Cancer*" Retrieved from http://www.ncbi.nlm.nih.gov/pubmed/20878101
[19] P A Ganz, J H Rowland, K Desmond, B E Meyerowitz& G E Wyatt (2005) "*Life after Breast Cancer: Understanding women's health-related quality of life and sexual functioning*" Retrieved from http://jco.ascopubs.org/content/16/2/501.short
[20] DursunDelen, Glenn Walker & Amit Kadam (2005). "*Predicting Breast Cancer Survivability: A comparsion of three data mining methods*" Retrieved from http://www.aiimjournal.com/article/S0933-3657(04)00101-0/abstract
[21]Herman Host M.D. &Eiliv Lund M.D. (2006). "*Age as a prognostic factor in breast cancer*" Retrieved from http://onlinelibrary.wiley.com/doi/10.1002/1097-0142(19860601)57:11%3C2217::AID-CNCR2820571124%3E3.0.CO;2-T/abstract
[22]S.E. Pinder, I.O. Ellis, M. Galea, S.O'Rouke, R.W. Blamey &C.W. Elston (2007). "*Pathological prognostic factors in breast cancer. III. Vascular Invasion: relationship with recurrence and survival in a large study with long-term follow up*" Retrieved from http://onlinelibrary.wiley.com/doi/10.1111/j.1365-2559.1994.tb01269.x/abstract
[23]Graham A. Colditz, MBBS; Walter C. Willett,MD; David J. Hunter, MBBS; Meir J. Stampfer, MD; Joann E. Manson,MD; Charles H. Hennekens,MD; Bernard A. Rosner,PhD& Frank E. Speizer,MD (1993). "*Family History, Age, and Risk of Breast Cancer – Prospective Data from the Nurses' Health Study*" Retrieved from http://jama.jamanetwork.com/article.aspx?articleid=407486
[24]SarojNiraula, Alberto Ocana, Marguerite Ennis &Pamela J. Goodwin (2012). "*Body size and breast cancer prognosis in relation to hormone receptor and menopausal status: a meta-analysis*" Retrieved from http://link.springer.com/article/10.1007/s10549-012-2073-x
[25] Walpole, Ronald E., Myers, Raymond H., Myers, Sharon L., & Ye, Keying (2007) "*Scientists & Engineers guide to probability and statistics*": Pearson Education, Inc.
[26] "*SPSS Tutorials: Binary Logistic Regression*" Retrieved from https://www.youtube.com/watch?v=Ak_t86zm_sQ
[27]"*Logistic Regression – Interpretation of Coefficients and Forecasting*" Retrieved from https://www.youtube.com/watch?v=eX2sY2La4Ew
[28]"*Logistic Regression – Introduction*" Retrieved from https://www.youtube.com/watch?v=gNhogKJ_q7U
[29]"*Logistic Regression – SPSS (Part 1)*" Retrieved from https://www.youtube.com/watch?v=OvQShzJ7Sns
[30] "*Logistic Regression – SPSS (Part 2)*" Retrieved from https://www.youtube.com/watch?v=zdJhydkcqv4
[31] "*Logistic Regression – SPSS (Part 3)*" Retrieved from https://www.youtube.com/watch?v=hxcDOoupB4Y

ESCO Project for Buildings of Government Agencies in Thailand

Monchai Prukvilailert [1], Prapat Wangskarn [1]

[1] Thammasat University, Thailand

Abstract. In Thailand, Department of Alternative Energy Development and Efficiency (DEDE) have organized the ESCO project to promote and encourage the use of machinery, materials and equipment having high efficiency for government buildings. ESCO company provides the invest and management for changing equipments in the buildings. In this paper, the evaluation of the project has been presented. The potential of electricity savings is about 77 million kwhr/year. It can reduce imports of crude oil about 6.58 thousand tons of crude oil (Ktoe/year). The budget to invest is BHT 1,504 million, with an average payback period of 4.85. However, we found that the establishment of the budget is the barriers. The recommendations and solutions using legal process have been presented to proceed the project in the future.

1 Introduction

On March 20, 2013 the Cabinet has approved the Government agency to implement measures to reduce energy consumption by at least 10% according to the regulation of energy development and promotion in Thailand[1]-[3]. Results from the survey data of the project and oversee the conservation of energy in the building sector about 800 buildings. The potential energy savings of buildings of government agencies that have the potential to reduce crude oil import up 75 ktoe per year (BH 1,800 million per year) by investing more than BH 6,300 million to solve the limitations of the government in the budget of up to 6,300 million baht in using high-performance equipment. Department of Alternative Energy Development and Efficiency DEDE, Ministry of Energy, have launched the project management in order to save energy in the building sector using Energy Service Company (ESCO) operating by the private sector. It offers a full range of energy conservation. For the cost of the operation, ESCO companies will be paid in amount of the cost savings that can be reduced.

Mechanism in Business Management (Energy Service Company: ESCO) would mean providing turnkey energy conservation by counselling, proposing project, project management, engineering design, energy Analysis, equipment installation and operation and find funding. The distinguished characters of the mechanism ESCO is to guarantee energy savings from the implementation of energy conservation projects. Compensation will be paid to the buildings if saving energy is not in accordance with agreed. Thus buildings confidence that ESCO mechanism is able to used to get energy saving worth all the money that were paid to the budget. It can reduce energy consumption and energy expenditures down, certainly in the future. However, there are policy barriers of ESCO for government buildings such as, the inability to establish the comprehensive budget package (Turnkey) and the implementation of commitments over the fiscal year.

In order to resolve such obstacles, National Energy Policy Council (NEPC) On March 23, 2013 has approved a pilot project to manage energy-saving (ESCO) in government buildings. The project management to save power by ESCO system in buildings will have to sign an agreement deal with the third parties including DEDE, Metropolitan Electricity Authority (MEA)/Provincial Electricity Authority (PEA) and government buildings. MEA and PEA act as ESCO companies that have to survey and design measures to conserve energy. Furthermore, MEA\PEA have the responsibility for defining hardware configuration, preparing an energy agreement, equipment installation as well as evaluation of the economy. MEA\PEA will be charged the managing cost adding to monthly electric bill of buildings. The pilot project was in demand. Therefore, more than 50 government buildings signed a memorandum on cooperation in the implementation of project management in order to save energy by ESCO especially for academic sector and Hospital.

2 Procedure

Energy Service Companies (ESCO) will act as turnkey implementation of energy conservation to the buildings to achieve the energy savings. ESCO acts for monitoring of energy (Energy Audit) before implementing, construction Management services, project financing, project guarantees, equipment maintenance and servicing and other services [11]-[12].

Figure 1 shows the allocation of an energy performance contract. The first action (the baseline year) buildings must pay all utility bills. During the implementation of the contract performance, energy costs decreases by the amount of the energy savings will pay to the Energy Service Companies (ESCO) for a fee to carry out investment. The remainder will be for buildings throughout the lifetime of the energy performance contract project. When end of the contract term, the total energy saving will return to building only.

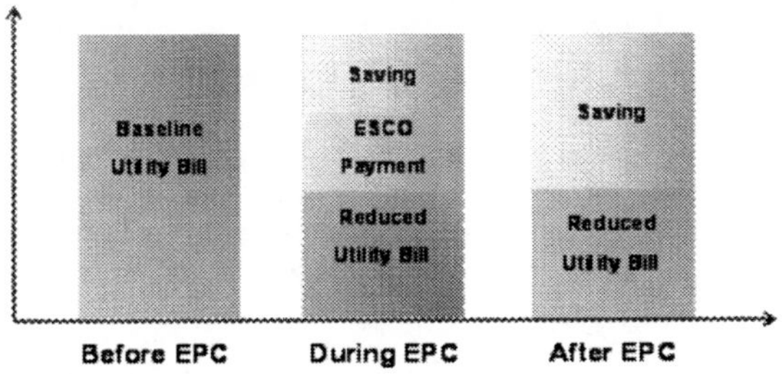

Figure 1 Allocation of the energy performance contract

In Figure 2, the process of implementation can be explained as follow:

1. ESCO selection (MEA or PEA)
2. Energy Audit

To determine the energy audit, ESCO company surveyed and monitored the power consumption of the buildings before defining the energy saving measures, especially the specific energy consumption (Specific Energy Consumption, SEC). After the survey and collect energy consumption data, ESCO analyzed to find ways to conserve energy. The process can be shown in detail as follow:

(1) Determine primary (Preliminary Audit) which assesses energy use scales to determine the potential of reducing the energy consumption. In addition, the preliminary examination would have cost for ESCO.

(2) Determine the specific type (single purpose or targeted Energy Audit) , which monitored only one device or system thoroughly such as changing light bulbs, a variable speed drive (VSD) and replacement of boilers (Boiler) and chiller (Chiller).

(3) Check to cover all (comprehensive audit), which details the use of energy and to prepare plans for energy conservation projects carefully. It will consider the impact of the various systems used to power the buildings according to weather, working hours and other variables associated with energy use. Advantages of the inspection are a comprehensive analysis of the costs and savings in detail. Including the impact of one system to another system, the disadvantage is the relatively high cost.

3. Energy savings proposal
4. Energy performance contract

In this project, the energy performance contract has been chosen based on guaranteed saving method. Therefore, buildings must perform financing for the project by themselves or borrow money from financial institutions. In this case, Energy Service Companies (ESCO) will guarantee the performance of the economy (saving performance) to the clients or agencies. Usually, the savings will amount, as well as the clients must pay back to the lender.

5. Project Implementation
   5.1 Construction Management services
   5.2 Project Financing
   5.3 Project guarantees
   5.4 Equipment maintenance and servicing
   5.5 Other services
6. Measurement and verification

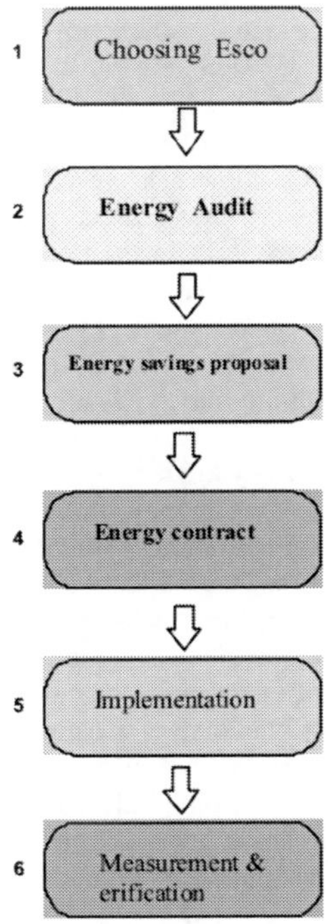

Figure 2. Operational processes using ESCO

There are 53 buildings (26 academic buildings, 26 Hospitals and 1 office buildings) joined the project. They have implemented energy saving measures using ESCO system.

3 Potential savings

The results showed that the building sector, signed a memorandum of number 53. This government buildings have the potential of electricity savings about 77 million kwhr/year. It can reduce imports of crude oil about 6.58 thousand tons of crude oil (Ktoe) per year. This represents the cost of BHT 310 million per year. The budget to invest is BHT 1,504 million, with an average payback period of 4.85 years as shown in Table 1.

For this project, four types of equipments are promoted to implement.

1. High efficiency boiler
2. High efficiency chillers

3. High efficiency air conditioning units
4. LED lamp

Table 1. Potential savings of 53 buildings.

Type	Electricity saving (kWh/y)	Thermal saving (MJ/y)	Energy saving (ktoe/y)
Hospital	30,152,669	1,227,937	2.59
Academic	46,407,574	0	3.95
Office	294,547	0	0.02
total	76,854,820	1,227,937	6.56

By tracking the status of the progress for all buildings in the project, we found that only one buildings (Rajamangala University, Lanna campus) can achieved the success of the implementation which is operated by budget of its own income. The results were unexpected comparing with the previous projects [6]–[12].

We have collected the information to fix the problems in the disbursement of funds. The building can be divided into two types of government. Type One, government building have their own income and budget. Type Two, government buildings received all budget from the government.

Building Type 1 (Academic buildings) can implement the project by spending money on their own. Buildings can implement immediately without waiting for the correct payment procedures. Most are academic buildings. The potential of electricity savings is BHT 46 million per year. Imports of crude oil can be reduced up to 3.96 tons of oil per year(ktoe), representing a cost of BHT 188 million per year. The budget to invest is BHT 1,064 million, with an average payback period of 5.67 years.

Building Type 2 (Hospital) only has an annual government statement of expenditure. This will have to wait to resolve disbursement procedures. However, the buildings have the potential of electricity savings is BHT 30 million units per year. Imports of crude oil can be reduced up to 2.63 tons of oil per year (Ktoe), representing a cost of BHT 122 million per year. The budget to invest is BHT 445 million, with an average payback period of 3.62 years. Table 2 show the number of equipments by type.

Table 2. Equipment by Type for Building Type 1 and Type 2.

Type of equipment	Building Type 1 (Academic) (Number)	Building Type 2 (Hospital) (Number)
1. Boiler	0	1
2. Chiller	5	6
3. Air conditioning (package unit)	0	10
3. Air conditioning (split type)	9,995	5,360
4. LED lamp	367,647	83,103

4 Example of successful project

Rajamangala University of Technology Lanna hired PEA acts to improve and increase energy efficiency by implementing measures to change the light bulbs to LED lamps and changing to use high efficiency air conditioning unit of 6 campuses. Total invest is BH 72,215,220 as detailed below.

1. Measure to change the light bulbs by LED lamps is implemented with the invested amount to BH 28,120,320. The payment period is 72 months, with monthly payment of BH 390,560.

2. Measure of using high efficiency air conditioning unit is implemented with the invested amounts to BH 44, 094,900. The payment period is 60 months, with monthly payment of BH 734,915. In summary, both measures are shown in Table 3-4.

5 Amendment Policy Issue

It is evident that more than 30 buildings of government buildings Type 2 cannot continue to implement the project. Therefore, we have investigated the legal information, including comments from all relevant government agencies. Then, we compiled and prepared the legislation for the implementation of projects to install high efficiency devices using ESCO system in the building sector. The legislation is used for the implementation by law in the future.

Table 3. Summarizes measure to change the light bulbs to LED

Description	unit	6 Campus
Number	tube	23,471
The cost of materials and installation equipment	Baht	27,067,414
The cost of materials and installation equipment	BH/Tubes	1,262.93
The cost of the operation.	BH/year	3,007,490
Loan rates	%	5
Total investments	Baht	30,074,904
Saving effect	BH/year	5,476,673
Payback time	year	5.49

Table 4. Summarizes Measure of using high efficiency air conditioning unit

Description	unit	6 Campus
Number	set	785
The cost of materials and installation equipment	Baht	39,685,410
The cost of the operation.	BH/year	4,409,490
Loan rates	%	5
Total investments	Baht	44,094,900
Saving effect	BH/year	10,053,457
Payback time	year	4.39

We have monitored the implementation of the project of building 53 of and received the information by coordinating with the Provincial Electricity Authority (PEA) and Metropolitan Electricity Authority (MEA) And monitor progress. We found that the establishment of the budget is the barriers. This is contrary to the rules of disbursement to the ordinary project. From the collection of problems such obstacles, we have analysed the recommendations and solutions to proceed the project in the future as follows:

1. Government buildings should establish a budget for the implementation of the project into two parts of data collection and preparation of proposals for the project of high-performance equipment modifications. If the building does not wish to continue the project, the party providers (MEA / PEA) would still have cost in this data collection.

2. The implementation of ESCO is the pattern of hiring a contractor (Lump Sum Turnkey), which according to the Prime Minister on Procurement 1992 Article 54 [4]. Therefore, the project management concept has to be approved by the State Council.

3. To solve the problem of the operational phase of the project within one year engagement according to a financial commitment 1991, the Cabinet has to approve the budget commitments over the year, covering the period of project until the end of the contract period for every government agency.

4. The energy performance contract has to be written by the duration of the contract covering the period of the project. The proposed form of contract also has to be approved by the Cabinet.

5. The budget for utilities should remain constant throughout the project duration. It should be presented to the Cabinet for approval.

6. The savings of the budget for utilities from the project are paid for the cost of energy management of ESCO company. Therefore, the Cabinet has to approve the estimated energy savings according to the authority of the Budget Procedures Act 1959.

7. Various government agencies also implemented the project in different ways. Therefore, there should be regulation of the office of the Prime Minister on the preparation of ESCO project management for energy savings in the building sector for all government agencies by authority of the Government Administration Act 1991[5].

6 Conclusion

The preliminary ESCO project management has been launched in order to save energy in the government building sector in 2015. The potential of electricity savings is about 77 million kwhr/year. It can reduce imports of crude oil about 6.58 thousand tons of crude oil (Ktoe) per year. This represents the cost of BHT 310 million per year. The budget to invest is BHT 1,504 million, with an average payback period of 4.85 The implementation of the project have been monitoring for 53 buildings of by coordinating with PEA and MEA acted as ESCO company. We found that only one building has the successful implementation of the project building. There are 44 buildings that are pending the implementation because of the barriers of budget payment for the cost of ESCO project management. The solutions and guidelines for future project have been analyzed and presented in this paper. The amendment policy in the way of ESCO system for government buildings should be approved by the Cabinet. Finally, the new regulation for establishing the ESCO system is the key to encourage the implementation of energy savings in the buildings.

Acknowledgements

The paper is based upon work supported by Department of Alternative Energy Development and Efficiency (DEDE) of Thailand.

References

1. Energy Development and Promotion Act B.E. 2535 (1992), Kingdom of Thailand.
2. Ministerial regulation B.E. 2539 (1996), Ministry of Energy, Kingdom of Thailand.
3. Energy Conservation Promotion Act (No. 2) B.E.2550, Kingdom of thailand.
4. Prime Minister on Procurement 1992 Article **54**, Kingdom of thailand.
5. Budget Procedures Act 1959, Kingdom of thailand.
6. C. Asavatesanupap, M. Santikunaporn, and P. Wangsakarn :Energy conservation through value-engineering for SMEs in Thailand, The 5th European Conference Economics and Management of Energy in Industry Proceedings, Vilamoura, Portugal (2009)
7. P. Wangskarn, and M. Sasananan: Comparison of performance-based VS cost-based energy saving project, 2nd WSEAS/IASME International Conference on Energy Planning, Energy Saving,

Environmental Education (EPESE'08) Corfu, Greece, October 26-28 (2008)
8. Thammasat University: Final Report on the Evaluation the Participatory Energy Conservation Project: Executive Summary (2011)
9. M. Prukvilailert and P. Wangskarn: Energy Conservation Potential in SMEs of Thailand 2011 ,IEEE International Conference on Smart Grid and Clean Energy Technologies (IEEE ICSGCE 2011) ,Chengdu, China (2011)
10. P. Wangskarn and M. Prukvilailert: Evaluation the Designated Government Buildings Promotion in Thailand 2011, IEEE International Conference on Smart Grid and Clean Energy Technologies (IEEE ICSGCE 2011) ,Chengdu, China (2011)
11. How to Hire an Energy Service Company, Energy Efficiency Project Management Handbook, California Energy Commission, Energy Efficiency Division, January 2000, P400-00-001D.
12. D. Gray: How to Hire an Energy Auditor to Identify Energy Efficiency Projects, California Energy Commission, January 2000, P400-00-001C.

Energy Conservation of the Designated Government Buildings in Thailand

Prapat Wangskarn [1], Monchai Prukvilailert [1]

[1] *Thammasat University, Thailand*

Abstract. The designated government buildings have implemented and administered energy program under the energy development and promotion Act 2007 for many years continuously until 2015. Appointment person responsible for energy, performing energy management and implementing the energy conservation work plan and measures are legal requirements for the designated buildings. Therefore, the ministry of Energy has launched the project to support the implementation of energy management. The aim of the project was to create the energy management system in the designated government buildings, and to reduce energy consumption. In this paper, the evaluation of the project has been presented from the achievements of 839 designated government buildings. The energy saving is more than 440 ktoe/year. This is about 3% of energy consumptions of buildings.

1 Introduction

The Strategic Energy has set a goal of national energy conservation with the aim to reduce at least 25% of energy consumption per unit of production (Energy Intensity) of manufacturing and services in 2033. By regulatory measures under the Energy Conservation Act 1992, the building sector (about 852 buildings) required to submit reports to Department of Alternative Energy Development and Efficiency (DEDE). However, buildings must arrange a report on the implementation of energy management for inspection and certification by the competent authority before sending the report to DEDE in March of each year. For promoting and overseeing the building sector, DEDE conducted to provide advice in support of the development of an energy management system to ensure compliance with legal requirements and also the event encourages awareness. The working Group on energy management of buildings has been established for preparation of energy management continuously. From the past experience, there is a need to support public sector building steadily. The examination and approval by the competent authorities or inspection and certification of energy management have been prepared.

In fiscal 2015, DEDE send team of consultants to carry out the inspection and certification of energy management in public buildings on the site to verify the actual plans and performance measures for energy conservation.

The government agency that owns the designated buildings has an order to implement energy management programs according to the energy development and promotion Act 2007 [1] in Thailand. The success of energy conservation promotion for the designated government buildings can induce the owner of the designated buildings of private-sector to implement the energy management program under the law [2]-[4].

In the past, we found that some government agencies of the designated buildings were unable to conduct the energy management system according to the law by the lack of personnel who has knowledge and experience [5], [6]. In order to conform to the law, appointment person responsible for energy, performing energy management and implementing the energy conservation work plan and measures must be executed.

Therefore, the project of promotion and conducting the energy conservation regulation and supervision of designated government buildings was launched continuously in 2015 [7].

The project provided the experts on energy management system according to the law to give guidance for the staff of buildings. The agency of the buildings has been forced to implement the energy conservation activities.

In this paper, evaluation of the project has been presented. The effectiveness of the project was assessed in according with energy management activities, number of persons responsible for energy and the energy savings potential.

2 Evaluation Procedure

There were 839 designated government buildings applied to the project. The buildings were categorized into 5 groups according as shown in Table 1.

Table 1. Number of buildings categorized.

No	Government agency	Number
1	Office	392
2	Academy	183
3	Hospital	140
4	Electrical	13
5	Other	111

The buildings were also categorized into 21 groups according to their government agencies (19 ministries, the Prime minister's office and independent public agencies) as shown in Table 2.

Table 2. Number of buildings categorized by the agency

No	Government agency	Number
1	Ministry of Agriculture and Cooperatives	21
2	Ministry of Information and Communication Technology	25
3	Ministry of Labour	20
4	Ministry of Defence	165
5	Ministry of Finance	19
6	Ministry of Foreign Affairs	2
7	Ministry of Tourism & sports	8
8	Ministry of Social Development and Human Security	2
9	Ministry of Transport	47
10	Ministry of Natural Resources and Environment	8
11	Ministry of Energy	26
12	Ministry of Commerce	2
13	Ministry of Interior	77
14	Ministry of Justice	7
15	Ministry of Culture Thailand	7
16	Ministry of Science and Technology	21
17	Ministry of Education	171
18	Ministry of Public Health	132
19	Ministry of Industry	12
20	The Prime Minister's Office	13
21	Independent public agencies	54

The objects of evaluation include:

1. Status of implementation of requirements according to the law;
2. Potential of energy saving by changing the significant equipments;
3. Energy savings.

The data and information were collected from the final reports of the participated designated buildings, questionnaires and direct interview in 2015. Then, the data collation and analysis was performed to evaluate the outcomes of the project.

3 Results

3.1. Status of implementation of requirements according to the law

According to the energy conservation Act 2007, the government agency of the designated building must submit the energy conservation work plan and having the person responsible for energy (PRE). The work plans for every buildings have been investigated in details.

The work plan has been written in form of a report. It shows the implementation and results of energy conservation activities of buildings. Eight process of energy management system have to be reported the implementation and evaluation as follows:

1. Designation and energy management committee
2. Evaluation of preliminary status of energy management
3. Energy conservation policy
4. Evaluation of energy conservation potential
5. Determination energy conservation plane and target
6. Implementation and operation.
7. Evaluation by internal audit.
8. Management

In the past few years, some buildings did not implement the work plan due to lack of knowledge. This project, therefore, provided expert persons in the field of energy management system to consult and prepare the work plan for all buildings. For the year 2015, Almost buildings were able to submit the work plans in the given period. However, due to the lack of knowledge and experience, some of work plan of buildings having the error of data and missing the important activities of energy management process were not approved.

The work plan that is completed and fully to be approved, it is necessary to show the contents of activities, results and evaluation for 8 steps of energy management process. Ministry of Education and Ministry of Defense have the most buildings in responsibility about 169 buildings and 163 buildings, respectively. Only few number of work plans of buildings that have not been approved. It can be seen that most of government buildings also want to create a culture of law-abidingness.

The results were shown in Table 3 for the investigation of the work plan for government agencies.

Table 3. The results of investigation the work plan categorized by the agency

No	Government agency	approved	Not approved
1	Ministry of Agriculture and Cooperatives	20	1
2	Ministry of Information and Communication Technology	25	0
3	Ministry of Labour	20	0
4	Ministry of Defence	163	2
5	Ministry of Finance	19	0
6	Ministry of Foreign Affairs	2	0
7	Ministry of Tourism & sports	7	1
8	Ministry of Social Development and Human Security	2	0
9	Ministry of Transport	46	1
10	Ministry of Natural Resources and Environment	8	0
11	Ministry of Energy	26	0
12	Ministry of Commerce	2	0
13	Ministry of Interior	77	0
14	Ministry of Justice	7	0
15	Ministry of Culture Thailand	7	0
16	Ministry of Science and Technology	21	0
17	Ministry of Education	169	2
18	Ministry of Public Health	132	0

No	Government agency	approved	Not approved
19	Ministry of Industry	12	0
20	The Prime Minister's Office	12	1
21	Independent public agencies	53	1

After the project period, from 839 designated buildings, the agencies of 759 designated buildings or more than 90% of the total number of buildings were able to prepare the completed work plan.

Moreover, more than 80% of the buildings of agencies have already appointed the person responsible for energy (PRE). It can be seen clearly that the project brought about an increased understanding of the implementation under the energy development and promotion Act. However, about 20% of designated buildings, the agencies were not able to complete the requirement of the number of PRE.

When checking the status of compliance with the laws of the government building in the status of the appointment of the person for responsible energy (PRE) of building we found that 685 buildings (80% of total number of buildings) can meet the requirements of the law. However, 116 buildings (13% of total number of buildings) still have not appointed PRE fully following the requirements of the law. Furthermore, 59 buildings have no any PRE. In summary, more than 2,500 PRE has been registered and has been appointed for buildings. The details are summarized below.

Table 4. The Number of person responsible for energy categorized by the agency

No	Government agency	Completed	Number of PRE
1	Ministry of Agriculture and Cooperatives	17	69
2	Ministry of Information and Communication Technology	23	86
3	Ministry of Labour	19	59
4	Ministry of Defence	108	492
5	Ministry of Finance	19	87
6	Ministry of Foreign Affairs	2	6
7	Ministry of Tourism & sports	6	18
8	Ministry of Social Development and Human Security	2	13
9	Ministry of Transport	45	183
10	Ministry of Natural Resources and Environment	7	23
11	Ministry of Energy	11	55
12	Ministry of Commerce	2	8
13	Ministry of Interior	62	192
14	Ministry of Justice	7	24
15	Ministry of Culture Thailand	4	18
16	Ministry of Science and Technology	15	62
17	Ministry of Education	153	776
18	Ministry of Public Health	117	443
19	Ministry of Industry	12	37
20	The Prime Minister's Office	11	38
21	Independent public agencies	43	191

3.2 Potential of energy saving by changing the significant equipments

Although the agency of the designated government buildings have already implemented energy management system, the government has the idea to subsidize for changing the significant equipments. That is the way to reduce the energy consumption in the buildings rapidly.
Therefore, the number of the significant equipments (such as Fluorescent lamp, chiller, and boiler) has been collected. Then, the Potential of energy saving by changing the significant equipments and using the high efficiency equipments has been evaluated.
For Fluorescent lamp, in Thailand, most buildings usually installed the Fluorescent lamp T8 (36 w). The total number of T8 is about 2 million lamps. In the case of changing T8 to LED (20w) which having the same luminous intensity, we can reduce energy consumption to 147,188,761 kwhr/year (about BHT 588 million per year) with the investment costs BHT 2,000 million.

For air conditioning unit (split type), this is the most equipment installing for air conditioning of every type of buildings in Thailand. In designated buildings of government agency, the total number 133,452 units of split type have been installed more than 10 years. In the case of changing to High efficiency equipment which having EER higher than 11.0, we can reduce energy consumption to 467,106,540 kwhr/year (about BHT 1,868 million per year) with the investment costs BHT 5,000 million. Table 5 shows potential of energy savings for various size of air conditioning unit.

Table 5. Potential of energy savings of air conditioning unit categorized by size

No	Size (Btu/h)	Number	Energy saving (kwh/year)
1	$\leq$12,000	13,528	19,916,359
2	13,000	7,932	14,213,539
3	15,000	3,622	6,951,479
4	18,000	18,889	43,688,097
5	24,000	18,064	57,752,346
6	30,000	27,298	102,813,384
7	36,000	25,704	117,473,344
8	40,000	8,561	44,560,704
9	44,000	2,953	17,267,279
10	48,000	5,751	35,094,669
11	50,000	1,150	7,375,340

For chiller, the equipment has been installed in the buildings usually more than 15 years with total number of 310 units and having the cooling capacity 73,000 Ton. In the case of changing to High efficiency chiller which having COP lower than 0.6 kw/Ton, we can reduce energy consumption to 238,634,704 kwhr/year (about BHT 1,454 million baht per year) with the investment costs BHT 1,500 million.

For boiler, this equipment is used for disinfection and sterilization of medical equipments in the hospital under control of Ministry of Public Health. Total number of boilers that have been installed more than 15 years is 77 units (thermal efficiency lower than 70%). In the case of changing to High efficiency boiler which having thermal efficiency higher than 85%, we can reduce energy consumption to 51,563,636 MJ/year (about BHT 37 million per year) with the investment costs BHT 250 million as shown in Table 6.

Table 6. Potential of energy saving by changing equipments.

No	Government agency	Saving ktoe /year
1	LED lamp	12.54
2	High efficiency air conditioning (split type)	39.80
3	High efficiency chiller	20.33
4	High efficiency boiler	1.22
	total	73.89

3.3 Energy savings

There were 839 designated buildings involved in the project. The total energy consumption was 438,940 tons of oil equivalent per year (toe/year) of which the total consumption consists of 4,846,583,072 kWh of electricity and 1096,618,329 MJ of thermal energy. During the project period, there were many energy conservation measures recommended by energy experts have been implemented. Energy savings are 169,124,575 kWh of electricity and 86,834,887 MJ of thermal energy. This savings is about 3% of energy consumptions of buildings.

3.4 Altitudes and Suggestions

Some agencies did not complete the requirements of energy management system according to the law due to problems during procedures. Staff also did not realize the importance of a person responsible for energy and energy conservation activities because of lack of support from administrators and budget. Therefore, administrator should focus on energy management system in their organization. The importance and knowledge of energy conservation should be instilled to staff at all levels.

Government buildings suggested to DEDE including many interesting issues for carrying out the project further in future as follows:

1. The budget of buildings for changing significant devices such as air conditioning units, electrical and lighting is very limited. Therefore, DEDE should establish a support project for energy saving equipment modification to the building sector.

2. To motivate and encourage the implementation of energy conservation building sector, DEDE should provide for national competitions and awards for outstanding public buildings participating.

3. DEDE should provide a person for responsible energy (PRE) for buildings directly. The provincial energy officials have to facilitate coordination and take part in the operations of the government building and shipping information via email directly to PRE of the building.

4. DEDE should provide a team of experts to advise constantly not only just during the preparation of energy management, but continue long-term.

5. DEDE should focus on the promotion of energy conservation thoroughly. This includes practical and pragmatic. Also, DEDE should provide a meeting or training specifically for executives of government buildings who have primacy.

3.5 Suggestions for the building sector

The results of the implementation of the program found that some government buildings cannot be fully implemented energy conservation. This partly stems from the internal administration, the attitude of the staff and administrators of public buildings. This conclusion is as follows.

1. Government buildings should have a meter installed in each of the separate entities to be able to monitor and track energy consumption in detail.

2. Large public Building including many departments is still lacking in operation such as gathering data in preparing the report, and the establishment of energy conservation measures. Therefore, the chief executive of the government building should define policy and the annual budget for the implementation of energy management in particular to reduce restrictions on operations.

4 Conclusion

The continuous project of promotion and conducting the energy conservation regulation and supervision of designated government buildings in 2015 has been evaluated in this paper. The development of energy management and energy conservation measures of 839 designated government buildings were evaluated using questionnaires, direct interview and the data from the report of energy management work plan and target. It can be concluded that buildings are able to implement energy management system with experiences and consistency. This influences to energy savings of buildings and energy conservation activities occurred in the period of project. The according result of potential of energy saving by changing to use high efficiency equipment is about 73.89 ktoe/year. Finally, energy savings occurred in period of project is about 438.94 ktoe/year. For the attitude of buildings, It can be seen that most of designated government buildings were able to provide the requirements of energy management system according to the law.

Acknowledgment

The paper is based upon work supported by Department of Alternative Energy Development and Efficiency (DEDE) of Thailand.

References

1. Energy Conservation Promotion Act (No. 2) B.E.2550, Kingdom of thailand

2. The Department of Alternative Energy Development and Efficiency, Enhancing the Promotion of Energy Management in Designated Factories and Designated Buildings, Final Report, (2009)
3. Energy Development and Promotion Act B.E. 2535 (1992), Kingdom of Thailand
4. Ministerial regulation B.E. 2539 (1996), Ministry of Energy, Kingdom of Thailand
5. C. Asavatesanupap, M. Santikunaporn, and P. Wangsakarn, "Energy conservation through value-engineering for SMEs in Thailand", The 5th European Conference Economics and Management of Energy in Industry Proceedings, April 2009, Vilamoura, Portugal
6. P.Wangskarn, and M.Sasananan,"Comparison of performance-based VS cost-based energy saving project", 2nd WSEAS/IASME International Conference on Energy Planning, Energy Saving, Environmental Education (EPESE'08) Corfu, Greece, October 26-28, (2008)
7. Thammasat University, The final report of promotion and conducting the energy conservation regulation and supervision of designated government buildings Project in 2015

Investigation of the DI Diesel Engine Performance using Ethanol-Diesel Fuel Blends

Malee Suntikunaporn[1], Snunkhaem Echaroj [1] and Channarong Asavatesanupap[2]

[1]*Department of Chemical Engineering, Faculty of Engineering, Thammasat University, Pathumthani, Thailand 12120*
[2] *Department of Mechanical Engineering, Faculty of Engineering, Thammasat University, Pathumthani, Thailand 12120*

Abstract. Ethanol-diesel blend is a promising candidate as a fuel for direct injection (DI) diesel engine. In this research, solubility of different compositions of ethanol-diesel blends from 2 to 15% (v/v) ethanol were tested for 20 days. Significant increases in solubility of the blends were observed after addition of 1% (v/v) *n*-butanol. The Kubota's RT140 diesel engine was operated using E7B1D92 blend at several engine speeds (1,000 to 1,600 rpm). The obtained results demonstrated that, when using E7B1D92 blend at an engine speed of 1,500 rpm, the engine power and torque of the RT140 engine were increased from 8.6 PS to 10.1 PS and 4.1 kg_f-m to 5.1 kg_f-m compared to pure diesel fuel. However, specific consumption fuel increased when E7B1D92 blend was used during the engine test. Additionally, analysis of exhaust gas revealed a decrease in smoke density when E7B1D92 was used instead of pure diesel.

1. Introduction

A clear deficit in the supply of fossil fuel diesel have driven millions of car-users to convert to vehicle powered by alternative sources of fuel such as biodiesel, gasohol (ethanol-gasohol blend) and even electricity. According to the World Oil Outlook, demand of diesel fuel is projected to increase from 26 million barrels per day (mb/d) to 36 mb/d by 2035 [1]. As a result, in order to reduce our dependency on fossil fuel, researches are now considering the possibility of mixing renewable fuel with fossil-derived diesel for diesel injection engine. In Thailand the Ministry of Energy planned to replace 44% of crude oil consumption with biofuels such as ethanol and biodiesel by 2021 [2]. Fortunately, ethanol is a renewable material that can be produced through fermentation of sugar cane [3], fruit peels and various lignocellulose materials [4]-[6].

Ethanol has a high octane number which makes it compatible for spark ignition (SI) engine. However, ethanol has a cetane number of approximately 5 which is very low compared to diesel oil with cetane number of 50. Additionally, the molecular weight of ethanol differs greatly from diesel molecule [7]. For these reasons, ethanol cannot be used directly for compression ignition diesel engine. Ethanol has to be blend with diesel to make a solution mixture which is usually referred to as diesohol.

The solubility of ethanol in diesel was observed in many researches in term of how long the blended solution remains in a homogenous solution after mixing. The ethanol-diesel blend consisting of 10 % (v/v) was found to remain in a homogenous phase for 72 hours, whereas the blends with 20 and 30 % (v/v) ethanol separated into two phases 24 hours and 5 minutes after agitation. Addition of 5% (v/v) *n*-butanol to the 20 and 30% (v/v) ethanol solution was found to lengthen duration at which the solutions remain in one phase to 14 days and 11 days [8]. Additionally, *n*-butanol was found to successfully enhanced solubility of the water-containing ethanol-biodiesel-diesel blends [9]. An improvement in solubility of the blend was due to the polarity of *n*-butanol's hydroxyl which facilitates ethanol's solubility in diesel. Another kind of additive that had been employed frequently was Sorbitan methyl- ester [10], [11]. A research reported that solvents that are capable of improving the solubility of ethanol-diesel blends have to contain a hydrogen donor group, a hydrogen acceptor group and a nonpolar group. The hydrogen donor groups (eg. –OH, –SH, =NH and $-NH_2$) and hydrogen acceptor groups (eg. =N–, =O, =S, –NH–) have high affinity with polar substance present in the blend. Additionally, nonpolar group interacts strongly with hydrocarbons in the blend through the vander-waal molecular interaction forces [12], [13].

Utilization of ethanol-diesel blend as a fuel for diesel engine has been investigated in many researches. Shadidi et al reported an increased in engine power of at least 3.17% after 99.6% ethanol was blended with pure diesel and tested with a diesel engine containing 6 cylinders. However it was also found that the specific consumption increased by approximately 7.3 to 15.8% [14]. Another experiment conducted using 6 cylinder diesel engine revealed that the specific fuel consumption of ethanol-diesel blend was higher than that of pure diesel [15], [16]. Apart from diesel fuel that is derived from refining fossil fuel, bio-oil from pyrolysis of wood has also been shown to improve during combustion in a dual injection diesel engine when mixed with ethanol. This is because bio-oil from pyrolysis of biomass alone is constrained by its poor physical properties, such as high acidity and low energy density.

Moreover, bio-oil from pyrolysis of biomass usually contains free-radical which promote polymerization during combustion at high temperature or even during storage at ambient temperature. Addition of ethanol to bio-oil has been found to help reduce polymerization activities [17]. There are more researches on the subject of testing the performance of ethanol-diesel blend using either 6 cylinders diesel engine or a modified diesel engine [18]-[21]. Unfortunately, so far very few researches have investigated how ethanol-diesel blends with and without addition of co-solvent perform on a 1 cylinder diesel engine.

In the present work, we report the solubility testing of blends or diesohol with different ethanol-to-diesel ratio from 2% to 15% (v/v) ethanol; E2B1D93, E4B1D95, E7B1D92, E10B1D89, and E15B1D84. Additionally, the effect of adding 1% (v/v) *n*-butanol on the solubility of the solution was also investigated. Performance of the 1-cylinder Kubota's RT140 diesel engine was evaluated according to the specific fuel consumption, engine power output, torque and exhaust density at different engine rotational speeds.

2. Experimental setup

2.1 Blending and solubility testing

Ethanol-diesel blends were prepared by mixing diesel and ethanol for 10 minutes at a rotation speed of 2000 rpm using an agitator. For some experiment 1%v/v *n*-butanol was added as co-solvent to help increase the solubility and stability of the blends. The additive was added immediately after the ethanol diesel solution was prepared and the agitator was used to blend the solution again for another 15 minutes at 2000 rpm. After the solution appeared to be homogenous each blend was transformed to a flask and place at ambient temperature for 20 days in order to examine the solubility of each blend. A picture of each blend was capture 30 minutes, 10 days and 20 days after the solutions' preparation.

2.2 Engine performance testing

Engine performance was examined using a single cylinder, four stroke Kubota RT140 DI diesel engine with specifications as outlined in Table 1. The engine was coupled by a shaft to an eddy current dynamometer executed by a Tokyokoki-Schenck controller device. The engine is cooled by water circulating from the cooling tower located outside the engine testing laboratory. An automated solenoid controlled type-flow meter from Tokyo Kokukeiki Company was used to measure the fuel consumption. A fuel tank was used to store 6L of fuel for the engine. The exhaust gas was captured by a nozzle and was delivered to be analyzed by Sokken GSM-3 smoke meter which utilize filter paper and light detecting technologies. The crank angle or fuel-delivery angle inside the engine was adjusted to 19.2° by combining 0.2 and 1° metal sheets. The controller monitor was used to observe the rate consumption of fuel, exhaust gas temperature (T1), oil temperature (T2) and smoke density of the exhaust gas. The controller was also used to adjust the engine speed and torque. A detail diagram of the system is shown in Figure 1.

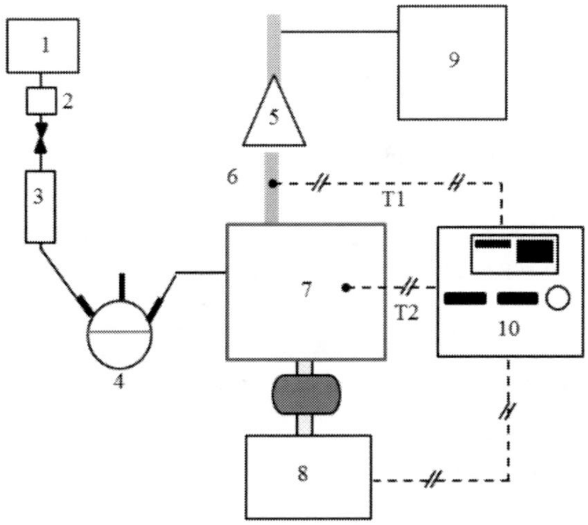

Figure 1 Experimental setup diagram of engine test facilities; 1. Fuel feed tank, 2. Filter, 3. Flow controller, 4. Feed modification, 5. Exhaust gas nozzle, 6. Exhaust gas, 7. Diesel engine, 8. Eddy dynamometer, 9. Smoke meter, and 10. Dynamometer controller device.

Table 1 RT140 DI Engine specifications

Parameter	Basic specification
Model	Kubota RT140 (Thailand)
Number of cylinder	1
Type	Direct injection diesel engine 4 stroke
Bore x stroke	97 x 96 mm
Displacement	709 cc
Compression ratio	18:1
Maximum power	14.0 HP /2400 rpm
Maximum torque	5.0 kg_f-m /1600 rpm
Injector pressure	200 kg/cm^2

The performance was recorded from a dynamometer and reported in terms of engine power (calculated by Equation 1), specific fuel consumption (calculated by Equation 2), and smoke density of the exhausted gas.

$$\text{Power} = \frac{Torque \times Speed}{716.2}(PS) \quad (1)$$

$$\text{Specific fuel consumption} = \frac{F_f \rho_f}{Power}\left(\frac{g}{PS \cdot h}\right) \quad (2)$$

Where F_f is the flow rate of the either ethanol-diesel blend or pure diesel fuel that is consumed by the diesel engine and ρ_f is the bulk density of the blend and fuel. The unit for engine speed is rpm, kg_f-m for engine toqure, g/PS-h for specific energy consumption, PS for engine power, and BSU for exhaust density.

3. Results and discussion

3.1 Blending and solubility testing

As shown in Figure 2, E15D85 blend separated into two phases after 30 minutes while the rest remained in a homogeneous solution. A two headed gray arrow was included in the figures beside the blend to highlight the separation of ethanol and diesel. A homogeneous solution appeared as a transparent clear yellowish color. It was found that E10D90 remained a homogeneous solution for roughly 30 minutes, but after 10 days it was observed to be separated into two phases. Ethanol content in E4D06 and E7D93 were observed to be completely miscible with diesel's hydrocarbon during the first 19 days. However, after 20 days E4D96 and E7D93 blends separated into two phases. E2D98 blend was still observed to remain in a homogeneous phase even after 20 days after agitation.

For the next part, an additive, 1%(v/v) *n*-butanol, was added after ethanol and diesel was agitated. Addition of *n*-butanol was observed to increase the solubility of E10B1D94 from 30 minutes to 10 days. It was found, as shown in Figure 3, that even after addition of additive E15B1D94 still separated into two phases almost immediate after agitation. However, after 20 days the E10B1D89 blend was separated into two phases. The blends that remain homogenous for at least 20 days are E2B1D97, E4B1D95 and E7B1D92 blends. An increase in the percentage of *n*-butanol can help increase the solubility of 15 (v/v%) ethanol in diesel fuel. It can be concluded that that additional of *n*-butanol was successful in improving the solubility of ethanol in diesel.

3.2 Engine performance testing

The ethanol-diesel blend (E7B1D92) was prepared and some properties such as density, viscosity and heating value were analysed. The significant properties of the blend compared to pure diesel are summarized in Table 2. According to Table 2, addition of ethanol to diesel was found to decrease the density, viscosity and heating value of the fuel.

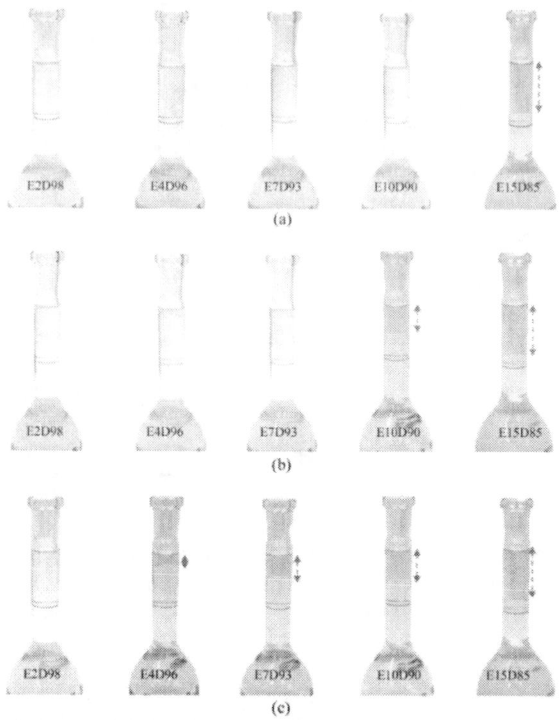

Figure 2 Solubility of different ethanol-diesel blends without *n*-butnaol 30 minutes a), 10 days b) and 20 days c) after agitation.

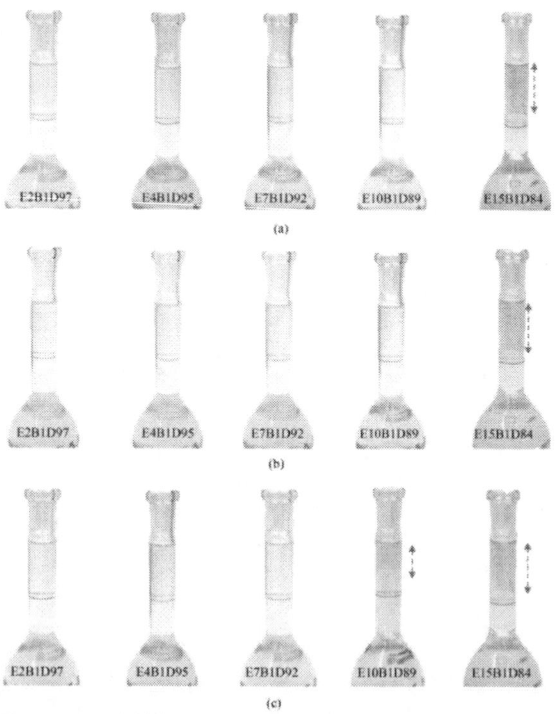

Figure 3 Solubility of different ethanol-diesel blends with *n*-butnaol 30 minutes a), 10 days b) and 20 days c) after agitation.

Table 2 Properties of E7B1D92 blends

Properties	Pure diesel	E7B1D92
Density 30°C,	837 kg/m^3	833 kg/m^3
Kinetic viscosity 40°C	2.6 cSt	2.5 cSt
Lower heating value	43.0 MJ/Kg	41.7 MJ/Kg

The relationship between engine power and engine speed for E7B1D92 was illustrated in Figure 4. It can found that the maximum engine power was attained at the engine speed of 1500 rpm. An increase in ethanol content clearly resulted in an increase in engine power. It can be observed that the maximum engine power of E7B1D93 blend was 10.1 PS, while pure diesel gave only 8.6 PS. Engine power for both ethanol-diesel blended fuel and pure diesel decrease to approximately 7 PS after the engine speed is reduced from 1500 rpm to 1000 rpm. Although the blended fuel has lower heating value compared to pure diesel, it gave significantly higher engine power. This is because the molecular formula for ethanol is C_2H_5OH which contain of an oxygen atom (approximately 35% oxygen). During combustion, fuel reacts with oxygen to release the desirable amount of energy. The engine will undergo a complete combustion when the content of oxygen in the system is sufficient. In contrary, if there is not enough oxygen then incomplete combustion will occur causing the diesel engine to perform below its capability. Therefore, the presence of ethanol molecule increases the supply of oxygen in the chamber, which results in a complete combustion and higher engine power.

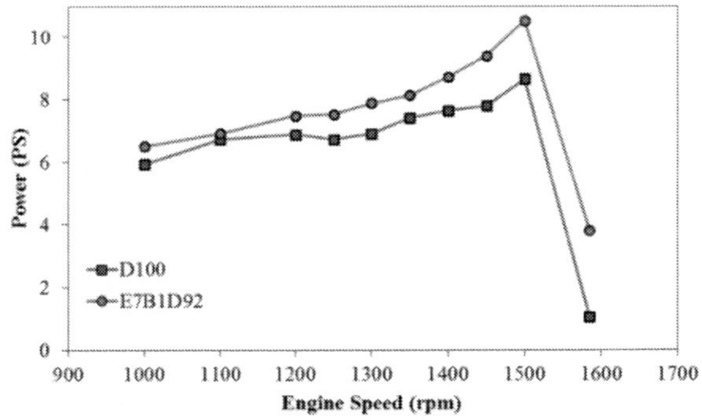

Figure 4. Relationship between engine power and engine speed of E7B1B92 blend and pure diesel

Variations of torque (kgf-m) with different engine speeds (1000 – 1600 rpm) were shown in the graph below. As shown in Figure 5, engine torque of the ethanol-diesel blend fuel is higher than that of the pure diesel. The result shows that at engine speed of 1,500 rpm E7B1D92 blend produce a torque of 5.1 kg_f-m. For pure diesel the maximum torque produced at 1500 rpm was only 4.1 kg_f-m. The increase in engine torque is again attributed to an increase in oxygen content due to combustion of ethanol molecules. Torque from ethanol-diesel blended fuel decreased to approximately 4.8 kg_f-m and remained constant from 1000 rpm to 1400 rpm. Similarly for pure diesel, torque remained constant at 4.2 kg_f-m.

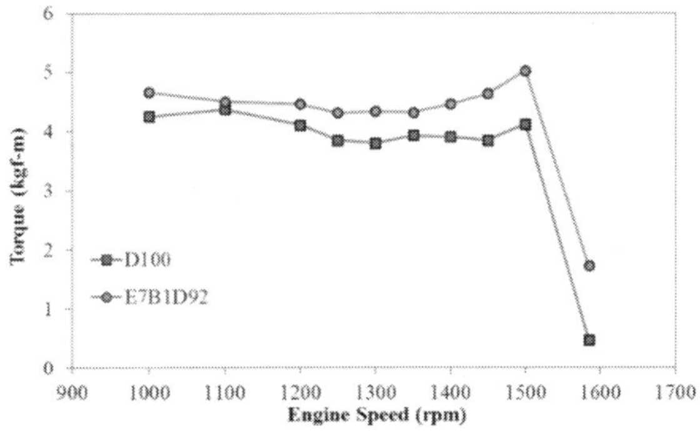

Figure 5. Relationship between torque and engine speed of E7B1B92 blends and pure diesel.

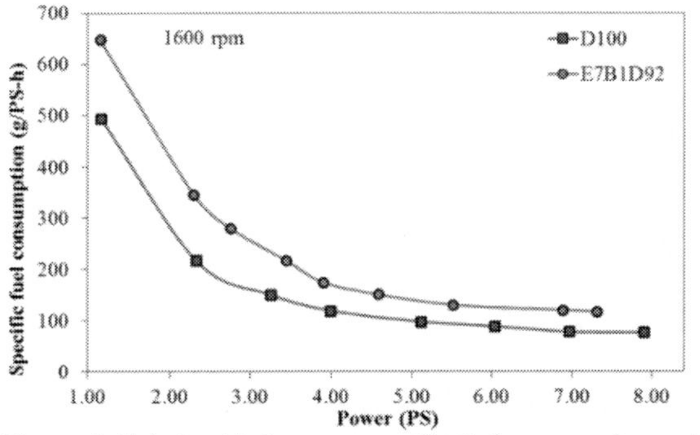

Figure 6. Relationship between specific fuel consumption and engine power of E7B1B92 blend and pure diesel for the engine speed of 1,600 rpm.

Variation of specific fuel consumption with engine speed is shown in Figure 6. The specific fuel consumption can be defined as the mass of fuel flow rate in g/h at one engine power output in PS. It is an important specification used to investigate the performance of the diesel engine. As shown in Figure 6, an increase in specific fuel consumption was observed as the content of ethanol was increased. At an engine power output of 1.2 PS the specific fuel consumption for E7B1D92 blend was found to increase by 31.1% compared to pure diesel. At 5.1 PS engine power output the specific fuel consumption for E7B1D92 blend was found to increase by 33.5% compared to pure diesel. The increase in the specific fuel consumption required for combustion of bended fuel is reasonable due to the lower heating values of ethanol-diesel blend compared to that of pure diesel. Lower heating value is defined as the heat release out of a combustion system given that the initial and final state of the products and reactants are at the same temperature. As shown in Table 2, the lower heating value for 7v/v% ethanol-diesel blends was only 41.7 MJ/Kg, while pure diesel gave a heating value of 43.0 MJ/Kg. Other studies also found that the fuel with lower heating value operated at a significant higher specific fuel consumption than fuel with higher heating value such as pure diesel [3]-[6].

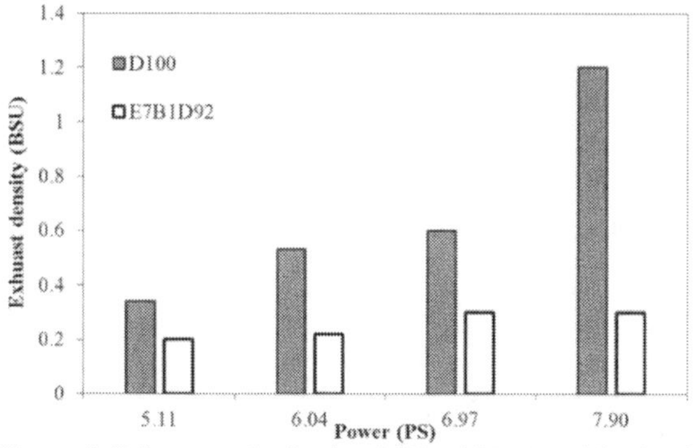

Figure 7. Exhaust smoke density (BSU) of E7B1B92 blend and pure diesel at four different loads for the engine speed of 1600 rpm.

Figure 7 shows that the soot or smoke produced by the ethanol-diesel blend is significantly lower than that produced from pure diesel. When the diesel engine is operated at 7.9 PS addition of ethanol caused the exhaust gas to reduce from 1.2 to as low as 0.3 BSU. When ethanol was added at the engine power of 5.1 PS, the exhausted density reduced from 0.38 to 0.2 BSU. This is also due to an increase in oxygen content during combustion of oxygen containing ethanol molecules which help improve the completeness of the combustion. Additionally it is demonstrated in Figure 7 that an increase in engine power from 5.11 PS to 7.9 PS results in an increase in the smoke density. For pure diesel the smoke density increases significantly from 0.38 to 1.2 BSU. However, for the ethanol-diesel blend the smoke density increases only slightly. Other researches investigating combustions of blends fuel using diesel engine yielded similar results. For instance, Chen et al. and Xingcai et al. revealed flame propagation and combustion duration by capturing in-cylinder photos during combustion of blends and pure diesel at different crank angles. The luminosity during combustion of pure diesel was observed to be significantly larger than that of the ethanol-diesel blend, which indicated higher formation of soot particulate in the chamber. Therefore, in their research it was also revealed that ethanol can help suppress soot formation [22]-[24].

4. Conclusions

In conclusion it was found that solubility of ethanol-diesel blends can be improved by addition of 1% (v/v) *n*-butanol. However, it was observed when the composition of ethanol was equalled to or higher than 10% (v/v) the blended solution separated into two layers after 20 days even after addition of *n*-butanol. Engine performance testing of the E7B1D92 blend revealed a significant increase in engine power and torque at 1,500 rpm compared to pure diesel fuel. However, specific fuel consumption also increase when E7B1D92 was tested and compared to pure diesel due to ethanol's significantly higher heating value than pure diesel fuel. Exhaust gas emitted during engine testing using E7B1D92 as a fuel was found to produce lower smoke density compared to that emitted from pure diesel fuel.

Acknowledgement

This work was supported by Faculty of Engineering, Thammasat University. Also, thanks to SIAM KUBOTA Corporation Co. Ltd for laboratory testing supports. One of us (S.E.) thanks Thammasat University for a scholarship.

References

1. L. Hermida, A.Z. Abdullah, A.R. Mohamed, Renew. Sustain. Ener. Rev. E **42**, 1223-33 (2015)
2. Thailand Board of Investment. *Bangkok Thailand Board of Investment* (2013).
3. S. Kumara, P. Dheerana, S.P. Singhe, I.M. Mishrab, D.K. Adhikari, Renew. Ener. E **83**, 850-8 (2015)
4. R.S. Orozco, Bioresources E **9**, 1873-85 (2014)
5. S. Brethauer, C.E. Wyman, Biores. Tech. E **101**, 4862–74 (2010)
6. S.R. Ajay Kumar Singh, Y. Kumar, H. Masih, J.K. Peter, J.C. Benjamin, P.K. Singh, Inter. J. Current Microbio. Appl. Sci. E **3**, 84-96 (2014)
7. DC. Rakopoulos, C.D. Rakopoulos, E.C. Kakaras, E.G. Giakoumis, Ener. Conv. Manage. E **49(11),** 3155-62 (2008)
8. J. Huang, Y. Wang, S. Li, A.P. Roskilly, H. Yu, H Li, App. Thermal Eng. E **29**, 2484-90 (2009)
9. W-J. Lee, Y-C. Liu, F.K. Mwangi, W-H. Chen, S-L. Lin, Y. Fukushima, Ener. E **36** , 5591-9 (2011)
10. A. Praptijanto, A. Muharam, A. Nur, Y. Putrasari,. Ener. Proced. E **68,** 345-54 (2015)
11. D. Hansdah, S. Murugan, L.M. Das, Alex. Eng. J. E **52**, 267-76 (2013)
12. SH. Yalkowsky, Amer.hem. Soc. 464 (1999)
13. L. Lešnik, B. Vajda,Z Žunič, C. Škerget, B. Kegl, Appl. Ener. E **111**, 558 – 70, (2013)
14. B. Shadidi, T. Yusaf, H.H.A. Alizadeh, B. Ghobadian, Appl. Ener. E **114**, 874-9 (2014)
15. E. Alptekin, M. Canakci, A.N. Ozsezen, A. Turkcan, H. Sanli, Fuel E **157**, 245-54 (2015)
16. MT. Cheng,HJ. Chen, LH. Young, HH. Yang, YI. Tsai, LC. Wang, J. Hazard. Mat. **E 297**, 234-40 (2015)
17. S. Lee, TY. Kim, Fuel E **162**, 65-73 (2015)
18. H. Chen , S. Shi-Jin, W. Jian-Xin, Proceedings Combust. Ins. E **31**, 2981-9 (2007)
19. L. Xingcai, H. Zhen, Z. Wugao, Li. Degang, Combust. Sci. Tech. E **176(8)**, 1309-29 (2004)
20. E. Alptekin, M. Canakci, A.N. Ozsezen, A. Turkcan, H. Sanli, Fuel **E 157**, 245-54 (2015)
21. Z. Chen, C. Yao, Q. Wang, G. Han, Z. Dou, H. Wei, Fuel **E 170**, 67-76 (2016)
22. U. Asad, R. Kumar, M. Zheng, J. Tjong, Appl. Ener. **E157**, 838-50 (2015)
23. R. Ballesteros, J. Guillen-Flores, J. Barba, Fuel E **157**, 191-201 (2015)
24. VB. Pedrozo, I. May, NM. Dalla, A. Cairns, H. Zhao, Appl. Ener. E **165**, 166-82 (2016)

MATEC Web of Conferences 60, 04004 (2016) DOI: 10.1051/matecconf/20166004004
ICCBS 2016

Study on the Pretreating Approaches for the Potato Straws

Yumin An , Jukui Wang, Ye Huang and Xiaomei Xu

Ningxia Normal University, Guyuan, Ningxia, China

Abstract. This paper proposes an approach to pretreat the potato straws. Specifically, potato straws are handled using various kinds of chemical solutions, including *HCl*, H_2SO_4, *NaOH* and $NaOH+H_2O_2$, under different concentrations. For each kind of solution, particular indicators, such as the cellulose content as well as scarification ratio of the treated straws, are studied in the paper. Based on orthogonal experiments, the best pretreatment effect is obtained by using the solution of 4% *NaOH* under temperature of 60° Celsius, solid-to-liquid ratio of 1:10, and processing time of 6d.

1 Introduction

In recent years, the resource comprehensive utilization becomes a hot topic both for the academia and the industry due to the excessive resource consumptions in the world. As a by-product of the crop production, straws are important and useful renewable resources. It is reported that the energy in 2-ton straws is equal to that in 1-ton coal. Utilization of straws is playing an increasingly more important role in agricultural production, natural resource development and environment protection. LOGEN in Canada [1] built a factory which made use of wheat straws to produce fuel ethanol in 2000. China has also made huge progress on utilizing cellulose to produce fuel ethanol in recent years. In the "863" Program of China [2], East China University of Science and Technology has built a factory on Fengxian, Shanghai, which made use of cellulose and could produce 600-ton fuel ethanol per year. Both of the two factories in Canada and China focused on using maize straws to produce fuel ethanol. However, there is no report of re-using potato straws as we know so far. In fact, potato production dominates the agriculture in Ningxia, China. There are over 1 million ton of potato straws produced per year in Ningxia, and traditionally, those straws are considered useless and thus stacked or even burned in the open air, resulting in amount of dust and fume, which takes greatly harm to the environment. Therefore, if we use those "useless" potato straws to produce fuel ethanol which is possible as verified by our experiments, natural resources would be re-used and environment could be protected. Thus, utilization of potato straws would take great advantages, especially be beneficial in Ningxia, China.

There is a technical bottleneck in the pretreatment of the potato straws, which is to enzymolyze the cellulose to get C_2H_6O efficiently. In the potato straws or other plants' straws, the lignin and the hemicellulose are combined by a covalent bond and the cellulose molecule is hidden inside, resulting in a natural barrier which prevents the cellulose molecule from touching the enzyme. Furthermore, considering that lignin is not water-soluble and has a complex chemical formation, straws are quite difficult to be degraded. All of these make it difficult to enzymolyze the cellulose in straws, which could only reach a ratio of 10%~20% [3].

Therefore, in order to enzymolyze the cellulose efficiently, straws must be pretreated with the problem of lignin degradation solved. The pretreatment process has a direct effect on the hydrolyzing and saccharifying result of the cellulose enzyme.

Lots of research have been done on the pretreatment problem for the maize straws. There are mainly three pretreatment approaches: physical approach [4], [5], chemical approach, and the combined approach of both [6]. So far, many papers focused on the dilute acid hydrolysis method, which has reached great achievements. In this paper, based on the efforts mentioned in the reference papers, we conduct a series of experiments using *HCl*, H_2SO_4, *NaOH*, $NaOH+H_2O_2$ to treat the potato straws and measure the resulting scarification ratios. Finally, we find the best pretreatment solution for the potato straws based on our experiments.

2 Proposed Approach

2.1 Experimental Materials

2.1.1 Potato Straws

The potato straws in the experiment were collected from Xiji and Guyuan region. They were crushed after insolation, and then sifted using 70 mesh sieves.

2.1.2 Experiment Instruments

Our experiment instruments include the electronic scales SE-202F, which is from the instrument factory in Dongguan, Guangdong; the air dry oven J2X-9076 from Shanghai BoXun Industrial Co., Ltd; the vacuum pump SHB-IV from Zhengzhou ChangCheng Industry and Trade Co., Ltd; the beater 79-2 from Changzhou GuoHua Electrical Appliance Co., Ltd; the spectrophotometer UV-1750 from Shimadzu Company; and the oscillator ZD-85 from Changzhou GuoHua Electrical Appliance Co., Ltd.

2.2 Proposed Approach

2.2.1 Pretreatment Using Different Solutions

HCI: We conducted three different experiments using *HCI*. Each of them would use 10 g straws which are released in 250 ml iodine flasks. The difference of those three experiments is the concentration (1%, 2% and 3%) of corresponding *HCI* solutions used. Note that these three experiments share the same solid-liquid ratio (1:10) of the *HCI*. The iodine flask, with *HCI* solution and straws in it, is covered by the plastic wrap and sealed by the elastic rubber band, then pretreated for 5 d under the temperature of 20° Celsius. After that, the solution will be neutralized using alkali for 1 d and then fully rinsed with water. Finally, the solution would be dried under the temperature of 80° Celsius, and we take the record of corresponding loss rate of cellulose, hemicellulose and lignin as well as scarification ratio.

We also conducted different experiments with solutions of H_2SO_4, $NaOH$ and $NaOH+H_2O_2$. The experimental process and configuration are same as applied in the *HCI* experiments.

2.2.2 Standard Glucose Curve Plotting

With this experiment, we record the absorption photometry of the standard glucose solution with DNS method [7] and then draw corresponding standard curve. In this experiment, we need 0.100 g solid glucose which has been dried for 2 hours under the temperature of 80° Celsius. The glucose then is dissolved with water in a flask and the solution is kept of 100 ml. Next, with 6 10-ml colorimetric flasks which have been washed and dried, we respectively put standard glucose solutions (1.0 mg/ml) of 0, 0.2 ml, 0.4 ml, 0.6 ml, 0.8 ml, 1 ml to corresponding flasks and then add distilled water to make sure finally the solution in each flask is 2 ml. Then we put 1.5 ml DNS to each flask, shake up, and heat the solution for 5 minutes. After cooling, we keep the solution at constant volume of 10 ml and record the absorption photometry with spectrophotometer in a 540 nm wave-length.

2.2.3 Experiment Process

Recording loss rate for the cellulose, hemicellulose and lignin: After the straws' pretreatment, straws are weighed and the amount of cellulose, hemicellulose as well as lignin are recorded. Based on those data, we can get corresponding loss rate. Note that the loss rate = (weight before pretreatment * amount of corresponding components – weight after pretreatment * amount of corresponding components) / weight before pretreatment * amount of corresponding components * 100.

Straws' enzymolysis and saccharization: we add 1 g straw, which has already been pretreated, 50 ml 0.1 mol/L HAc-NaAc (pH=4.8), 1 ml 2% cellulose solution, and 2 drops of toluene to a 100 ml trigonal tank with stopper. Then the tank is put to an oscillator (60 r/min) with constant temperature of 50° Celsius for 36 hours. Next we put 0.5 ml glucose solution to a 10 ml colorimetric tube, add 1.5 ml DNS, and shake up. The colorimetric tube is heated using boiling water for 5 minutes. When the solution is cooled, we keep it at a constant volume of 10 ml. Then we record the absorption photometry with spectrophotometer in a 540 nm wave-length. Besides, we also record corresponding absorption photometry for the straws' solution without cellulose under the same experimental configuration. Then the scarification ratio could be calculated using this equation: (absorption photometry with cellulose – absorption photometry without cellulose) * 0.9 * 10 * 51 * 10-3 / amount of the cellulose in straws after pretreatment * sample mass * 0.5 * 14.638.

Table 1. The element rate and loss rate in the sample after pretreatment.

Solution	**The weight, element rate and loss rate in the sample after treatment (loss rate -> LR)**						
	Weigh-t (g)	**Cel-Lul-ose**	**LR**	**Hemi-cellul-ose**	**LR**	**Lig-nin**	**LR**
1%HCI	8.99	36.2	15.1	16.39	32.2	5.36	42.1
2%HCI	8.44	36.6	14.2	15.38	36.4	5.20	43.8
3%HCI	8.28	37.0	13.3	15.10	37.5	5.13	44.5
1% H_2SO_4	8.65	39.5	7.36	17.96	25.7	7.12	23.0
2% H_2SO_4	9.28	38.6	9.50	16.80	30.5	7.23	21.8
3% H_2SO_4	8.51	38.4	9.96	15.90	34.2	7.40	20.0
1% NaOH	8.00	39.1	8.30	17.34	28.3	4.70	49.2
2% NaOH	8.21	39.0	8.65	16.92	30.0	4.20	54.6
3% NaOH	7.68	38.8	9.03	16.5	31.7	3.9	57.8
1% NaOH + H_2O_2	8.31	35.1	17.6	17.3	28.4	7.30	21.1
2% NaOH + H_2O_2	8.10	34.3	19.6	16.5	31.7	7.20	22.2
3% NaOH + H_2O_2	7.55	33.9	20.5	16.2	33.0	6.9	25.4

3 Experiment Results

3.1. The Amount and Loss Rate of the Cellulose, Hemicellulose as well as Lignin in the Sample

The amount and loss rate of the cellulose, hemicellulose as well as lignin in the sample ,which has been pretreated, is shown in Table 1, the weights of the samples before treatment are all 10 g.

According to Table 1, the sample pretreated by $NaOH+H_2O_2$ suffers from the greatest loss of cellulose, *HCl* for the loss of hemicellulose and *NaOH* for the loss of lignin.

3.2 The Standard Glucose Curve

The absorption photometry of the standard glucose solution is shown in Table 2. The weights of samples before treatment are all 10 g.

Table 2. Absorption photometry of the standard glucose solution.

concentration (mg/ml)	0	0.02	0.04	0.06	0.08	0.10
absorption photometry	0	0.302	0.595	0.888	1.209	1.438

Based on the data in table 2, we can get the standard glucose curve which is shown in Figure 1.

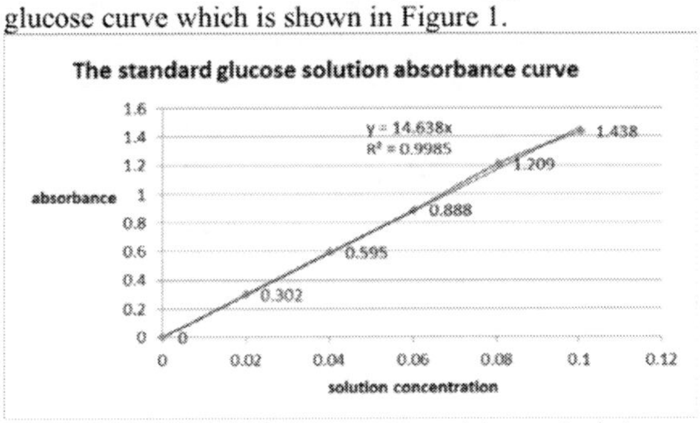

Figure 1. Absorption photometry curve of the standard glucose solution

3.3 The Scarification Ratio of Samples

The absorption photometry and scarification ratio of samples which have been pretreated with cellulose or without cellulose are shown in table 3, where the weight of sample before treatment is always 1 g. Based on data shown in table 3, we get the Figure 2.

Table 3. Scarification ratio of the samples after pretreatment.

solution	**absorption photometry with cellulose**	**absorption photometry without cellulose**	**Scarific-ation ratio**
1% HCl	0.428	0.057	6.43
2% HCl	0.580	0.014	9.70
3% HCl	0.693	0.004	11.68
1% H_2SO_4	0.645	0.034	9.70
2% H_2SO_4	0.487	0.049	7.12
3% H_2SO_4	0.404	0.002	6.60
1% NaOH	1.129	0.061	17.12
2% NaOH	1.192	0.023	18.82
3% NaOH	1.402	0.083	21.32
1%NaOH+ H_2O_2	0.617	0.066	9.83
2% NaOH+H_2O_2	0.570	0.023	10.0
3% NaOH+H_2O_2	0.797	0.154	11.9

According to Table 3 and Figure 2, the sample which is handled using *NaOH* has the largest scarification ratio.

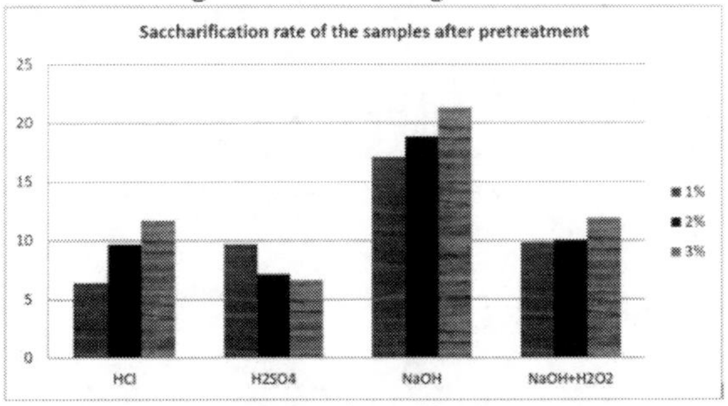

Figure 2. Sample scarification rate

Based on previous experiments, we conduct another orthogonal experiment with potato straw pretreated by *NaOH*. The control parameters in the experiment are the concentration of the *NaOH* solution, pretreating temperature, pretreating duration, solid-to-liquid ratio, and mixing speed. Results are shown in Table 4 and Table 5.

Table 4. Control parameters in the orthogonal experiment.

level	**A: NaOH (%)**	**B: tempe-rature**	**C: Duration (d)**	**D: Solid-to-liquid ratio**	**E: Mixing speed (r/min)**
1	1	40	3	8	60
2	2	50	4	10	70
3	3	60	5	12	80
4	4	70	6	14	90

According to Table 4 and Table 5, the concentration is the most important parameter. Then comes temperature. Followed by solid-to-liquid ratio, duration and mixing speed.

4 Conclusion

From the view of the economic cost and influence on the environment, this paper suggests an efficient pretreating approach for potato straws after comparing the experimental results from different pretreating parameters. The suggestive method is applying 4% *NaOH* solution with a 1:10 solid-to-liquid ratio and 70 r/min mixing speed under 60° for 6d. Using the proposed approach, the scarification ratio of the straws could be more than 19%.

Table 5. Orthogonal experimental results(Absorption phtometry --> AP, Amount of glucose --> AOG, Scarification rate --> SR).

No	**A**	**B**	**C**	**D**	**E**	**AP**	**AOG**	**SR**

.							(%)	
1	1	40	3	8	60	0.317	5.38	4.104
2	1	50	4	10	70	0.756	13.16	11.12
3	1	60	5	12	80	0.650	11.13	9.424
4	1	70	6	14	90	0.571	9.90	8.16
5	2	40	4	12	90	0.387	6.73	5.86
6	2	50	3	14	80	0.584	9.89	9.03
7	2	60	6	8	70	0.924	16.09	14.51
8	2	70	5	10	60	0.693	11.75	10.79
9	3	40	5	14	70	0.540	9.22	7.40
10	3	50	6	12	60	0.567	9.84	7.84
11	3	60	3	10	90	0.829	14.36	12.09
12	3	70	4	8	80	0.588	10.25	8.181
13	4	40	6	10	80	0.771	13.44	15.53
14	4	50	5	8	90	0.762	12.45	14.39
15	4	60	4	14	60	0.930	16.22	18.47
16	4	70	3	12	70	0.730	12.23	14.13

This research is funded by 2014 Guyuan Science-Technology Support Project, together with 2014 Ningxia Nomal University School-Level Scientific Research Program.

References

1. Sharon Boddy. http://www.grist.org/news/-maindish/2006/12/12/boddy. (2007)
2. Zhang Suping, Yan Yongjie, Technology of Rsing Cellulose to Make Ethanol [J]. Progress in chemistry, **7** (8): 1130-1132, (2007)
3. YANG X X, CHEN H Z. Bioconversion of Corn Straw by Coupling Ensiling and Solid State Fermentation [J]. J Bioresourse Tech., **78** (3): 277-280, (2001)
4. Wu Kun, Zhang Shimin, Zhu Xianfeng. Lignin Biodegradation Research Progress [J]. Journal of Henan Agri. Univ., **34** (4): 349-354, (2000)
5. Higuchit. Lignin Biochemistry, Biosynthesis and Biodegradation [J]. J Wood science and technology, **24**: 23-63, (1990)
6. Ye Hong, Li Jiacheng, Distillation and Its Application in Plant Cellulose Resources [J]. **14** (11): 8-10, (2000)
7. Li Riqiang, Xin Xiaoyun, Liu Jiqing. The Separation of Natural Straw Cellulose Decomposing Bacteria Breeding [J]. Shanghai Environ. Sci., **21** (1): 8-1, (2002)

Authors' background

Your Name	Title*	Research Field	Personal website
Yumin An	professor	inorganic chemistry	None
Jukui Wang	associate professor	inorganic chemistry	None
Ye Huang	associate professor	inorganic chemistry	None
Xiaomei Xu	associate professor	inorganic chemistry	None

Economic and energy efficiency of the solid biofuels produced from digested pulp

Wojciech Czekała [1], Sylwia Bartnikowska[1], Aleksandra Lewicka[2], Artur Bugała[3], Zbyszek Zbytek[4], Andrzej Lewicki[1].

[1] Institute of Biosystems Engineering, Poznan University of Life Sciences, ul. Wojska Polskiego 28, 60-637 Poznan, Poland
[2] Institute of Environmental Engineering, Poznan University of Technology, Piotrowo 3a, 60-965 Poznan, Poland
[3] Institute of Electrical Engineering and Electronics, Poznan University of Technology, ul. Piotrowo 3a, 60-965 Poland
[4] Industrial Institute of Agricultural Engineering, ul. Starolecka 31 60-963 Poznan, Poland

Abstract. One of the methods of renewable energy production is biogas plant. Products of anaerobic digestion are biogas and residual material called digested pulp or digestate. This material is commonly used as a fertilizer. One of a new possibility of digested pulp management is solid biofuels production after separation. This paper presents the calculation of cost for production of briquettes from digested pulp. The performed economic analysis proves that the production of briquettes from digested solid fraction (DSF) can be profitable and provide additional income for the biogas plant.

1 Introduction

Nowadays rational waste management is one of the most important environmental issues. The reason of this situation is an upward trend of the generated waste. On the other hand it is a consequence of the increase of human population and higher standards of living, especially in developed countries. Among the waste materials, the specific group includes organic waste and residues, which are characterized by high content of moisture and organic matter. This group includes animal faeces [1], sewage sludge [2-4], residues from the forestry sector [5] and agri-food waste [6].

Due to the properties of above mentioned waste, the most suitable methods of management are biological processes, in particular composting [7-9]. Composting carried out under proper conditions allow to obtain a valuable organic fertilizer. Another advantage of composting is the reduction of weight and volume of waste [10,11]. Except from composting more and more attention is paid to use organic waste as substrate in biogas plants [12,13]. In those installations anaerobic microorganisms can degrade organic matter to the final product i.e. biogas [14]. Biogas is a mixture of various gases and the most important one, in terms of energy, is methane [15].

The by-product of this process is digested pulp which is removed each day from reactor chambers. Due to chemical composition digested pulp is generally used in agriculture for fertilizing purposes [16,17]. However, digestate which has specific properties can be processed into other products. For example, the solid fraction after the separation - referred to as digested solid fraction (DSF) - is a suitable material for composting or producing solid biofuels [18-20]. Taking into account relatively low public support for renewable energy sources in Poland, the proper management of digested pulp can have a direct significant impact on improvement of the economic balance of the operating biogas plants.

2 Aim and the range of research

The aim of the study was to analyze the economical aspect of production of briquettes, prepared from solid fraction of digested pulp obtained from a selected agricultural biogas plant. The scope of research included the selection of the appropriate elements of the installation and energy analysis of produced solid fuel.

3 Materials and methods of research

The object of the study was digested pulp from one of the most modern agricultural biogas plant in Poland. Power of discussed installation is about 1 MW of electricity and 1 MW of thermal energy, at daily production of biogas over 13 000 m^3.

In the analyzed plant the daily production of DSF was about 10 Mg with a dry matter content of about 25%. According to the Polish legal norms, the use of waste for energy is classified as the recovery method R1 - "principal use as a fuel or other means to generate energy" [21]. The analysis of the physical and chemical properties of digested pulp and prepared solid biofuel taken directly from the production line of the biogas plant (fig. 1) was performed in the Laboratory of Ecotechnology at the Institute of Biosystems Engineering at Poznan University of Life Sciences. According to the norm [22] the dry matter content was determined and following the standard [23] the dry matter of briquette from digested pulp according to the norm [24] the heat of combustion and calorific value of the tested briquettes was determined using the calorimeter method. Measurement of the heat of combustion was based on complete combustion of the substrate sample placed in a calorimeter bomb immersed in water. Measurement was performed with clean oxygen conditions.

Based on the obtained results the available information about the installation and using the presented assumptions, the economic analysis of the production of briquettes from the solid fraction of digestate was performed.

Figure 1. Briquettes from the solid fraction of digestate

4 Discussion of results

4.1 The properties of the tested material

The research results show that the dry matter of the produced briquettes was 87%, while the dry organic matter was 88.64%. In turn, the heat of combustion of the analyzed sample had a value of 18.22 MJ•kg^{-1} and the calorific value was 16.58 MJ•kg^{-1} (tab. 1).

Table 1. The parameters of the tested material

Substrate	Briquette from digestate
Dry matter [%]	87.00
Organic dry matter [%]	88.64
Heat of combustion [MJ·kg^{-1}]	18.22
Heating value [MJ·kg^{-1}]	16.58

4.2 Economic analysis

The unit price of briquettes production from DSF is determined based on many factors. The most important is the cost of the substrate which is often times is high. In this study the cost of substrate was gained for free using the digestate, i.e. by-product of the fermentation process. Other elements of the balance were the following costs: depreciation of equipment, the cost of electricity, salary for employees, the cost of repair and maintenance of machines, cost of building the hall dedicated to drying and the production of briquettes as well. The balance sheet also included the costs of bags used for briquettes packing. Currency exchange rate was 1 € = 4.2249 PLN

4.3 Depreciation costs

In this project the separator of AgriKomp company was used. The obtained DSF was transported to the belt dryer (PSD 500 Airtechno) that used the energy from generated in the biogas plant. To produce biofuel, the briquetting machine BIOMASSER DUO with efficiency 355.2 kg•h^{-1} (ASKET) was used. In the calculation, the depreciation cost of the machine included the total cost of the machine, years of work and the amount of substrate processed per year by the device. The analysis assumes that the operating time for all our used machinery was 10 years. In order to estimate the depreciation costs of briquetting machine, its efficiency (355.2 kg•h^{-1}), was taken into account. And it allowed to calculate the annual production of briquette. Concerning calculation of depreciation of the separator and the dryer, additional information about humidity of digestate solid fraction was required. This allowed to calculate that from one Mg of fresh DSF can be obtained 287 kg of briquettes. The total cost of depreciation of equipment used in the analysis is 12.81 €•Mg^{-1} (tab. 2). For dryer depreciation cost half of the total sum of machine has been adopted, because the device is also used for drying other substrates in the agricultural farm.

Table 2. The total cost of depreciation

	Separator	Dryer belt	Briquetting machine
The price for a machine [€]	52072.24	69000	12966.98
Amount of briquettes [€•Mg^{-1}]	1047.55	1047.55	1037.18
The total cost of depreciation	12.81 €·Mg^{-1}		

4.4 Electricity costs

The cost of electricity is a ratio of the following components: hours needed to produce one Mg of briquettes, the current price of industrial electricity fixed by the distributor of the power grid and power consumption of the device. The industrial price of electricity from the ENEA power supplier and distributor was assumed to be 0.11€•kWh. The analysis assumed that the cost of electricity for the separator was 26,04 €·Mg^{-1}. This listing also includes salary for an employee. On the other hand, energy costs for drying have zero value, because recovery heat on biogas plant is free. In case of briquetting machine, the energy input published on the producer website (70 kWh•Mg^{-1}) was used in calculations, along with other components of the ratio, i.e. time required to produce 1 Mg of briquettes - 2.82 h and the price of electricity. The calculation showed that the total cost of electricity for briquetting machine was 7.46 €•Mg^{-1}.

4.5 Costs of repairs and maintenance of machines

The analysis assumes that the costs of repairs and maintenance of machines constitute 30% of the amortization cost for each device. Therefore, it has been proved that the total costs amounted to 2.86 €•Mg^{-1}, and the structure was diverse (table 3). The largest share of the costs relating to repairs and maintenance (52.1%) was in case of separator, on drying (34.6%) and briquetting (13.3%) consecutively.

Table 3. The structure of repair and maintenance cost of machines

Device	**€·Mg^{-1}**
Separator	1.49
Dryer belt	0.99
Briquetting machine	0.38

4.6 Salary for employees

Calculating the salary costs for employees one-shift (8 hours) labour was assumed, with a rate of 3.55 €•h^{-1}. Whereas taking into account that briquetting can be operated by one person, the cost was calculated as 10.01 €•Mg^{-1} (tab. 4). It is worth to highlight that the costs of separator servicing are included in the electricity costs.

Table 4. The salary for the employee briquetting

Number of Employees [-]	1
Number of work shift [-]	1
Number of hours per one shift [h]	8
Hourly rate [€/h]	3.55
Number of hours which is need to produce 1 Mg of briquettes	2.82
Salary for employee per 1 Mg of biofuel [€·Mg^{-1}]	10.01

4.7 Additional costs

The additional cost was related to the construction of 300 m^2 hall intended for drying purposes and briquettes production. The cost of this construction was 66.365 €. Calculated per 1 Mg of briquette the obtained cost was 1.62 €.The total balance sheet also included the cost of packing bags for briquettes. Assuming the price of 1,000 bags with a capacity of 15 kg at the level of 0.14 € it was calculated that the cost of the bags for 1 mg briquettes was 6.94 €.

Economic analysis shows that there are three main components of the costs for the analyzed case. The first group includes the costs associated with the exploitation of the separator which in total amounted to 32.50 € (Tab. 5). Within this group, the highest share (80.1%) was assigned to the electricity associated with the concerns amortization and approx. 4.6% are the costs of repair and maintenance work.

Table 5. The unit costs for DSF briquettes

	1 Mg DSF briquettes cost	**€·Mg^{-1}**
	DIGESTED SOLID FRACTION	
1.	Purchase of raw materials	0
	SEPARATOR	
2.	Depreciation of separator	4.97
3.	Electricity for separator	26.04
4.	Repair and maintenance of machinery	1.49
	The total cost for the group - separation	32.5
	BELT DRIER	
5.	Depreciation of dryer	6.159
6.	Electricity for dryer	0
7.	Hall dedicated to drying	1.62
8.	Repair and maintenance of machinery	0.99
	The total cost for the group - drying	9.20
	BRIQUETTING MACHINE	
9.	Depreciation of briquetting machine	1.25
10.	Electricity for briquetting machine	7.46
11.	Repair and maintenance of machinery	0.38
12.	Salary for employee	10.00
13.	Bags for packing briquettes	6.94
	The total cost for the group – briquetting machine	26.03
	TOTAL COST OF 1 Mg DSF BRICQUETTES PRODUCTION	**67.71**

The second component represents the expenses spent on the exploitation of the dryer, totally amounted to 9.20 €. The largest share of the costs (71.6%) was the dryer depreciation cost. The 17.6% of the costs was related to the construction of the hall, and 10.8% were the costs for repair and maintenance (tab. 5).

The last component was the costs associated with the exploitation of briquetting machine which accounted to 26.03 €. Distribution of costs in this group was more diverse compared to the two previous ones (tab. 5). The largest costs (38.4%) were related to the salary of employees and electric energy (28.7%). In turn the smallest share (1.5%) were the costs of repair, maintenance and amortization of the machine (4.8%).

In the presented analysis, the cost of production of 1 Mg of briquettes from DSF was 67.73 € (Tab. 5). The 48.0% of this amount were the costs for the separator, 38.4% were related to the briquetting machine and 13.6% were drying costs.

5 Summary and conclusions

In the analyzed agricultural biogas plant the daily production of fresh DSF is 10 Mg. The digested pulp can be used by the R10 method - as organic fertilizer, by R3 – for compost production, R1 – as a fuel [21], or by R1 - using the digestate as a fuel.

The performed economic analysis proves that the production of briquettes from DSF can be profitable. The right argument is the low cost of production of 1 Mg of briquette which is 67.71€ at the market price of solid biofuels ranging from 100 € to 170 € per Mg. This means that the sale of briquettes can provide additional income for the biogas plant. Moreover, the production of briquettes allows the management of the DSF, which is perceived as a problematic by-product in the process of methane fermentation. Because the calorific value of DSF briquettes is similar to other ones, it determined the tested material an attractive biofuel.

References

1. W. Czekała, J. Dach, A. Ludwiczak, A. Przybylak, P. Boniecki, K. Koszela, M. Zaborowicz, K. Przybył, D. Wojcieszak. The use of image analysis to investigate C:N ratio in the mixture of chicken manure and straw. Proc. SPIE. 9631, Seventh International Conference on Digital Image Processing (ICDIP 2015), 963117. (July 06, 2015) doi: 10.1117/12.2197041 (2015)
2. I. Białobrzewski, M. Mikš-Krajnik, J. Dach, M. Markowski., W. Czekała, K. Głuchowska. Model of the sewage sludge-straw composting process integrating different heat generation capacities of mesophilic and thermophilic microorganisms. Waste Management **43**, 72-83. DOI:10.1016/j.wasman.2015.05.036 (2015)
3. K. Malińska, M. Zabochnicka-Świątek, J. Dach. Effects of biochar amendment on ammonia emission during composting of sewage sludge. Ecological Engineering **71**, 474-478 (2014)
4. S. Kujawa, R.J. Tomczak, T. Kluza, J. Weres, P. Boniecki. A stand for the image acquisition of composted material based on the sewage sludge. 4th International Conference on Digital Image Processing. Kuala Lumpur (Malaysia), proceedings of SPIE, Vol. 8334, Article Number: 83341R, DOI: 10.1117/12.949899 (2012)
5. J. Starzyk, W. Czekała. The Influence of Admixtures Accelerating the Pine Bark Composting Process on Variation in the Bacteriological State of Composts. Archives of Environmental Protection **40**, 4, 125–135, DOI: 10.2478/aep-2014-0044 (2014)
6. W. Czekała, K. Kozłowski, J. Dach, P. Boniecki, A. Lewicki, D. Janczak, K. Jóźwiakowski, T. Piechota. Energy Conversion from Biomass to Hydrogen and Methane. 4th International Conference on Materials Engineering for Advanced Technologies (ICMEAT) Location: London, ENGLAND Date: JUN 27-28, 2015 Pages: 654-657 (2015)
7. W. Czekała, K. Malińska, R. Cáceres, D. Janczak, J. Dach, A. Lewicki. Co-composting of poultry manure mixtures amended with biochar – The effect of biochar on temperature and C-CO2 emission. Bioresource Technology **200**, 921–927 (2016)
8. K. Malińska, M. Zabochnicka-Świątek. Selection of bulking agents for composting of sewage sludge. Environment Protection Engineering **39**, 2, 91-103 (2013)
9. A. Piotrowska-Cyplik, Ł. Chrzanowski, P. Cyplik, J. Dach, A. Olejnik, J. Staninska, J. Czarny, A. Lewicki, R. Marecik, J. Powierska-Czarny. Composting of oiled bleaching earth: Fatty acids degradation, phytotoxicity and mutagenicity changes. International Biodeterioration & Biodegradation. Vol. 78, March 2013: 49-57 (2013)
10. A. Lewicki, J. Dach, P. Boniecki, W. Czekała, K. Witaszek. The Control of Air Humidity and Temperature in Relationship with a Biowaste Composting Process. Advanced Materials Research Vol. 909 pp 455-462. DOI 10.4028/www.scientific.net/AMR.909.455 (2014)
11. D. Janczak, M. Marciniak, A. Lewicki, W. Czekała, K. Witaszek, P.C. Rodríguez Carmona, M. Cieślik, J. Dach. Bioreactor Internet System for Experimental Data Monitoring and Measurement, Procedia Technology **8**, 209 – 214 (2013)
12. Dach J., Koszela K., Boniecki P., Zaborowicz M., Lewicki A., Czekała W., Skwarcz J.,Wei Q. Piekarska-Boniecka H., Białobrzewski I. The use of neural modelling to estimate the methane production from slurry fermentation processes. Renewable and Sustainable Energy Reviews **56**, 603–610. doi:10.1016/j.rser.2015.11.093 (2016)
13. A. Lewicki, J. Dach, D. Janczak, W. Czekala. The experimental macro photoreactor for microalgae production, Procedia Technology **8**, 622-627 (2013)
14. A. Lewicki, P. Boniecki, J. Dach, D. Janczak, W. Czekała, Z. Zbytek, M. Cieślik. Neural Estimation of Biomethane Efficiency from Maize Silage, Maize Straw and Cattle Slurry. 4th International Conference on Materials Engineering for Advanced Technologies (ICMEAT) Location: London, ENGLAND Date: JUN 27-28, 2015. Pages: 649-653 (2015)
15. J. Dach, W. Czekała, P. Boniecki, A. Lewicki, T. Piechota. Specialised Internet Tool for Biogas Plant Modelling and Marked Analysing. Advanced Materials

Research Vol. 909 pp 305-310. 7 pkt. DOI 10.4028/www.scientific.net/AMR.909.305 (2014)
16. D. Zirkler, A. Peters, M. Kaupenjohann. Elemental composition of biogas residues: Variability and alteration during anaerobic digestion. Biomass and bioenergy **67**, 89-98; (2014)
17. H.I. Owamah, S.O. Dahunsi, U.S. Oranusi, M.I. Alfa. Fertilizer and sanitary quality of digestate biofertilizer from the co-digestion of food waste and human excreta. Waste Management **34**, 747–752 (2014)
18. S. Pedrazzi, G. Allesina, T. Belló, C.A. Rinaldini, Tartarini P. Digestate as bio-fuel in domestic furnaces. Fuel Processing Technology **130**, 172–178; Tambone F., Terruzzi L., Scaglia B., Adani F. 2015. Composting of the solid fraction of digestate derived from pig slurry: Biological processes and compost properties. Waste Management 35, 55–61 (2015)
19. M.A. Bustamante, J.A. Alburquerque, A.P. Restrepo, C. Fuente, C. Paredes, R. Moral, M.P. Bernal. Co-composting of the solid fraction of anaerobic digestates, to obtain added-value materials for use in agriculture. Biomass and bioenergy **43**, 26-35 (2012)
20. M. Kratzeisen, N. Starcevic, M. Martinov, C. Maurer, J. Müller. Applicability of biogas digestate as solid fuel. Fuel **89**, 2544–2548 (2010)
21. Polish Act of 14 December 2012 on Waste
22. PN-75 C-04616/01 – Water and wastewater. Special research of sludge. Determination of water content, dry matter, organic matter and mineral substances in sewage sludge.
23. PN-Z-15011-3 :2001 - Compost from municipal solid waste. Determination of pH, organic matter content, organic carbon, nitrogen, phosphorus and potassium.
24. PN-81/G-04513 –"Solid fuels. Determination of gross calorifie value and calculation of net calorific value".

MATEC Web of Conferences **60**, 04006 (2016) DOI: 10.1051/matecconf/20166004006
ICCBS 2016

Influence of carbon dioxide, temperature, medium kind and light intensity on the growth of algae *Chlamydomonas reinhardtii* and *Scenedesmus obliquus*

Przemysław Piotr Olejnik[1],a, Andrzej Lewicki[2], Piotr Boniecki[2], Aleksandra Lewicka[3], Damian Janczak[2], Wojciech Czekała[2]

[1]*Department of Biotechnology and Food Microbiology, Poznan University of Life Sciences, ul. WojskaPolskiego 48, 60-627 Poznan, Poland*
[2]*Institute of Biosystem Engineering, Poznan University of Life Sciences, ul. WojskaPolskiego50, 60-627 Poznan, Poland*
[3]*Institue of Environmental Engineering, Poznan University of Technology, Piotrowo 3a, 60-965 Poznan, Poland*

Abstract. Microalgae attracts the attention of scientists because of the possibility of using in the energy industry as one of the substrates for the production of renewable energy. So far, the greatest emphasis was put on attempts to obtain strains, and technologies of their culturing, in order to efficiently acquire fat from cells and its further conversion to biodiesel using transesterification reaction. Increasingly, algae are considered also as an efficient biomass producer, which can be used as a substrate for methane production in biogas plants. In this study the influence of different physical and chemical conditions, on the growth of two algae species: *Chlamydomonasreinhardtii* and *Scenedesmus obliquus* was investigated. Based on the literature and the data obtained for the algae growth on the standard medium and the digestate remaining after fermentation, one may suggest further investigations on the use of other liquid waste from agriculture and industry for algae breeding, including chemical. analysis and supplementation of these mediums so as to provide the best conditions for their growth.

1 Introduction

The continuous development of the technologies and ongoing improvement of the financial situation of the society causes the consumption increase. This phenomenon increase demand of fuel and affects the amount of waste. Nowadays more and more emphasis is placed on issues related to waste management and widely understood environmental protection [1,2,3]. European Union inclines its members to invest in renewable energy sources. One of the most stable and promising is biogas production [4,5,6], which like in case of composting process[1,7], not only solves the problem of waste, but also produce energetic fuel. Because the problem is so current the even latest achievements of computer analyzes using Artificial Neural Network have been engaged to bring solutions [8,7]

Algae are a very promising alternative for biomass production. They are characterized by the ability to proceed photosynthesis, which is determined by possession of chlorophyll a and other additional photosynthetic pigments, whose presence depend of organis [10]. In addition, they do not produce specialized organs such roots, stems or flowers. These organisms are present mainly in water environment, as well as on a land, in places with high humidity. It is estimated that about 50% of oxygen is produced by algae, mainly in the oceans, which cover about 71% of the Earth [11].

Algae are usually divided on macro- and microalgae. The first group can form a variety of structures, such as leaves or roots, but the cells forming them are not specialized and do not perform specific functions [10]. To this group of organisms belong certain algae species that can form fronds with a length up to 100 meters. Microalgae are, however, single-celled organisms which sometimes are present in a loose cell groups, chains or diplococci. In addition, these organisms are divided to two kingdoms: prokaryotes or eukaryotes. Prokaryotic algae can be called cyanobacteria or blue-green algae. Them excessive growth in water bodies result in water blooms, during which some species can produce cyanotoxins. These compounds are highly toxic and harmful to humans and animals [12]. Some species of microalgae can be used in human nutrition, for example *Spirulina sp.*, *Chlorella sp.* or *Nanochloropsis sp.* are commonly used in human nutrition, cosmetic and chemical industry [13].

Microalgae focused attention of scientists because of the potential use of the energy industry as one of the substrates in renewable energy production. So far, the greatest emphasis was placed on trying to obtain strains and technologies of their culturing to raise content of neutral fats, which can be convert to biodiesel using transesterification reaction [11,14,15]. Futhermore, algae are considered as an efficient producer of the biomass, which can then be used as substrate for methane production [16,17].

Microalgae present higher degree of utilization solar energy in photosynthesis, compared to plants. They need to grow only inorganic compounds, organic compounds are produced in the process of photosynthesis [18]. That is why scientist are working to make use of its potential [19, 20]. An important advantage of algae is can be cultured in many systems: from closed bioreactors, in which is possible to control bioprocess conditions, up to open pools, which are the cheapest solution in algae production. However, there are problems with microbiological contaminations, bacteria and others can freely go into culture from air. Therefore, sometimes it is not required to produce sterile culture, for example in the energy industry with open pools, which are the cheapest solution. Unlike the first-generation fuels, where the substrate are products which can be successfully used in the food industry, algae do not compete with human nutrition, and can be grown on land useless for agriculture. Fuels from algae are cold third generation biofuels [11,18].

2 Materials and methods

In this study two strains of microalgae: *Chlamydomonasreinhardtii* and *Scenedesmus obliquus* were used. The aim of this study was to determine the effects of a few factors: temperature, carbon dioxide concentration, light intensity and type of used medium on growth rate of used strains.

2.1 Culturing media

Standard TAP medium was used as control [21] and liquid fraction of digestate from biogas plant as alternative medium. Digestate pulp was centrifugated (4 000 rpm for 10 minutes). For further research were used supernatant, which were dilluted by distilled water to final concentration 0,5%. All media were sterilized by autoclaving. Cultures were grown in Erlenmeyer flasks of 500 ml capacity, in which were put 100 ml of medium.

2.2 Inoculate preparation

Cultures were inoculated by prepared cells suspensions from *Ch. reinhardtii* and*S. obliquus* cultures from raised in a light chamber (20 ° C, 72 mmol/m^{2}s, 2% CO_2) in 500 ml flasks containing 100 ml of culture, growth on TAP medium. Then, when the cultures raised an absorbance of light with a length of 684 nm of 1-1.5 for *Ch.reinhardtii* and 2-2.5 for *S. obliquus*, cultures were centrifuged (3 000 rpm for 10 min) and resuspended, respectively TAP medium and 0.5% fermentation pulp, in such a volume to concentrate cells ten times. The procedure was carried out again to wash the rest of the pollutants and suspended in the same volume of fluid as after first step. Prepared inoculates were added to relevant flask in a volume of 1 ml.

2.3 Culturing conditions

The cultures were conducted in light chambers that allow you to control the following parameters: light intensity, temperature, relative humidity and carbon dioxide concentration. Cultures were conducted in various combinations of conditions for each combination three times. The study of the following conditions: temperature 10°C and 20°C; the concentration of carbon dioxide: atmospheric (about 0.03%) and 2%, the exposure to light 72 and 10 µmol/m^{2}s; all the time kept constant relative humidity of 70%.

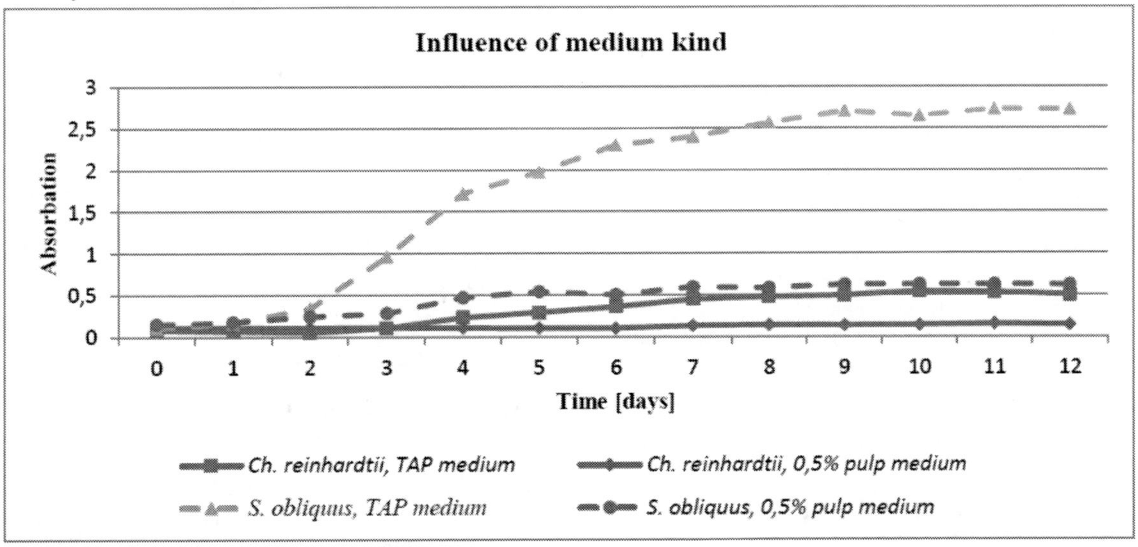

Figure 1. Effect of different medium kind (TAP and 0,5% fermentation pulp) on the cultures of *Ch. reinhardtii* and *S. obliquus* conducted in 2% CO_2 concentration, 72 µmol/m2s light intensity and 20°C.

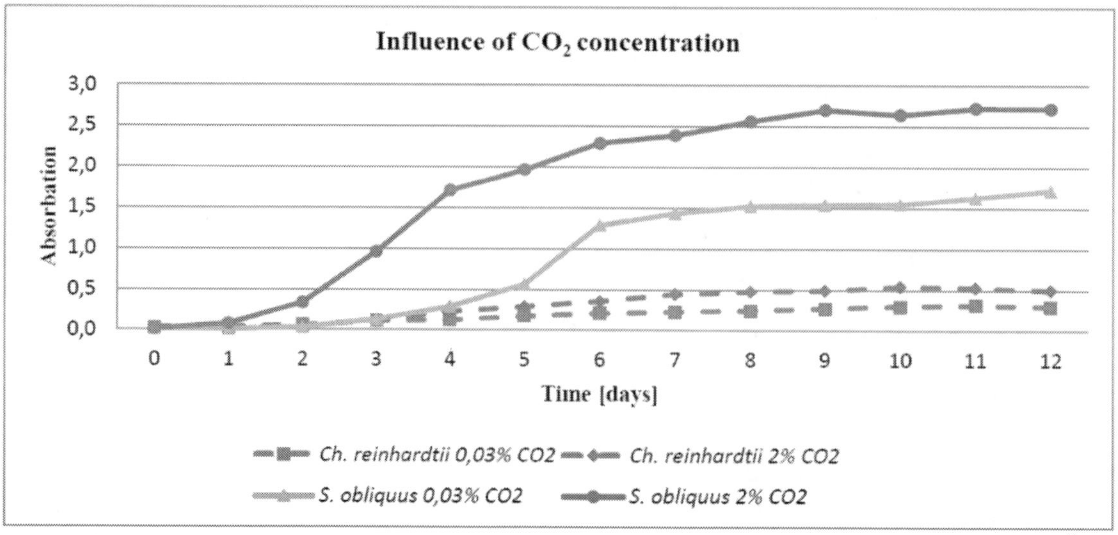

Figure 2. Effect of different CO_2(0,03% and 2%) on the cultures of *Ch. reinhardtii* and *S. obliquus* conducted on TAP medium in 72 µmol/m2s light intensity and 20°C.

The prepared cultures were carried out in the above-mentioned conditions for 12 days, and sample were taken at 24-hour intervals and analyzed using a spectrophotometer Specord PC 205 (Analytikjena company). Measurement of absorbance of 684 nm wave length light was measured in the range of 0 to 0.3, otherwise the sample was diluted [22].

3 Results

3.1 Influence of medium type

Both in the case of *Ch. reinhardtii* and *S. obliquus* replacing the TAP medium, which is dedicated to the cultivation of algae, to 0.5% fermentation pulp had a negative impact on the growth of algae. In the case of *Ch. reinhardtii* growth has slowed down up to four times, and in the case of *S. obliguus* five times.

In the literature, there are reports of the use of other liquid wastes as the substrate for growing algae [23]. For some substrates it was observed efficient biomass productivity, as well as highlights the possibility of using algae for the treatment of a variety of liquid wastes, mainly by conversion of inorganic nitrogen compounds and incorporation to algal biomass, which can be further used as starting material in biome than plants.

Based on these results it is concluded that it is necessary to analyze a larger number of substrates from different places, as well as analysis of received materials and testing their supplementation with certain chemicals to create a suitable environment for algae to develop.

3.2 The impact of the carbon dioxide concentration

Supplementation of the culture using carbon dioxide had positive impact to the growth of both the substrates a standard (TAP) and 0.5% digestate pulp. These results are confirmed in the literature (Anderson, 2005; Singh and Singh 2014). Supplementing cultures with carbon dioxide is a commonly used technique to increase the rate of cell growth and to increase their final concentrations in relation to the volume of the culture.

Supplementation withcarbon dioxide have also aspects related to environmental protection - the use of waste gas containing a high concentration of carbon dioxide in breeding algae can have a positive impact on reducing emissions of this gas by industrial plants. This possibility is also part of the European Union's current policy aimed to reducing carbon dioxide emissions by member states [24].

3.3 Effect of light intensity

It was found that the lighting at the level of 72 µmol/m^{2}s greatly accelerates the growth of both species of algae.

Interestingly, the comparison of the growth curves for *S. obliquus* culture, carried out at 20 °C on TAP medium with light intensity of 12 µmol / m^{2}s and different concentrations of carbon dioxide (0.03 and 2%), were not shown any major difference in speed growth. Probably light intensity is limiting factor for cell grow. The data obtained correspond to the information literature on the impact of light on the growth of algae [25,26].

3.4 Effect of temperature

In the case of temperature impact on the growth of algae were not shown any differences which might be different from the data in the literature [18,25,26]. Temperatures closer to the optimum temperature for the used strain (about 20 °C), caused the faster growth. However, the low temperature turns out to be a strong factor limiting the growth of tested species, even when others remain at an optimum level (high light intensity and carbon dioxide concentration and growth in the medium TAP).

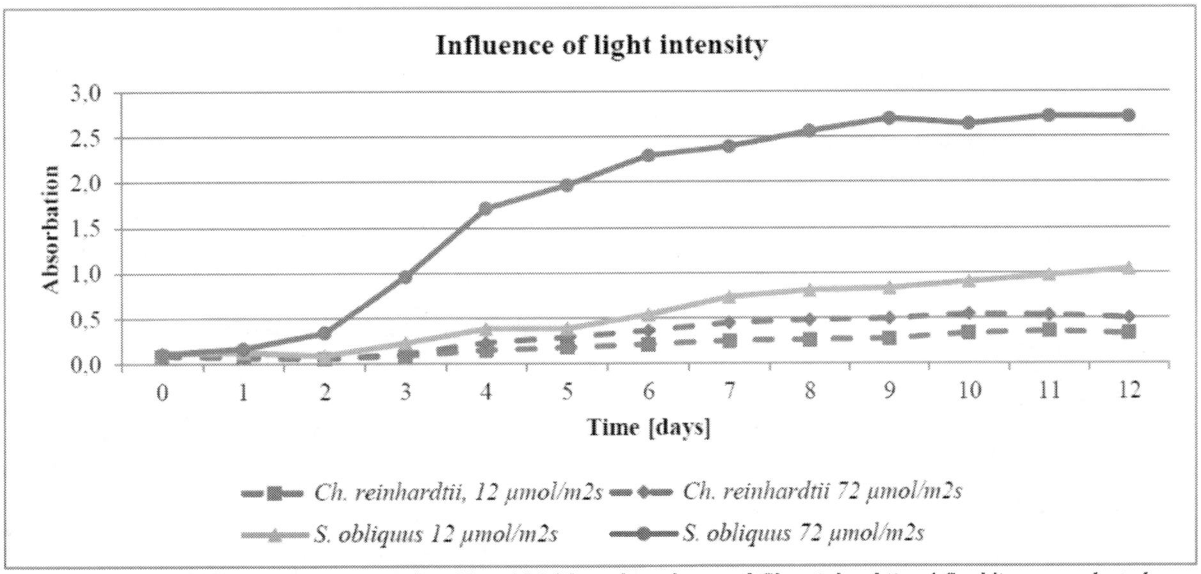

Figure 3. Effect of different temperature values10 and 20 °C on the cultures of *Ch. reinhardtii* and *S. obliquus* conducted on TAP medium with 2% CO_2 concentration and 72 μmol/m^{2}s light intensity.

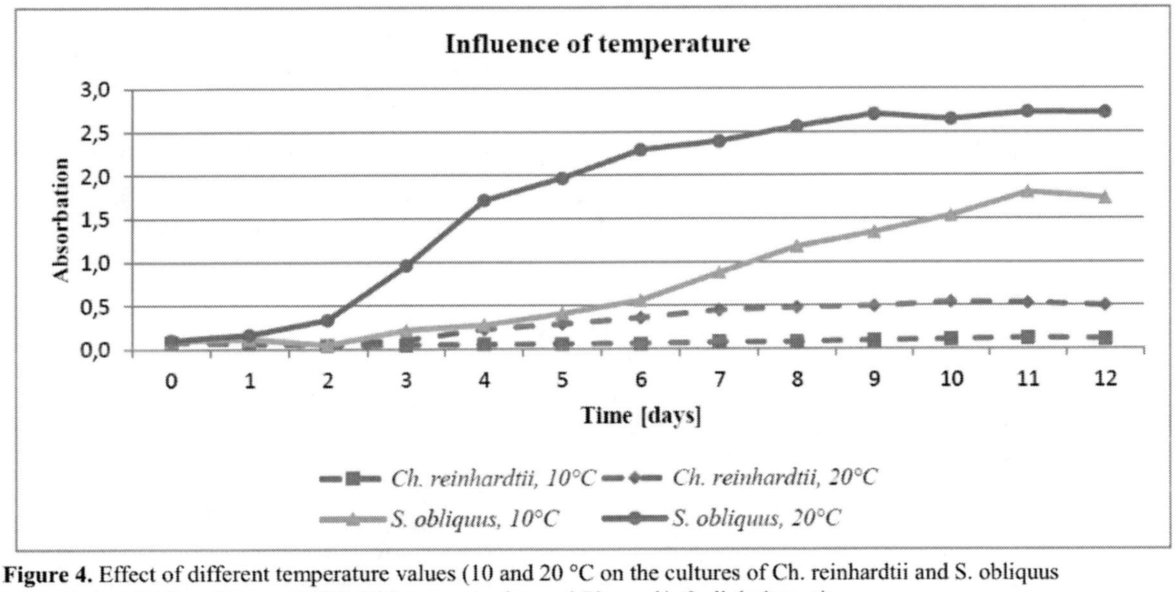

Figure 4. Effect of different temperature values (10 and 20 °C on the cultures of Ch. reinhardtii and S. obliquus conducted on TAP medium with 2% CO2 concentration and 72 μmol/m2s light intensity.

4 Discussion

In the above study the effect of various physical and chemical parameters on the growth of microalgae from species *Ch. reinhardtii* and *S. obliquus*were investigated. Based on the literature and obtained data aboutthe growth of algae on standard TAP medium and diluted digestate pulp can suggest needed of further studies on the use other liquid wastes from agriculture and industry to breeding algae. It can be needed to characterized chemical composition of used wastes and its supplementation in appropriate substances.

Furthermore, it seems reasonable to investigate the effect of feeding cultures by wastes gas reach in carbon dioxide from different processes, such as combustion or various types of fermentation. The development of efficient procedures of this procedure can significantly increase the rate of algae growth and also allows minimizing the process costs. Additionally use of such procedures seems to be beneficial to the environment. Nowadays when research on alternative energy sources seems to be more urgent and different scientists get involved using various techniques from Internet to process control [27] to Artificial Neuron Networks [28] research on algae as energetic biomass of the future seems to be a promising solution for energetic stability of the World.

Acknowledgement

This paper was created as a result of the realization of the project entitled "The innovative energy production in biogas plant through utilization of poultry manure with the alternation of a plant substrate for algae" financed by The National Centre for Research and Development (contract number: GEKON1/O5/214528/34/2015).

References

1. K. Malińska, M. Zabochnicka-Świątek. 2013. Environment Protection Engineering **39**(2), 91-103
2. A. Piotrowska-Cyplik, Ł. Chrzanowski, P. Cyplik, J. Dach, A. Olejnik, J. Staninska, J. Czarny,. A. Lewicki, R. Marecik, J. Powierska-Czarny. 2013. International Biodeterioration& Biodegradation. **78**, March 2013
3. A. Wolna-Maruwka, J. Dach. 2009. Archives of Environmental Protection. **35**(3), 87-100.
4. J. Dach, W. Czekała, P. Boniecki, A. Lewicki, T. Piechota. 2014. *Advanced Materials Research* **909**, 305-310. (A)
5. J. Dach, P. Boniecki, J. Przybył, D. Janczak, A. Lewicki, W. Czekała, K. Witaszek, P. C. RodríguezCarmona and M. Cieślik, 2014. Energy **69**, 34-38. (B)
6. W. Czekała, K. Kozłowski, J. Dach, P. Boniecki, A. Lewicki, D. Janczak, K. Jóźwiakowski, Piechota T. 2015. *4th International Conference on Materials* Engineering for Advanced Technologies (ICMEAT), 2015 Pages: 654-657.
7. J. Starzyk, W. Czekała. 2014. Archives of Environmental Protection **40**(4), 125–135.
8. D. Janczak, A. Lewicki, R. Mazur, P. Boniecki, J. Dach, J. Przybyl, M. Pawlak, K. Pilarski, W. Czekala, 2013. *5th International Conference on Digital Image Processing. Beijing (China), proceedings of SPIE*, Vol. **8878**, Article Number: 88783R.
9. A. Lewicki, P. Boniecki, J. Dach, D. Janczak, W. Czekała, Z. Zbytek, M. Cieślik. 2015. *4th International Conference on Materials Engineering for Advanced Technologies (ICMEAT)* Pages: 649-653.
10. Lee, Re. 2008. *Phycology Basic characteristics of the algae.*
11. A. Richmont. 2004. *Handbook of Microalgal Culture Biotechnology and Applied Phycology.*
12. H. W. Paerl, R. S. Fulton, P. H. Moisander, J. Dyble 2001. The Scientific World Journal**1**, 76–113
13. L. Ding-mei, Q. Yu-zao. 1997. Journal of Applied Phycology 25–28.
14. S. Mandal, N. Mallick. 2009. Applied Microbiology and Biotechnology**84**(2): 281–9.
15. T. M. Mata, A. A. Martins, N. S. Caetano. 2010. Renewable and Sustainable Energy Reviews**14**(1): 217–32.
16. M. Dębowski, M. Zieliński, A. Grala, M. Dudek. 2013. Renewable and Sustainable Energy Reviews**27**: 596–604.
17. A. Xia, et al.. 2015. Renewable and Sustainable Energy Reviews**51**: 209–30
18. R. A. Andersen. 2005. Algal Culturing Techniques. Elsevier Academic Press.
19. S. Cerbin, K. Nowakowski, J. Dach, K. Pilarski, P. Boniecki, J. Przybyl, A. Lewicki, 2012. *4th International Conference On Digital Image Processing (ICDIP 2012). Proceedings of SPIE.* Vol.: **8334** Article Number: 83342A.
20. A. Lewicki, J. Dach, D. Janczak, W. Czekala, 2013. *Procedia Technology* Volume: **8** Pages: 622-627
21. D. S. Gorman, R. P. Levine. 1965. *Proceedings of the National Academy of Sciences of the United States of America***54**(6): 1665–69
22. D. U. Santos-Ballardo, et all. 2015. *Aquaculture***448**: 87–92.
23. H. Azarpira, D. Kondiram, G. Pondhe. 2014. Advances in Environmental Biology**8**(7): 2419–23.
24. L. Więcław-Solny, et all. 2012. Polityka Energetyczna**15**(4): 111–123.
25. S. P. Singh, P. Singh. 2014. Renewable and Sustainable Energy Reviews**38**: 172–79.
26. C. Sorokin, R. W. i Krauss. 1958. *Plant physiology***33**(2): 109–13.
27. D. Janczak, M. Marciniak, A. Lewicki, W. Czekała, K. Witaszek, P. C. RodríguezCarmona, M. Cieślik, J. Dach, 2013 *Procedia Technology*, vol. **8**, 209-214. DOI: 10.1016/j.protcy.2013.11.029.
28. J. Dach, K. Koszela, P. Boniecki, M. Zaborowicz, A. Lewicki, W. Czekała, J. Skwarcz, W. Qiao, H. Piekarska-Boniecka, I. Białobrzewski. 2016 Renewable andSustainable Energy Reviews**56**, 603–610.

MATEC Web of Conferences **60**, 04007 (2016) DOI: 10.1051/matecconf/20166004007
ICCBS 2016

Optimization of the selection process of the co-substrates for chicken manure fermentation using neural modeling

Andrzej Lewicki[1], Artur Bugała[2], Hanna Piekarska-Boniecka[3],Aleksandra Lewicka[4], Piotr Boniecki[1], Marta Cieślik[1]

[1] *Poznan University of Life Sciences, Institute of Biosystem Engineering, ul. WojskaPolskiego 28, 60-637 Poznan, Poland*
[2]*Poznan University of Technology, Institute of Electrical Engineering and Electronics, ul. Piotrowo 3a, 60-965Poznan, Poland*
[3]*Poznan University of Life Sciences, Department of Entomology and Environmental Protection, ul. WojskaPolskiego 28, 60-637 Poznan, Poland*
[4]*Poznan University of Technology,Institute of Environmental Engineering,ul. Piotrowo 3a, 60-965 Poznan, Poland*

Abstract. Intense development of research equipment leads directly to increasing cognitive abilities. However, along with the raising amount of data generated, the development of the techniques allowing the analysis is also essential. Currently, one of the most dynamically developing branch of computer science and mathematics are the Artificial Neural Networks (ANN). Their main advantage is very high ability to solve the regression and approximation issues. This paper presents the possibility of application of artificial intelligence methods to optimize the selection of co-substrates intended for methane fermentation of chicken manure. 4-layer MLP network has proven to be the optimal structure modeling the obtained empirical data.

1.Introduction

ArtificialNeural Networks(ANN) are an importantapplicationof cognitivemethodsusedin the areaof empirical researchcarried out inthe area of widelyunderstoodartificial intelligence[1-4].In particular,promising results are related to an application one of the importantcharacteristics ofANN, i.e. the ability to solve the problems concerning regressionandapproximation issues. It givesthe possibility of using of efficientANNsimulators,among others, in order toshape up the predictiveissues. ANNcanbesuccessfully used practically in anysituation, where themain objective is toestimatethe value ofa variablebased onacquired(e.g. experimentally) characteristics(i.e. descriptors) [5-7].
However, inavailable literature related to the discussed topic there is little informationconcerning the use ofANNtomodelthe process of biomethane emission[8, 9].It seems highly appropriateto tryto buildthe regressionneural model, generatedbased onempirical datacollectedduringresearch conductedunder laboratory conditions.Developed andtestedneural modelcanserve as a toolsupporting thedecision-making processesthat occur during operationof biogas plants[10].Itscorrect applicationimproves and rationalizes thesystem forselecting theproper mixture ofinput substratesso that themethane fermentation processtakes placeunderpossibly optimal conditions.

The objective of the paper was todevelopthe neuralestimatorintendedto predictthe amount ofemittedbiomethanefrom the fermentation process ofchicken manurewithadditions of othersubstrates.As atraining setwas usedthe databaseof the biogasefficiency ofthe substrates andtheirmixtures obtainedin the Laboratory ofEcotechnologyat the PoznanUniversity of Life Sciencesunder the scientific supervisionof Prof.JacekDach.

The Laboratory ofEcotechnologyis the largestPolishbiogas laboratory, withmore than250fermentersoperating accordingto DIN38414/S8[11].

2.Methodology

In order to accomplishthe aforementioned task, the followingworkflow has beenproposedandthen implemented(Fig. 1).

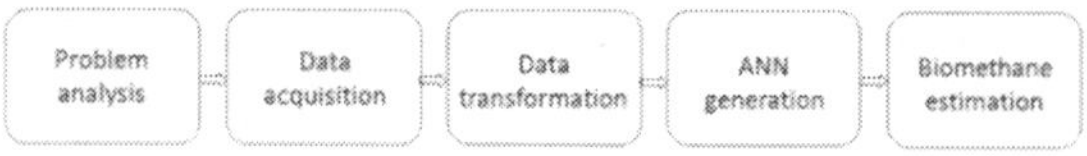

Figure 1. Action scheme

The most important step in ANN generating is to produce an adequate training set containing in its structure encoded empirical data. For this purpose, the Authors have defined the numeric input variables and predicted output variable resulted from the structure of the formulated scientific problem. Five input variables have been adopted as a share of respondents substrates:

- GRASS - grass [g Fresh Mass (FM)/reactor],
- STRAW - straw [g FM/reactor],
- CHICKEN - chicken manure [g FM/reactor],
- MAIZE - maize silage [g FM/reactor],
- INOCULU - inoculum [g FM/reactor].

Jako 1 zmienną wyjściową przyjęto:

- BIOMETH - biomethaneproduction efficiency [m3/Mg FM].

Using theobtainedexperimental results,set of empirical dataconsistingof 156measurement cases has been generated. The above mentioned set has beenused tocreateANNand, therefore, has been dividedinto 3subsets:

- training fileincluding 78 cases,
- validationfileincluding39 cases,
- test fileincluding39 cases.

Some part oftraining file(cases:33to 39) intended for ANN generatoris showninTable 1.

Table 1. Fragment of the training file for the ANN simulator (BIOMETH – output variable)

No.	GRASS	STRAW	CHICKEN	MAIZE	INOCULU	BIOMETH
33	0.00	13.55	0.00	0.00	1192	**180.89**
34	0.00	13.55	0.00	0.00	1210	**175.45**
35	000	0.00	38.85	0.00	1164	**70.51**
36	0.00	0.00	38.95	0.00	1162	**68.19**
37	0.00	0.00	38.80	0.00	1161	**69.35**
38	0.00	0.00	0.00	43.	1156	**109.19**
39	0.00	0.00	0.00	43.	1156	**113.14**

The simulator of artificial neural networksimplementedin a commercialStatisticapackagehas been usedfor the design ofneural models. The following types ofneural networkshave been subjected totesting:

- linear networks.
- MLP networks (MultiLayer Perceptron).
- RBF networks (Radial Basic Function).
- GRNN networks (Generalized Regression Neural Network – regressive networks).

The development of neural modelswas a 2-stage procedure.At the beginning,an effectiveoption supporting the process of ANN designinghas been used, i.e. *AutomatedWebDesigner*implementedin Statistica computer system. This toolallowed toautomateandsimplify theprocedures of preliminaryseeking of a set with predictiveneural networksmodeling the investigated process.

In the second stage of neural models development another useful device has been used i.e. *User Networks Designer*. It offers the possibility of advanced interference in the parameters and training methods of generated neural networks. This tool has been activated frequently in order to modify both the initial settings of parameters, learning algorithms and the ANN structure itself.

3.Results

3.1 Artificial neural networks

From 100of generatedneural models,a fileof10isolatedneuraltopologies has been extracted as it showsTable 2 (ANN no. 10is the bestnetwork).

Table 2. File of 10 generated ANN

	Type	RMS error	Imputs	Layer1	Layer2	Quality
1	RBF	11.953	5	9	-	0.29790
2	MLP	11.406	5	1	-	0.26561
3	RBF	9.8235	5	10	-	0.24572
4	RBF	9.1699	5	11	-	0.22898
5	MLP	8.5263	5	2	-	0.21456
6	RBF	5.3826	5	13	-	0.13413
7	MLP	4.4734	5	3	-	0.11134
8	MLP	3.7940	5	6	-	0.09538
9	RBF	3.5926	5	14	-	0.08896
10	**MLP**	**2.655757**	**5**	**8**	**4**	**0.0633**

where:

- RMS (Root Mean Square) error - it is a total error made by the network on a data set (learning, test or validation). It is determined throughout summing the squared individual errors, then dividing the obtained sum by the number of included values and determining the square root of the quotient obtained. RMS error is the most convenient single value to be interpreted that describes the total ANN error.
- Qality – the value of root mean square error regression.

4-layer MLP networkwith structureshown in Figure 1turned out to be the optimalstructure modelingobtainedempirical data. The input layeris composed offiveneuronswitha linearPSP(Postsynaptic Function)andactivation function. The firsthidden layeris composed oftwosigmoidalneurons, i.e. with linearPSPfunction andlogisticactivation function. The secondhidden layeris built fromthreeneuronswith structureidentical to theneurons from thefirst layer. The networkoutputwasonesigmoidal neuron. The developedneural modelwastrainedusingBP(BackPropagation) method in 3 cyclesfor 50epochsandoptimized with algorithmCG(Conjugate Gradients) for 196epochs. The following parameters have been adopted during the training process with the algorithm of BP error:

- decreasing training coefficient: $\eta = 0.1$ do $\eta = 0.01$,
- momentumfactor: $\alpha = 0.22$.

The structure ofgeneratedANN, typeMLP: 5-8-4-1,is shown in Fig.2.

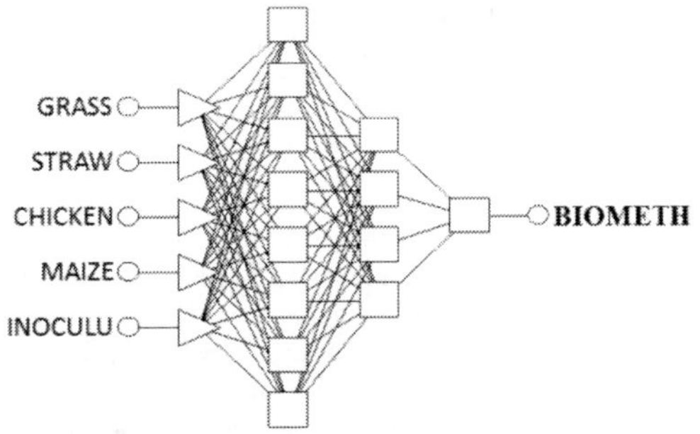

Figure 2.Structure ofthe optimalANN, typeMLP: 5-8-4-1, with2hidden layers.

Unidirectional MLP neural networks are commonly used in ANN topology practice. Multilayer Perceptron represents the so-called category of parametric neural models. Where characteristic is that the number of neurons constituting its structure is considerably less than the number of cases of the training set. The basic characteristics of MLP network include the following features:

- MLP is a unidirectional network,
- MLP is trained by „with-a-teacher" method,
- has a multi-layer structure, with the following layers: input, hidden, output
- architecture of the connections within the network allows communication only between the neurons located in contiguous layers,
- neurons being a part of ANN, MLP type,aggregate the inputdataby defining theinputs weightedsums(using the linearformula ofaggregation),
- activation functionof the inputneuronsis linear, hidden neurons-non-linear, whilethe nature ofoutput neuronsis generallynonlinear,
- due to thesaturation level present(in the sigmoidactivationfunctions), all the data processed by the networkrequire an appropriaterescaling(preprocessing andpost processing).

The quality of the generated MLP network, as a predictive tool, is identified by the so-called statistics of regression issues, which are shown in Table 3.

Table 3. Regression statistics of generated model for the files: training, validation and test

	biomethane for training file	**biomethane** for validation file	**biomethane** for test file
Data mean	70,81500	79,05513	89,66154
Data S.D.	45,39954	52,06174	40,76000
Error mean	0,442888	-0,07916	0,304094
Error S.D.	3,655942	4,18808	3,908609
Abs E. Mean	2,672447	2,971817	2,857903
S.D. Ratio	**0,080530**	**0,080440**	**0,095890**
Correlation	**0,996799**	**0,996759**	**0,995508**

where:

- Data mean - the mean value of the output variable, calculated on the basis of the preset values of this variable, collected (respectively) in the training, validation or testing set. The regression statistics are calculated independently for the training, validation and test set.
- Data S.D. - the standard deviation calculated for the specified (as above) values of the output variable.
- Error mean - mean error (module of the difference between the reference value and the value obtained at the output) for the output variable.
- Error S.D. - the standard deviation of the errors for the output variable.
- Abs E. Mean - mean absolute error (the difference between the reference value and the value obtained at the output) for the output variable.
- S.D. Ratio - quotient of the standard deviations both for errors and data. This is the main indicator of the quality of the regression model developed by the network.
- Correlation - standard R.Pearson correlation coefficient for the setpoint value and the value obtained at the output.

Table 3shows thatthe correlation isat the level of0.99for the following files:training, validationandtest one, while the quotient of standard deviations for errorsand thedatarangesfrom 0.08in case ofvalidation file up to0.09forthe test file.

The assessment of sensitivity of developed MLP on individual input variables has been performed in order to determine the level of significance of representative parameters used to build the neural model. The procedure of sensitivity analysis is implemented in the Statistica package as a tool for assessing the impact of the various input variables on the quality performance of generated neural model. The sensitivity analysis provides an insight into the usefulness of particular input variables. Moreover, it indicates high-ranking variables that without any loss of quality of the network can be omitted. Furthermore it also points the key variables (low rank value), which must not be ignored.

DOI: 10.1051/matecconf/20166004007

Table 4. The values of quotients of errors and rank for the five input variables

	grass	straw	chicken	maize
Rank	3	1	4	2
Error	17.69924	67.48457	17.64613	32.38631
Ratio	4.614127	17.59298	4.600281	8.442992

gdzie:

- Rank - significance level of input variable, organizes the variables by importance: 1 is the dominant variable.
- Error - the network quality in the absence of a variable: the lower the rank number of the ANN input variable, the higher error of reduced network (without the input variable).
- Ratio - quotient of the error of the reduced network by the ANN error obtained using all variables. If the Ratio is less than one, than removal of the variable improves the quality of the network.

The sensitivity analysis of the MLPneural model5-8-4-1onthe input variables of analyzed processshowed thatthe most importantparametersin the process of neural estimation of the amount ofgeneratedbiomethaneare(in order):

- Rank 1: STRAW
- Rank 2: MAIZE
- Rank 3: GRASS

In order to visualize the behavior of the generated neural model, depending on the values of the main descriptors (STRAW, MAIZE, GRASS) Figure 3 shows the three surfaces of ANN responds illustrating the biomethane efficiency in a function of key input variables of developed neural model.

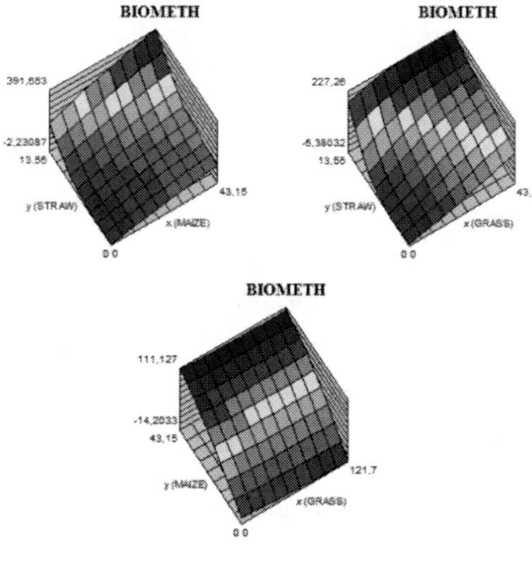

Figure 3. Answer surfacesfor network MLP:5-8-4-1

4.Discussion

The analysis of the empirical resultsobtainedduring the testsor receivedin the course of industrial processes controlsometimesis very complex.Usuallyit is related to a largeamount ofdata obtained.

Duringcontinuous processessuch asindustrial production ofbiogasthroughoutmethane fermentationwe have to dealwith a specific case.In the aforementioned situationit is possible toobtain a veryextensivetraining setbased ona number of variablesanalyzedover a long periodof time[12,13].In order to properlyinterpret the multifactorialresultsit is necessary torefer toadvancedstatistical methodsbased onArtificial Intelligence.According to[14; 15] application of thistype of analysis enables defining the trends, dependencies and propercontrol of the processes,not only in thelaboratory scale butmainlyin the industrial one.

5. Conclusions

The resultsanalyzedusingANNhave shownthat the optimumstructuremodelingobtainedempirical dataisa 4-layer MLP network. The receivedvisualizationsand statementsare consistentwith the experience ofa biogas plantstaffandscientists studyingthe efficiencyof biogassubstrates.It proves that chosen method can besuccessfully implemented into the plannedapplication-optimizingthe selection of the co-substratesfor fermentation of the chicken manurein order to obtainthe highest possibleof biomethane production.

Acknowledgement

This paper was created as a result of the realization of the project entitled "The innovative energy production in biogas plant through utilization of poultry manure with the alternation of a plant substrate for algae"financed by The National Centre for Research and Development (contract no.: GEKON1/O5/214528/34/2015).

References

1. Czekała W., Dach J., Ludwiczak A., Przybylak A., Boniecki P., Koszela K., Zaborowicz M., Przybył K., Wojcieszak D., Witaszek K.. *Proc. SPIE. 9631, Seventh International Conference on Digital Image Processing (ICDIP 2015),* **963117**. (doi: 10.1117/12.2197041 (2015)
2. Przybył K., Koszela K., Boniecki P., Mueller W., Raba B., Lewicki A. Proc. SPIE **9159**, *Sixth International Conference on Digital Image Processing (ICDIP 2014),* **91590W**; doi:10.1117/12.2064243 (2014)
3. Raba B., Nowakowski K., Lewicki A., Przybył K., Zaborowicz M., Koszela K., Boniecki P. SPIE 9159,

*Sixth International Conference on Digital Image Processing (ICDIP 2014),***91590H** (April 16, 2014); doi:10.1117/12.2064091 (2014)

4. Dach J., Koszela K., Boniecki P., Zaborowicz M., Lewicki A.,., Czekała W., Skwarcz J., Qiao W., Piekarska-Boniecka H., Białobrzewski I. Renewable andSustainable Energy Reviews**56**, 603–610 (2016)

5. Boniecki P., Piekarska-Boniecka H., Koszela K., Zaborowicz M., Przybył K., Wojcieszak D., Zbytek Z., Ludwiczak. A., Przybylak A. , Lewicki A. *Proc. SPIE. 9631, Seventh International Conference on Digital Image Processing* (ICDIP 2015), **963115**. (July 06, 2015) doi: 10.1117/12.2197029 (2015)

6. Koszela K; J. Otrząsek; M. Zaborowicz; P. Boniecki; W. Mueller; B. Raba; A. Lewicki; K. Przybył. Proc. SPIE 9159, *Sixth International Conference on Digital Image Processing (ICDIP 2014),***915913**; doi:10.1117/12.2064274 (2014)

7. Kujawa S., Tomczak R.J., Kluza T., Weres J., Boniecki P.,. *4th International Conference on Digital Image Processing. Kuala Lumpur (Malaysia), proceedings of SPIE,* Vol. **8334**, Article Number: 83341R, DOI: 10.1117/12.949899 (2012)

8. GueguimKanaa E.B., Olokeb J.K., Lateefb A., Adesiyanb M.O. Renewable Energy, Volume **46**, Pages 276–281

9. Lewicki A., Boniecki P., Dach J., Janczak D., Czekała W., Zbytek Z., Cieślik M.. *4TH International Conference on Materials Engineering for Advanced Technologies (ICMEAT 2015)* Pages: 649-653 (2015)

10. Cerbin S., Nowakowski K., Dach J., Pilarski K., Boniecki P., Przybyl J., Lewicki A., *4th International Conference On Digital Image Processing (ICDIP 2012). Proceedings of SPIE.* Vol.: **8334** Article Number: 83342A Doi: 10.1117/12.954164 (2012)

11. Janczak D., Marciniak M., Lewicki A., Czekała W., Witaszek K., Rodríguez Carmona P. C., Cieślik M., Dach J., *Procedia Technology,* Vol. **8**, 209–214. DOI: 10.1016/j.protcy.2013.11.029 (2013)

12. Li J., Wei L., Duan Q., Hu G., Zhang G. Bioresource Technology, Volume **156**, Pages 307–313 (2014)

13. Li J., Kong C., Duan Q., Luo T., Mei Z., Lei Y. Bioresource Technology, Volume **193**, Pages 62–67 (2015)

14. Ozkaya B., Demir A., Sinan Bilgili M. Environmental Modelling & Software, Volume 22, Issue **6**, Pages 815–822 (2007)

15. Abu Qdaisa H., BaniHanib K., Shatnawia N. Resources, Conservation and Recycling, Volume **54**, Issue 6, Pages 359–363 (2010)

MATEC Web of Conferences **60**, 04008 (2016) DOI: 10.1051/matecconf/20166004008
ICCBS 2016

Energy and economic potential of maize straw used for biofuels production

Zbyszek Zbytek[1], Jacek Dach[2], Tadeusz Pawłowski[1], Anna Smurzyńska[2], Wojciech Czekała[2], Damian Janczak[2]

[1] *Industrial Institute of Agricultural Engineering, ul. Starolecka 31 60-963 Poznan, Poland*
[2] *Institute of Biosystems Engineering, Poznan University of Life Sciences, ul. Wojska Polskiego 28 60-637 Poznan, Poland*

Abstract. The paper presents the energy and economic comparison of two technologies of maize straw utilization: solid biofuel production (briquettes) and methane fermentation. The research experiments have shown that maize straw is the material which can be efficiently implemented in both technologies. Maize straw usage as briquettes can generate more energy (10.956 GJ Mg^{-1}) than methane fermentation (9.74). In Europe, biogas is used in co-generation units for production of electric and heat energy. Due to higher price of electricity, economic profitability of maize straw usage for biogas production is over twice higher (182 USD) than in case of briquettes production (96 USD).

1 Introduction

Nowadays, the plant biomass is considered to be the oldest source of energy. It can be obtained both from conventional agricultural sector, as well as from forest crops, however, more often comes straight from the energy crops used in order to produce electricity and heat [1]. It is obtained from the herbaceous plants with high yields (mallow, Jerusalem artichokes), woody plants (willow, poplar), grasses (i.e. miscanthus) or from plant production (straw, hay). The structure of existing sources of biomass from crop production mainly includes the use of wheat and rye straw. And this resulted directly from a decline in cattle numbers in Poland. Currently, the straw for energy demands is applied in the unprocessed form i.e. in the form it has been harvested or processed into the form of briquettes and pellets. However, new possibility of energy production from plant biomass is a maize straw. Maize has a high yield potential ratio, 12-15 Mg of dry mass of the whole plants from 1 ha. This factor makes possible to use it not only for biogas production (fresh mass, silage) or bio-ethanol (grain) but also for direct combustion [2].

Maize straw (the main subject of this paper) is characterized by a variable humidity resulting from various development phases and harvest deadlines. In order to improve the maize straw energy properties it should be additionally dried. Niedziółka showed [3] that the average calorific value of the maize straw was in fresh state 6.5 MJ kg^{-1}, whereas dried reached 15.5 MJ kg^{-1}. Furthermore, raw and unprocessed straw is a material with low energy value, due to decreased concentration of energy per unit volume. Therefore, the most common method of maize straw management in Poland is production of pellets intended for thermal energy production. This process allows to obtain the pellets with considerably better density compared with initial one. Thus, the pellets provide longer combustion time in comparison with incompact biomass [4]. Another advantageous feature of use of processed biomass from plant production for energy purposes is reduction of transportation costs, and moreover the ability of automatization of the combustion process.

Maize is a widely cultivated plant. Except for being a good feed for animals, it can be considered as a raw material for food or chemical industry. Along with an increase of energy production from the plant biomass, the interest in maize cultivation intended for energy purposes including the biogas production, has also raised [5]. Until now, commonly used feedstock for biogas plants was the maize silage as a substitute improving the energy efficiency of biogas plants [6]. Thus, it became kind of stimulator for fermentation process and facilitated the utilization of sewage sludge [7] and slurry [8]. Due to the production costs of the substrate, presence of crop monocultures and providing the food for domestic animals, the importance of maize silage as a feedstock for biogas plants gradually decreases. Therefore, we are looking for new directions of maize use or application of new herbaceous plants. More recently, there have been attempts to use a maize straw as a structural material for composting of the sewage sludge [9-10]. A new trend of management in agricultural biogas plants is application of the maize straw, thus eliminating the methane production from the maize silage. In this case there is lack of conclusive determination of costs of the total energy production from the maize silage and straw.

Within last years, it has been observed also dynamic development of advanced technologies of artificial intelligence (neural networks) for optimization of production and treatment of agricultural biomass and biowaste. Those activities were usually close to energetic and environmental problems, following often the idea promoted by European Union: "waste-to-energy". The development of IT systems and implementation of large computing power of the computers has significantly contributed to neural networks implementation. High influence on the development of artificial neural networks had a technical evolution in the field of neuro-processors construction, which are analog hardware applications. In recent years, the artificial neural networks have been used in agricultural engineering [11-12], eco-energetics [13-14] and environmental engineering [15-16]. The obtained rewarding results in terms of classification and prediction were used to construct the expert systems working in real-time, effectively supporting the decision-making processes in many sectors of agriculture, food processing and biofuel production [17-18]. This even resulted in development of the first agricultural equipment and machinery based on the methods of artificial intelligence, and therefore, without constant supervision and manual. Moreover, also in terms of application of maize straw, the appropriate measures have been taken in order to use the neural networks to produce biogas from maize silage and straw [19].

2 Aim and the range of research

The aim of this study was to analyze the energy potential and economic balance of maize straw used as new biofuels source (biogas and briquettes – the most popular methods of energetic biomass utilization in Central Europe). The analyzes have included use of maize straw as a substrate for biogas and briquettes production. One of the important analytical aspect was to compare the energetic efficiency as well as price of raw material with other kinds of substrates, typically used for biogas and solid biofuel production.

3 Materials and methods

The research and basic physical analyzes were made in the Industrial Institute of Agricultural Engineering. The biogas experiment for estimation of energetic value of maize straw during fermentation was made in Laboratory of Ecotechnologies, at the Institute of Biosystems Engineering, Poznan University of Life Sciences (PULS). Heat of combustion tests of maize straw was estimated in the Institute of Technology and Life Sciences.

The research material (maize straw) was collected from PULS experimental farm located in Złotniki, near Poznań, in autumn 2014, during maize harvest for grain. The dry matter content was made according to Polish standard norm PN-75C-04616/01 (drying by 24 h in 105°C), whereas organic dry matter according to PN-Z-15011-3 norm (samples combustion in 525°C for 3 hours) [20-21].

The heat of combustion was made on the dried straw, initially crushed by rotor mill. For analysis, the calorimeter was used according to the Polish standard and European norm PN-EN ISO 9831. The analysis consisted of the total combustion of 1 g sample in pure oxygen atmosphere with 2.8 MPa pressure in calorimeter submerged in water of 2.7 dm^3 volume. The combustion heat was calculated automatically by internal calorimeter calculation program.

The biogas production tests were made in Laboratory of Ecotechnologies, the biggest Polish biogas laboratory with over 250 different fermenters. The analyzes were done according to the German norms DIN 38414/S8 and VDI 4630 - standardized biogas guidance issued by the Association of German Engineers in Dresden.

The biogas production experiments (both in meso- and thermophilic fermentation) were conducted in the set of multi-chamber biofermenter (Fig. 1) constructed in the Laboratory of Ecotechnologies. On the basis of obtained results, available information concerning installation, presented assumptions, economic analysis of briquettes production from the solid fraction of digestate, was performed.

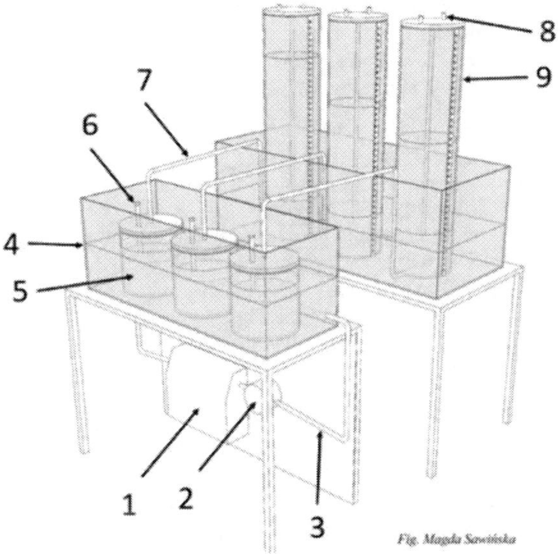

Figure 1. Scheme of biofermentor for biogas production research (3-chamber section): 1 – water heater with temperature regulator, 2 – water pump, 3 – insulated conductors of calefaction liquid, 4 – water coat, 5 – biofermentor with charge capacity 2 dm3, 6 – sampling tubes, 7 – biogas transporting tube, 8 - gas sampling valve, 9 – biogas volume-scale reservoir

The biogas production was conducted in the glass biofermenters (capacity of 2 dm^3). The maize straw samples were fermented in inoculum i.e. liquid fraction of post-fermented pulp from the typical biogas plant working in Działyń, near Poznań. The temperatures of fermentation were: 39°C±1 (mesophilic fermentation) and 55°C±1 (thermophilic fermentation) in order to link this experiment with working conditions of methane fermentation process occurring on typical biogas plants in Europe. The volume of produced biogas (as well as composition: CH_4, CO_2, H_2S and NH_3) was checked once a day (each 24 hours) and introduced to the calculation program. This program automatically calculated the total production of biogas and biomethane per ton of fresh or dry matter.

The energy value of biogas produced from maize straw fermentation bases on methane content in the gas (biogas is the mixture of CH_4 and CO_2). The calculation of maize straw energy value in methane fermentation process was made by the following formula:

$$E = V_{CH4} \cdot We_{CH4}$$

where:
E – amount of energy produced from material [MWh];
V_{CH4} - volume of methane produced from 1 Mg of substrate [m^3];
We_{CH4} – energy value of methane [0,00917 MWh $\cdot m^{-3}$]

In case of combustion of biogas in co-generation unit for production of electric and heat energy (situation present in over 97% all biogas plants in Europe), the calculated amount of energy has to be recalculated by coefficients:

$\eta_e = 0,4$ (electric efficiency of co-generation unit);
$\eta_t = 0,48$ (thermal efficiency);

The price of maize, straw as well as other substrates for briquettes and biogas production were taken from the actual market data (September-October 2015) published by Polish agricultural agencies. The price for electric and heat energy was taken from the Polish Energy Regulatory Office (www.ure.gov.pl).

4 Results and Discussion

4.1 The energy efficiency

The research results showed that the dry matter (TS) of the fresh maize straw amounted 77.8%, while the dry organic matter was 89.8%. In turn, the heat of combustion of the analyzed briquettes sample had value of 16.142 MJ•kg^{-1}, while the calorific value amounted 14.245 MJ•kg^{-1} (tab. 1).

Table 1. The parameters of the tested material (maize straw and briquettes)

Substrate	Values
Total Solids (maize straw) [%]	77.8
Volatile Solids (maize straw) [%]	89.8
Heat of combustion (briquettes) [MJ·kg^{-1}]	16.142
Heating value (briquettes) [MJ·kg^{-1}]	14.245

The analysis of basic maize straw parameters shows that this material has high content of organic dry matter (over 89.5%) which is favorable for energetic use as solid biofuel (i.e. briquettes) and promising for methane fermentation. The heat of combustion was also high (16.142 MJ kg^{-1}) and real heating value amounted 14.245 MJ kg^{-1}. This value is similar to other popular species of straw.

Table 2. The biogas and biomethane production from 1 Mg of maize straw

Sample	CH_4 content [%]	Biogas yield [m^3 Mg^{-1}]	Methane yield [m^3 Mg^{-1}]
Maize straw	51.9	442.1	229.6

The experiment with maize straw fermentation has proved very high biogas production efficiency. This efficiency is over twice higher than maize silage – the most popular substrate for biogas plants in Europe. However, high biogas and biomethane potential is related to much higher dry matter content (tab. 1) comparing to typical maize silage (32%).

The analysis of maize straw energy transformation, made using two technologies (briquettes production and biogas fermentation), let us to develop simplified energy balance for those methods. This balance is shown in table 3.

Table 3. Energetic balance of methane fermentation and briquette production of 1 Mg of maize straw dry matter

Energy type	MWh Mg^{-1}	GJ Mg^{-1}
Methane fermentation of maize straw		
Total content in biogas, including:	2.71	**9.74***
- electric energy (for sale)	0.97	-
- heat energy (for sale)	-	4.21
Briquettes production from maize straw		
Heat of combustion of maize straw	-	14.245
Energy of maize straw drying	-	0.55
Energy for briquettes production	0.162	-
Energy in briquettes	-	13.695
Losses during combustion (furnace efficiency 80%)	-	2.739
Net energy from maize straw briquettes	-	**10.956**

* calculated by coefficient that 1 MWh = 3.6 GJ

The total amount of energy content in biogas produced from 1 Mg of maize straw is 9.74 GJ and this value is a little bit smaller than net energy which can be used by combustion of briquettes. However, it has to be highlighted that over 97% of European biogas plants use biogas for production of electric and heat energy in cogeneration units. Thus, the real energy value from energy transformation of maize straw by combustion of produced biogas is 0.97 MWh of electric energy and 4.21 of heat. About 15% of energy is lost, mainly in the form of heat.

4.2 Economic analysis

The real economic profitability from energetic use of maize straw in both analyzed technologies is presented in tables 4 and 5.

Table 2 presents the results of biogas and biomethane production from 1 Mg of maize straw by fermentation process. After biogas production, it is combusted in co-generation unit with 40% of electric and 48% of thermal efficiency (as shown in Materials and Methods chapter). The amount of electric energy production from 229.6 m^3 of CH_4 was 0.97 MWh, however heat production was 1.17 MWh (4.21 GJ). In order to obtain the profits from the energy sold, those values were multiplied by Polish electric energy price (150 USD MWh^{-1}) and heat energy price (12.5 USD GJ^{-1}).

Table 4. Profits of transformation of 1 Mg maize straw dry matter by methane fermentation process into electric and thermal energy, including cost of material

Parameter	Value (USD)
Profit from energy sold:	
- electric	146
- heat	53
Total profit	199
Cost of 1 Mg of maize straw	17
Net total profit	**182**

The economic calculations of maize straw energetic usage profitability show that net total profit is 182 USD for 1 Mg of used material. This high value is related to the level of electric energy price in Poland. Due to the European Union climate policy and decarbonization of European economies (CO_2 taxation) electric energy is expensive.

Table 5. Profits of transformation of 1 Mg maize straw dry matter by briquettes production (electric energy price = 150 USD MWh^{-1} and heat energy price = 12.5 USD GJ^{-1})

Parameter	Value (USD)
Energy costs for briquettes production	24
Profit from briquettes combustion	137
Total profit	113
Cost of 1 Mg of maize straw	17
Net total profit	**96**

The economic analysis for briquettes production shows much lower profit comparing to maize straw usage as substrate for biogas production. This is related mainly to the fact that heat energy price in Poland is several times lower comparing to electric energy price. In consequence, even transformation of maize straw for solid biofuel (briquettes) can generate more energy from 1 Mg of biomass, this is heat energy, cheaper than electric energy. That is why transformation of maize straw by methane fermentation is more profitable under Polish conditions, which are similar to these present in the whole Central Europe.

5 Conclusions

Based on the aforementioned analysis, the following conclusions were made:

1. Maize straw is the material which can be efficiently transformed into solid biofuel and biogas.

2. Use of maize straw as briquettes can generate more energy (10.956 GJ Mg^{-1}) than methane fermentation (9.74).

3. In Europe, biogas is used in co-generation units for production of electric and heat energy. Due to higher price of electricity, economic profitability of maize straw usage for biogas production is over twice higher than in case of briquettes production.

References

1. A. Kowalczyk-Juśko, B. Kościk, K. Jóźwiakowski, A. Marczuk, J. Zarajczyk, J. Kowalczuk, M. Szmigielski, Sagan, A. Przem. Chem. **94**, 10 (2015a).

2. A. Kowalczyk-Juśko, A. Marczuk, J. Dach, M. Szmigielski, J. Zarajczyk, K. Jóźwiakowski, J. Kowalczuk, D. Andrejko, B. Ślaska-Grzywna, N. Leszczyński, Przem. Chem. **94**, 2 (2015c)

3. I. Niedziółka, M. Szymanek, A. Zuchniarz, Agric. Eng. **7**, 95 (2007)

4. A. Kowalczyk-Juśko, J. Kowalczuk, M. Szmigielski, A. Marczuk, K. Jóźwiakowski, J. Zarajczyk, A. Masłowski, B. Ślaska-Grzywna, A. Sagan, J. Zarajczyk, Przem. Chem. **94**, 10 (2015b)

5. J. Dach, P. Boniecki, J. Przybył, D. Janczak, A. Lewicki, W. Czekała, K. Witaszek, P. C. Rodríguez Carmona, M. Cieślik, Energy, **69** (2014) DOI: 10.1016/j.energy.2014.02.013

6. W. Czekała, K. Kozłowski, J. Dach, P. Boniecki, A. Lewicki, D. Janczak, K. Jóźwiakowski, T. Piechota, 4th International Conference on Materials Engineering for Advanced Technologies (ICMEAT) Location: London, ENGLAND Date: JUN 27-28 (2015)

7. I. Białobrzewski, M. Mikš-Krajnik, J. Dach, M. Markowski, W. Czekała, K. Głuchowska, Waste Management. **43**, (2015). DOI: 10.1016/j.wasman.2015.05.036

8. D. Janczak, M. Marciniak, A. Lewicki, W. Czekała, K. Witaszek, P. C. Rodríguez Carmona, M. Cieślik, J. Dach. Procedia Technology **8**: (2013) DOI: 10.1016/j.protcy.2013.11.029

9. K. Malińska, M. Zabochnicka-Świątek.. Environment Protection Engineering **39**, 2 (2013)

10. K. Malińska, M. Zabochnicka-Świątek, J. Dach. Ecological Engineering **71** (2014) DOI: 10.1016/j.ecoleng.2014.07.012

11. K. Nowakowski, P. Boniecki, R. J. Tomczak, S. Kujawa, B. Raba. 4th International Conference on Digital Image Processing. Proceedings of SPIE, **8334**, 833425 (2012) DOI: 10.1117/12.954155.

12. K. Koszela, J. Otrząsek, M. Zaborowicz, P. Boniecki, W. Mueller, B. Raba, A. Lewicki, K. Przybył. Proc. SPIE 9159, Sixth International Conference on Digital Image Processing (ICDIP 2014), **915913** (2014) DOI: 10.1117/12.2064274

13. S. Cerbin, K. Nowakowski, J. Dach, K. Pilarski, P. Boniecki, J. Przybyl, A. Lewicki, 4th International Conference On Digital Image Processing (ICDIP 2012). Proceedings of SPIE. **8334**, 83342A (2012) DOI: 10.1117/12.954164

14. J. Dach, K. Koszela, P. Boniecki, M. Zaborowicz, A. Lewicki, W. Czekała, J. Skwarcz, Wei Qiao, H. Piekarska-Boniecka, I. Białobrzewski, Renewable and Sustainable Energy Reviews **56** (2016) DOI: 10.1016/j.rser.2015.11.093

15. S. Kujawa, R. J. Tomczak, T. Kluza, J. Weres, P. Boniecki 4th International Conference on Digital Image Processing. Proceedings of SPIE, **8334**, 83341R (2012) DOI: 10.1117/12.949899

16. W. Czekała, J. Dach, A. Ludwiczak, A. Przybylak, P. Boniecki, K. Koszela, M. Zaborowicz, K. Przybył, D.

Wojcieszak, K. Witaszek, Proc. SPIE. 9631, Seventh International Conference on Digital Image Processing (ICDIP 2015), **963117** (2015) DOI: 10.1117/12.2197041

17. J. Dach, W. Czekała, P. Boniecki, A. Lewicki, T. Piechota. Advanced Materials Research **909** (2014) DOI 10.4028/www.scientific.net/AMR.909.305

18. P. Boniecki, H. Piekarska-Boniecka, K. Koszela, M. Zaborowicz, K. Przybył, D. Wojcieszak, Z. Zbytek, A. Ludwiczak, A. Przybylak, A. Lewicki. Proc. SPIE. 9631, Seventh International Conference on Digital Image Processing (ICDIP 2015), **963115**. (2015) DOI: 10.1117/12.2197029

19. A. Lewicki, P. Boniecki, J. Dach, D. Janczak, W. Czekała, Z. Zbytek, M. Cieślik. 4th International Conference on Materials Engineering for Advanced Technologies (ICMEAT) (2015)

20. A. Wolna-Maruwka, J. Dach Archives of Environmental Protection. **35**, 3 (2009)

21. A. Piotrowska-Cyplik, Ł. Chrzanowski, P. Cyplik, J. Dach, A. Olejnik, J. Staninska, J. Czarny, A. Lewicki, R. Marecik, J. Powierska-Czarny. International Biodeterioration & Biodegradation **78** (2013) DOI: 10.1016/j.ibiod.2012.12.007

Isolation of Nanocrystalline Cellulose from oil palm empty fruit bunch – A response surface methodology study

Yee Kai Song[1,2], Irene Mei Leng Chew[3], Thomas Shean Yaw Choong[1], Jully Tan[2] and Khang Wei Tan[2]

[1]Faculty of Engineering, University Putra Malaysia, 43400 UPM, Serdang, Selangor Darul Ehsan, Malaysia
[2]Faculty of Engineering, Technology and Built Environment, UCSI University, Cheras, 56000 Kuala Lumpur, Malaysia.
[3]School of Engineering, Monash University, 46150 Bandar Sunway, Selangor Darul Ehsan, Malaysia.

Abstract. The research work studied the extraction of NanoCrystalline Cellulose (NCC) from oil palm empty fruit bunch (EFB), with aid of Response Surface Methodology (RSM). Particle size analysis using Malvern Zetasizer had confirmed the extracted NCC fall within the desired nano scaled range. The impact of three input parameters, namely concentration of NaOH solution during alkaline treatment, concentration of H_2SO_4 solution during acid hydrolysis, and duration for acid hydrolysis on NCC particle were investigated. From ANOVA study, it had suggested that the current RSM model is significant to interpret the interaction among the all three input parameters.

1 Introduction

Back to 1951, Rånby (1951) demonstrated a fundamental piece of work which stated that native cellulose fibers can be degraded using sulphuric acid as hydrolysis agent [1]. In 50 years apart into 21st century, Rånby's work in acid hydrolysis on native cellulose had been a solid guidance on blossoming numerous cellulose particle extraction from biomass sources, to name a few, tunicate [2], ramie [3], and wood [4].

Empty fruit bunch (EFB) is a by-product from mills to obtain valuable oil. In most tropical countries, especially Malaysia, EFB from palm oil tree is one of the abundantly available fibrous biomass materials, which is always regarded as waste and extremely low in commercial value. Some general applications on EFB are pulp production, used as stabilizing agent for soil, horticulture, or used as fu.el for power generation [5]. Applications can be critically many; however the commercial viability is trivial. Perhaps, by hydrolyzing EFB into fermented saccharides, which can be further developed into valuable ethanol [6] seems to be a good idea to increase ultilization value of EFB. Though, the high handling cost, large inventory spaces needed due to its bulkiness, purity of ethanol produced, and dependency of ethanol price to the market could jeopardize its commercial viability. These factors are leading to high demand of cost effectiveness in NCC extraction from any form of biomass in order to have high commercial viability in the future.

As aforementioned, sulphuric acid hydrolysis is the most fundamental chemical method to isolate crystalline region in any form of cellulosic material [1]. With such specific method to produce NanoCrystalline Cellulose (NCC) from cellulose, it imposed a setback to the practicality of NCC on industry that required high processing temperature, which is caused by reduced thermal stability, or thermal degradation prior to acid hydrolysis. Poletto et al. [7] reported that lesser the percentage of crystalline extracted from wood with association with smaller crystalline size could lead to accelerated thermal degradation. This finding is essential to justify the necessity of searching an alternative method to extract NCC with lower particle size. The study on the role of acid particularly on its impact on the isolation of crystalline particle is crucial, because from the magnitude of degradation it basically determine the size of NCC extracted. Response surface methodology (RSM) was commonly used as an optimization tool due to its well establishment in examining experiment variables, deriving model, and factors used in the optimization parameters back in 1983 [8]. In the current study, a foundation was laid down for further research by studying the significance and interactions among the three input parameters, i.e. concentration of NaOH solution during alkaline treatment, concentration of H_2SO_4 solution during acid hydrolysis, and duration for acid hydrolysis, on their impact to determine the NCC particle size. It provided an insight of how to reduce the involvement of acid during NCC isolation to obtain finer NCC particles.

2 Methodology

2.1 Isolation of NCC

DOI: 10.1051/matecconf/20166004009

2.1.1 Preparation of empty fruit bunches

Empty fruit bunches (EFB) were grinded to a size of approximately 10 mm. The grinded EFB were washed for 10 min to remove dust, dirt, wax, and impurities on the surface of EFB. 10 g of grinded EFB were dispersed in 800 mL distilled water on a stirring hot plate at a constant temperature of 50 °C for 2 h to strengthen the outcome of this pretreatment. The suspension was then sieved through a 300 µm mesh size sieve. Followed by that, the sieved EFB was treated with 800 mL of 10 - 30 % NaOH solution at a temperature of 80°C for 6 h, and then the EFB was washed with distilled water until pH 7 was obtained. This process is generally known as the alkaline treatment to remove hemicelluloses along with partial lignin on the EFB. The alkaline treated EFB was dried in the oven for 24 h at a temperature of 70 °C to remove moisture content. Next, bleaching process was done to remove the lignin, i.e. 3 g of the alkaline-treated EFB was dispersed in 350 mL of distilled water that is mixed with 10 mL of hydrogen peroxide (H_2O_2) and 20 - 22 drops of glacial acetic acid (CH_3COOH). The mixture was stirred at medium temperature of 70 °C for 7 h, followed up by sieving in a 60 µm mesh size sieve. Then, the mixture was washed with distilled water until pH 7 was obtained, and dried in oven at 70°C for 24 h for further process.

2.1.2 Extraction of NCC

Sulphuric acid hydrolysis was conducted to extract NCC from the treated EFB. 0.3 g of bleached EFB was dispersed in 4 mL consist of 56 - 64 % weight of sulphuric acid (H_2SO_4) solution at a constant temperature of 45°C for 1 - 3 h. The hydrolysed samples were washed using distilled water until pH 5 is obtained. After washing, centrifugal was used to separate the NCC from the heavy component at 3000 rpm for 10 min.

2.2 Characterization and Analysis

2.2.1 Particle Size Analysis

The extracted NCC was dispersed in distilled water and analyzed using Malvern Zetasizer, Nano ZS series. The measurement was repeated twice and z-average was generally regarded as particle average size in this study.

2.2.2 Response Surface Methodology (RSM)

Response surface methodology (RSM) is statistical-based software which is useful to optimize the isolation process of NCC in this study. From preliminary experiment, three (3) independent variables which are concentration of alkaline used (NaOH), concentration of acid used (H_2SO_4), and duration of sulphuric acid hydrolysis were being investigated. Box-Behnken Design (BBD) was employed to reduce the complexity which only requires few significant points from three level factorial arrangements. Hence, the total number of experiments needed is 17, and the z-average is the only response in this study. The boundary limit for the three input variables are 10 – 30 %, 56 – 64 % and 1 h to 3 h, for concentration of alkaline used (NaOH), concentration of acid used (H_2SO_4), and duration of sulphuric acid hydrolysis, respectively. The resulted response surface model is validated statistically with analysis of variance, or in abbreviation ANOVA.

3 Result and Discussion

3.1 Particle Size Analysis

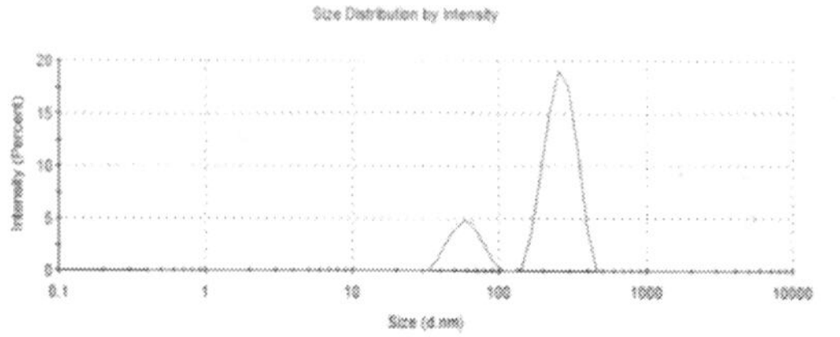

Figure 1. Particle Size Analysis by Malvern Zetasizer, Nano ZS Series.

For the first run of experiment, NCC was obtained with the following condition i.e. 1 M NaOH solution, 60 % H_2SO_4 solution and 2 h hydrolysis time. As shown in **Fig. 1**, the analysis indicated that the size distribution by intensity has two peaks, recorded as 59.79 nm for peak 1, and 264.9 nm for peak 2. As a whole, the z-average was recorded as 420.6nm. It was reported that NCC is a typical elongated rod-like shape with diameter ranging from 10nm to 100 nm while length ranging from 100nm to 1000nm [9]. Polydispersity index (PDI) value of 0.528 was obtained due to non-uniformity of the particle size. It was theorized as in the effect of the orientation of NCC across the dynamic light scattering during the measurement. According to Stoke-Einstein theory [10], the larger particle diffuses in a slower rate compared to the smaller particle. For instance, when NCC diffuses across the aqueous medium in horizontal orientation against the light source, the size interpretation by Zetasizer is smaller, due to a faster speed detected for its difussion. Similarly, if NCC diffuses across the aqueous medium in vertical orientation against the light source, the size interpretation by Zetasizer will be bigger. It is theorized that the larger surface area contribute to higher resistances in the direction of diffusion, resulting in lower speed of NCC diffusion. This phenomenon is illustrated in **Fig. 2**. In recognition of this assumption, particle size measurement was repeated for samples in RSM study.

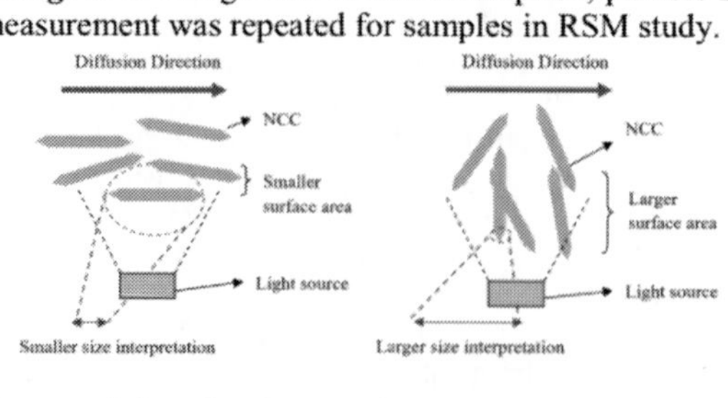

Figure 2. Effect of surface area in the direction of diffusion leading to size interpretation by Zetasizer.

3.2 Response Surface Methodology

With exactly defined domain of independent experimentation variables, a much simpler illustration (**Fig. 3**) can be done to understand vividly on how these variables could be fitted into Box-Behnken Design (BBD) model which required only 17 sets of experiment data for process optimization. The box is the boundary limit controlled at only 3 level figures as in -1, 0, and +1 to the experimental domain that represent the minimum and maximum value obtained from the 17 sets of experiment data. Three-dimensional (3D) contour plot were generated by RSM. In general, the 3D graph determined the overall contribution of the input variables to the z-average. Green region signifies that input variables had less impact to the output variables. The significance of input variables increases with intensification of colorization from green region to dark blue region progressively. The stationary point, which is denoted by dark blue region on lower part of wedge-like 3D graph is the optimized point. Hence, it can be justified that both concentration of NaOH and H_2SO_4 are critical parameters in altering z-average. At low concentration of NaOH and low concentration of H_2SO_4, the z-average diameter is of the highest point throughout the 3D contour plot. This result is valid as both of NaOH and H_2SO_4 had to perform their function in isolating of NCC. Without utilization of NaOH, the removal of hemicelluloses layer and partial lignin would had become a bottleneck process to proceed further into inner layer of EFB. Without sufficient of H_2SO_4, the extraction of NCC will be affected and hence resulted in larger particle size.

Further steps are taken to analyze the effect of duration of sulphuric acid hydrolysis on the z-average. **Fig. 4** showed pattern changes as the duration of acid hydrolysis increases. Starting from the left to the right, the high-significant dark blue region expanded in elongation form across 3D contour plot as the duration of acid hydrolysis increases. It also signifies that duration of acid hydrolysis is able to consolidate larger range of alkaline concentration and acid concentration in order to produce lower z-average diameter.

Table 1 tabulated data generated by RSM, which is known as ANOVA. It measure the significance and correlation between each studied input parameters. P-value is a numerical reference to validate the interactions among the three input variables. Statistically, it is desirable to have a P-value below 0.05 to justify a significant interactions. Following this "golden rule", interaction between B and C, which are acid concentration and acid hydrolysis duration, owning a p-value of 0.0042 that is much smaller than 0.05. This has statistically-proven that the interaction between these two variables is the most critical parameters to determine the z-average.

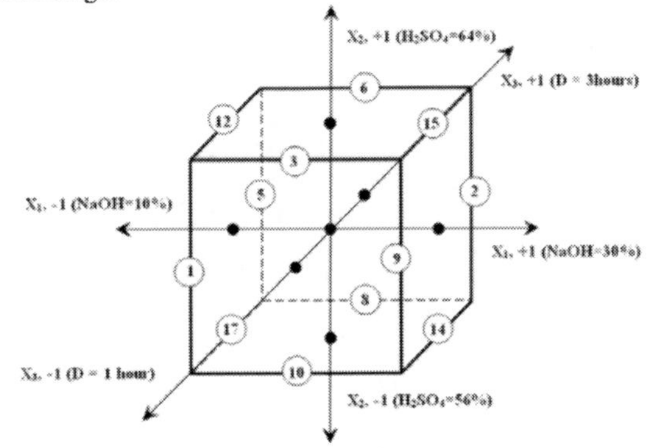

Figure 3. Illustration of three independent variables (X1: NaOH concentration, X2: H2SO4 concentration, X3: Duration of H2SO4 hydrolysis) fitted in Box-Behnken Design (BBD).

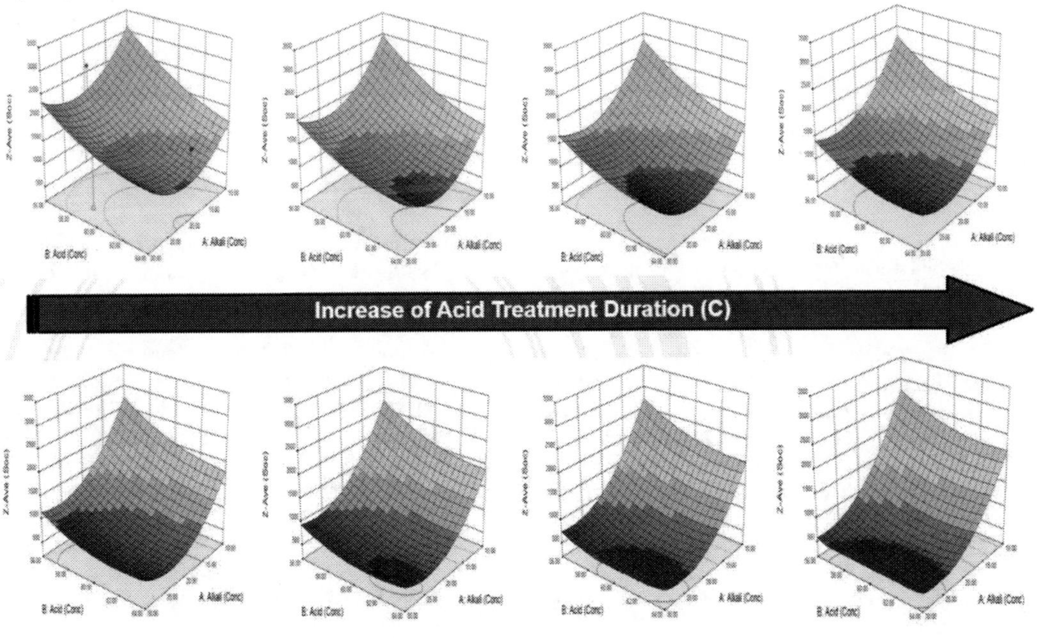

Figure 4. Effect of duration of H_2SO_4 hydrolysis on the z-average progressively.

DOI: 10.1051/matecconf/20166004009

Table 1. Analysis of Variance (ANOVA)

	Particle Size (nm)			
	Sum of Squares	Deg. of Freedom	F-value	P-value
Model	2.45 x 10^5	9	8.71	0.0141
Lack of fit	14817.13	3	12.35	0.0759
Pure Error	800.06	2	- NIL -	- NIL -
Residual	15617.19	5	- NIL -	- NIL -
Total	2.606 x 10^5	14	- NIL -	- NIL -
Linear				
A - Alkali Conc.	534.65	1	0.17	0.6962
B - Acid Conc.	2949.12	1	0.94	0.3758
C - Duration	7638.48	1	2.45	0.1786
Interaction				
AB	1879.22	1	0.6	0.473
AC	3186.6	1	1.02	0.3588
BC	77423.06	1	24.79	0.0042
Quadric				
A^2	1.105 x 10^5	1	35.39	0.0019
B^2	25983.19	1	8.32	0.0344
C^2	3981.42	1	1.27	0.3101

It was observed that the quadric relation of A^2 resulted in value of 0.0019, which is lower than 0.05. When alkaline concentration is decreased, z-average diameter is increased. It is due to the higher alkaline concentration, the higher the ability to dissolve hemicelluloses/lignin to a greater extent, hence exposing more cellulose crystalline region for acid hydrolysis.

Fig. 5 illustrated that the parity plot to display the closeness of actual value to the predicted value. The less deviation from the straight line indicates less deviation from the predicted value. It is a satisfactory correlation between experimental data and predictive data.

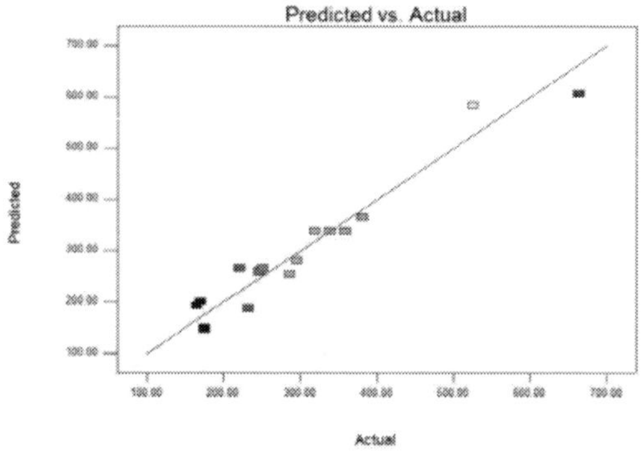

Figure 5. Parity plot displaying the distribution of experimental data against predicted data of the resulted y value.

However, it can observe that the increase duration of sulphuric acid hydrolysis would impose a major setback to be used as a deciding parameter in process optimization. Wang et al. [11] reported that at 64 % wt of H_2SO_4, insignificant amount of cellulose remained. This is attributed to complete hydrolysis of cellulose to soluble sugars and various by-products. The finding proved that excessive treatment using H_2SO_4 either in higher concentration or prolonged treatment duration, NCC yield percentage would not improved but it will be degraded into other products.

From the past decades, there are a lot of studies had been conducted in attempt to optimize the yield of NCC. Rånby [1] reported as high as 40 % NCC yield at 22 % wt of H_2SO_4 of prolonged treatment time. Not long later a study shown that 65 % wt of H_2SO_4 are capable of extracting NCC but it is only limited to maximum of 30 % yield [12]. Most recent advancement is that Bondeson and Oksman (2006) attempted to extract NCC straight from microcrystalline cellulose (MCC) via acid hydrolysis [13]. However, this approach was reported to have a wide range of yield percentage, from 0 to 95 %. Apart from unstable production of NCC, agglomeration phenomenon is also reported for above 47 % yield of NCC. To authors' knowledge, development of NCC production via acid hydrolysis did not take off a significant leap forward since more than 6 decades ago. Hence, for sulphuric acid hydrolysis, this range is proposed to be the optimized condition for NCC extraction from EFB.

4 Conclusion

NCC was successfully isolated from EFB using a conventional sulphuric acid hydrolysis method. Zetasizer on particle size analysis using Zetasizer is a quick and feasible option which suggested the isolated NCC has particle size below 300 nm. ANOVA suggested the conventional chemical approach, especially for NaOH concentration, H_2SO_4 concentration, and duration of H_2SO_4 hydrolysis, played a crucial part to determine the final particle size. This is significant as it provides a fundamental for future research to reduce involvement of acid, as well as other chemical involved.

Acknowledgement

The authors are thankful to Ministry of Higher Education (MOHE) for the financial support through ERGS (ERGS/1/2013/TK07/UCSI/03/1), and CERVIE for RGS (PROJ-IN-FETBE-015). The authors also thankful to the Institution of Engineers Singapore (IES) for the funding to develop part of the prototype.

References

1. Rånby, B. G., (1951). The Colloidal Properties of Cellulose Micelles.*Discussions of the Faraday Society*. 11, pp. 158 – 164.
2. Kimura, F., Kimura, T., Tamura, M., Hirai, A., Ikuno, M. and Horii, F. (2005).Magnetic Alignment of the Chiral Nematic Phase of a Cellulose Microfibril Suspension. *Langmuir*. **21**, pp. 2034 – 2037.
3. De Menezes, A. Jr., Siqueira, G., Curvelo,A. A. S., and Dufresne,A., (2009). Extrusion and Characterisation of Functionalised Cellulose Whiskers Reinforced Polyethylene Nanocomposites. *Polymer*. **50**, pp. 4552 – 4563.
4. Beck-Candanedo, S., Roman, M. and Gray, D. G. (2005). Effect of Reaction Conditions on the Properties and Behaviour of Wood Cellulose

Nanocrystal Suspensions.*Biomacromolecules*.**6**,pp 1048–1054.
5. Diego Piarpuzan, Julian A. Quintero, Carlos A. Cardona (2011) Empty fruit bunches from oil palm as a potential raw material for fuel ethanol production, *Biomass and Bioenergy* **35**, 1130-1137.
6. Ohara, S., Kato, T., Fukushima, Y.and Sakoda, A., (2013). Selective Ethanol Production from Reducing Sugars in a Saccharide Mixture.*J BiosciBioeng*. **115(5)**, pp. 540 – 543.
7. Poletto, M., Zattera, A. J., Forte, M. M. C., and Santana, R. M. C. (2012). Thermal Decomposition of Wood: Influence of Wood Components and Cellulose Crystallite Size. *Bioresource Technology*. **109**, pp. 148 – 153.
8. Cheynier, V., Feinberg, M., Chararas, C and Ducauze, C. (1983). Application of Response Surface Methodology to Evaluation of Bioconversion Experimental Conditions. *Appl. Env. Microbiol*. **45**, pp. 634-639.
9. Ruiz, M. M., Cavaillé, J. Y., Dufresne, A., Gérard, J. F., and Graillat, C. (2000).Processing and Characterization of New Thermoset Nanocomposites Based On Cellulose Whiskers. *Compos Interfaces*. **7(2)**, pp. 117-131.
10. Sharma, M., and Yashonath, S. (2007). Size Dependence of Solute Diffusity and Stoke-Einstein Relationship: Effect of Van Der Waals Interactions. *Diffusion Fundamentals*, pp. 11.1-11.5.
11. Wang, Q. Q., Zhu, J. Y. and Reiner, R. S. (2012). Approaching Zero Cellulose Loss in Cellulose Nanocrystal (CNC) production: Recovery and Characterization of Cellulosic Solid Residues (CSR) and CNC. *Cellulose*. **19**, pp. 2033-2047.
12. Mukherjee, S. M. and Woods, H. J. (1953). X-ray and Electron Microscope Studies of the Degradation of Cellulose by Sulphuric Acid. *Biochim Biophys Acta*. **10**, pp. 499-511.
13. Bondeson, D. A. M. and Oksman, K. (2006). Optimization of the Isolation of Nanocrystals From Microcrystalline Cellulose by Acid Hydrolysis. *Cellulose*. **13**, pp. 171-180.

The experimental study of heat extraction of supercritical CO_2 in the geothermal reservoir

Cyun-Jie Huang[1], Jui-Ching Hsieh[2],David T. W. Lin[1],Jenn-Kun Kuo[3] andYuh-Chung Hu[4]

[1]*Institute of Mechatronic System Engineering, National University of Tainan,No. 33, Sec. 2, Shu-lin St., Tainan City 700, Taiwan (R.O.C.)*
[2]*Industrial Technology Research Institute, Green Energy and Environment Research Laboratories,No.195, Sec. 4, Chung Hsing Rd., Chutung, Hsinchu, 31040, Taiwan (R.O.C.)*
[3]*Department ofGreenergy, National University of Tainan,No. 33, Sec. 2, Shu-lin St., Tainan City 700,Taiwan (R.O.C.)*
[4]*Department of Mechanical and Electromechanical Engineering,National Ilan University, No.1, Sec. 1, Shen-Lung Rd., I-LanCity 260,Taiwan(R.O.C.)*

Abstract. The heat transfer phenomena of supercritical CO_2 are experimentally investigated in a horizontal tube for improving the efficiency of CO_2-EGS.This studydiscuss the experimental verification of the numerical simulations. The experiment is conducted for the pressure, the flow rate, and particle size 1.54mm.In addition, theexperiment and simulation that the maximum heat extractionis occurred at the 9MPa pressure and mass flow rate of 0.00109 kg/s. Themaximumspecific heat extraction at 9MPa and flow rate of 0.00082 kg/s.The results show that thenumerical model has been experimentally verified of the feasibility. Furthermore, the pseudo-critical point had a significant influence on the heat extraction, temperature difference and specific heat extraction.

1 Introduction

The global warming of the earth becomes more and more serious. The use of green energy is a good choice. Geothermal energy is a renewable energy, it can be sustainable developed the new type energy [1]. Many researchers have rated it highly.

The geothermal system is divided into traditional geothermal system and non-traditional geothermal system. Non-traditional geothermal systems are normally called enhanced geothermal system (EGS). The operated process of EGS exists two stages. First, the injection well shall be at least drilled three kilometersdeep and the temperature shall be more than 250°C. Second, the high pressure working fluid composed of chemical component (depend on geologic condition) is used to create the fracture in the reservoir though the injection well. The high pressure water is used to absorb the heat from the fracture of the reservoir and then flow out from the production well. Finally, the extracted heat from the working fluid can be used to generate power. In the past, water is used as working fluids, but previous studies have suggested that CO_2can bethe possible EGS working fluid forits chemistry property and density. It is a new breakthrough in the geothermal development. Brown first proposes CO_2-EGS in 2000 [2].

In recent years, there are many researchers to conduct the study of CO_2. The CO_2exists unique feature, in areas with scarce water resources using CO_2 as an alternative to water as a working fluid [3]. The thermal siphon power generation and analysis between water and carbon dioxide is presented byAtrens et al. [3].In 2014, Xu et al. investigated fluid and rock chemical interaction in CO_2-EGS [4].Hsieh et al. [5]discussed the heat transfer phenomena of supercritical CO_2 are experimentally investigated inupward flow vertical tube with silica-based porous media. For the availability of CO_2-EGS, researchers have discussed several topics, including CO_2 mineralization by injecting CO_2 into granite and sandstone [6], carbon capture utilization and sequestration[7],geothermal production and CO_2 sequestration in CO_2-EGS [8].

Researchers have investigated heat transfer phenomena related to CO_2-EGS,determine that buoyancy effects are significant for all flow orientations [9, 10, 11]. Most of these studies show that the flow rate and pressure are the most important parameters in the EGS. However, all of these studies have been limited by the absence of an experimental system for investigating the performance of supercritical CO_2 in a reservoir.

This study is to determine the efficiency of EGS by using the supercritical CO_2 to perform the heat extraction ofthe silica-based porous media in the experimental scale.

2 Experiment

2.1. Experimental apparatus

The schematic diagram of the experimental system is shown in figure 1. The experimental system consists of a CO_2 pressure vessel, high pressure pump, preheated water bath, cooling water bath, heating unit, test section, differential pressure gage, flow meter, and data logger. The photograph of experimental system is shown in figure2. Here, CO_2 pumped into the preheated water bath and reaches the supercritical status. Then, the supercritical CO_2 flows into the test section and extracts the heat to simulatethegeothermal reservoir. The temperature profile, pressure, and mass flow rate are measured and recorded. The test section composes of stainless tube, and ten calibrated T-type thermocouples and differential pressure gage inserted into the test section to record experimental data. The inside, outside diameter and length of test sections is 29.9,55and 235mm, respectively. The test section is filled with silica-base particles. To determine the behavior of supercritical CO_2, various experimental parameters are applied during the experiment including the system pressure (7.5MPa, 9MPa, ,10MPa, 11MPa and12.5MPa), mass flow rate (0.00082and 0.00109 kg/s), and the particle size1.54mm. In addition, the initial test section temperature is pre-set at 200°C by a constant heat flux to simulate the actual reservoir temperature.

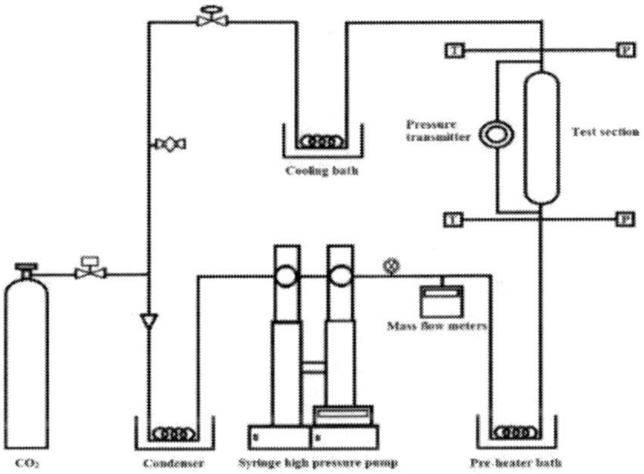

Figure 1. Schematic diagram of experimental apparatus.

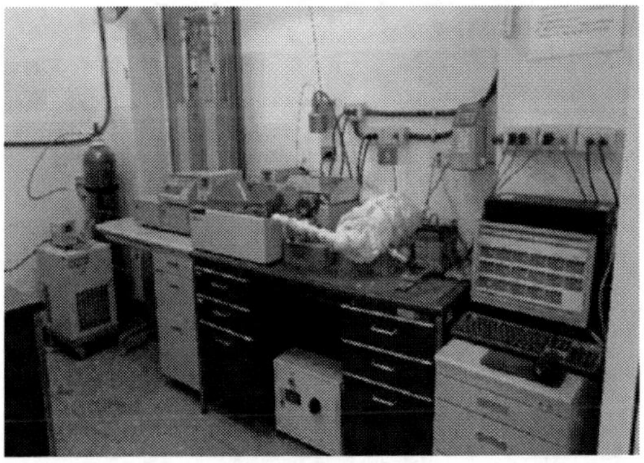

Figure 2. The photograph of experimental apparatus.

2.2Experimental parameters

This studyinvestigates the heat extractionof thesupercritical CO_2at different pressure and mass flow rate.

The heat transfer rate from the tube wall to CO_2iscalculated in a steady state as follows:

$$\dot{Q} = \dot{m}(h_{out} - h_{in}) \quad (1)$$

where $\dot{m}$ indicates the mass flow rate, h_{in} is the inlet enthalpy, h_{out} is the outlet enthalpy of supercritical CO_2.

$$Q_{in}(W) = \dot{Q}(W) + Q_{LOSS}(W)(2)$$

where$\dot{Q}$ indicates the heat extraction rate of supercritical carbon dioxide, Q_{loss} is heat loss.

$$q = \frac{\dot{Q}}{\dot{m}}(3)$$

where $\dot{Q}$ indicates the heat extraction rateof supercritical carbon dioxide, $\dot{m}$indicates the mass flow rate.

3Experimental results and discussion

In the present study, the experiments are carried out with various supercritical pressures from 7.5 to 12.5MPa, and various mass flow rates 0.00082and 0.00109 kg/s at the initial wall temperature 200 °C . As the initial wall temperature reached a steady state, the supercritical $CO_2$is poured into the test section. All the experimental data reported in this study isprocessed based on the physical properties of CO_2 provided by the NIST Refrigerants Database [12]. Figure 3 shows the relationship of specific heat with temperature in the critical region under different pressure by using the NIST Refrigerants Database. As shown in this figure, the pseudo critical temperatures are 31.7°C (7.5 MPa), 40°C (9 MPa), 45.02°C (10 MPa) 49.7°C (11 MPa) and 55.95 °C (12.5 MPa) for CO_2 at these pressures.

The profilesof temperature difference are shown in figure 4. The temperature difference with different pressure (7.5, 9, 10, 11 and 12.5 MPa), mass flow rate (0.00082and0.00109 kg/s) are presented. Through figure 4, we can observe that thetemperature difference decrease as the mass flow rate increases. The reason is thatthe lower velocity of supercritical $CO_2$is benefit to heat extraction. In addition, the temperature difference is nonlinear with the pressure variation. It is dependent on the complex supercritical thermal properties of CO_2. We observe that the maximum temperature difference is happened at 7.5MPa pressure. Theminimumtemperature difference is happened at 9MPa pressure.It corresponds the maximum specific heat is occurred at the same pressure, it`s pseudo critical temperatures are 31.7°Cand pressure at 7.5MPa. The temperature difference in heat extraction is one of the important factors.The larger temperature difference is, the better heat extraction is.

For saving the experimental cost, the simulated model is built to fit the experimental system. The optimization of simulated model can provide the range of the operated parameters. Then, the best operated parameters can be found based on the experiment. First, the comparison of heat extraction between experiment and simulationwith different pressure (7.5, 9, 10, 11 and 12.5 MPa), mass flow rate (0.00082 and0.00109 kg/s)are shown in figure 5.

 DOI: 10.1051/matecconf/20166004010

Through figure 5, we can observe that the heat extraction increases as the mass flow rate. The results show that the maximum heat extraction is occurred at the 9MPa pressure and 0.00109 kg/s. The error of heat extraction between the numerical model and experiment is 0.61%.In addition,simulation model has been experimentally verified.It proofs that the numerical model is suitable forsimulating the supercritical CO_2 porous flow tostudy the complex phenomena. Therefore, it find the combination of pressure, flow rate, and particle size for the optimal heat extraction.

Second, the heat extraction of numerical model are shown in figure 6.The heat extraction of numerical model with different pressure(7.5, 9, 10, 11 and 12.5 MPa) and mass flow rate(0.00082 and0.00109 kg/s) at particle size 1.54mmis presented.Through figure 6,we canobserve thatthe maximum heat extraction is occurred at the 9MPa pressure.Finally, theoptimal heat extraction profiles of experiment are shown in figure 7. Theoptimal heat extraction with different pressure (8.9, 8.95, 9, 9.05 and 9.1 MPa), mass flow rate0.00109 kg/s is presented.Through figure 7, we can observe that themaximum heat extractionis happened on 9 MPa.

For the consideration of cost of realistic practice, the heat extraction of working fluid per kilogram is necessary to study. The specific heat extraction is defined as the heat extraction per kilogram supercritical CO_2.The specific heat extraction profiles are shown in figure 8. The specific heat extraction with different pressure (7.5, 9, 10, 11 and 12.5 MPa), mass flow rate (0.00082and0.00109 kg/s) atparticle size 1.54mmis presented. Through figure 8, we can observe that the specific heat extraction increases as the mass flow rate decrease. The results show the maximumspecific heat extraction is occurred at the 9MPa pressure and 0.00082 kg/s.Moreover, the specific heat extraction at 9MPaof 0.00082 kg/sis approximately 13.2% higher than that of 0.00109kg/s.The specific heat extraction can pointout what kind of pressure and flow rate to approach economic efficiency better.Better the specific heat extraction, theeconomic efficiency is.

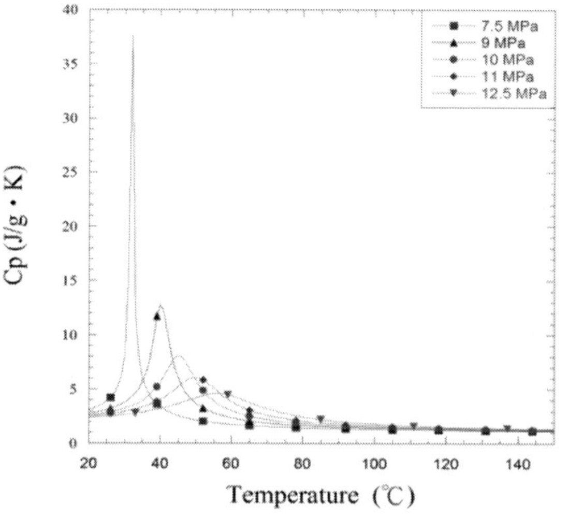

Figure 3. Specific heat of supercritical CO_2 with various temperature and pressure in the critical region [12].

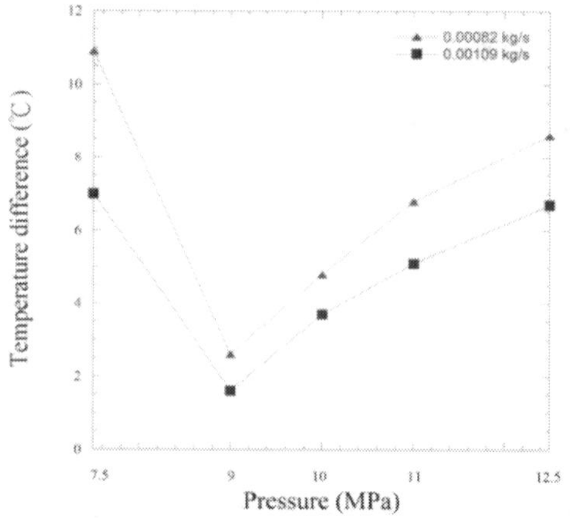

Figure4. The profiles of temperature difference with different pressure and mass flow rate.

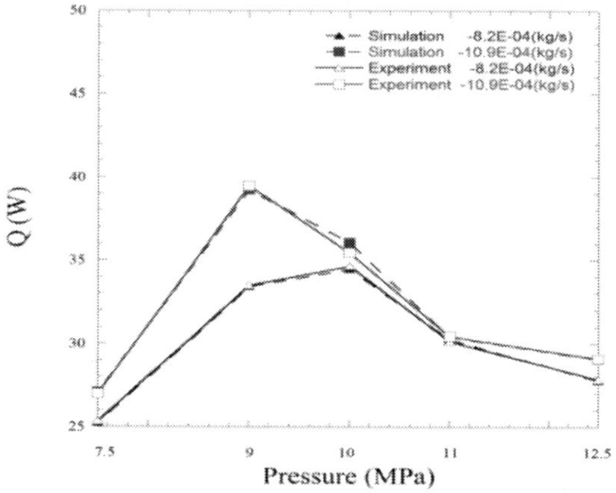

Figure 5. The comparison of heat extraction betweenexperiment and simulation.

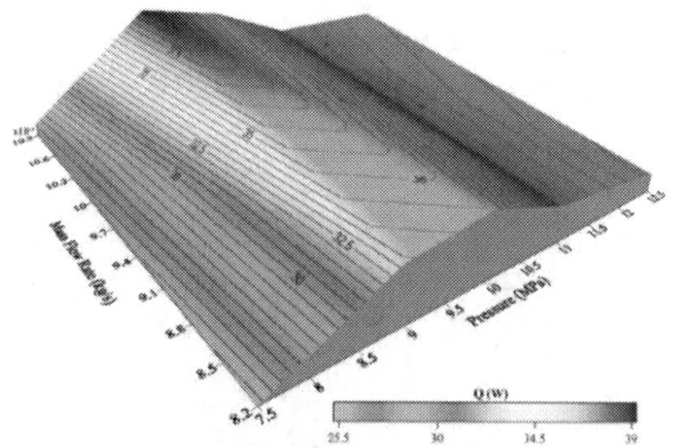

Figure6. The optimal heat extraction with various pressure and mass flow rate.

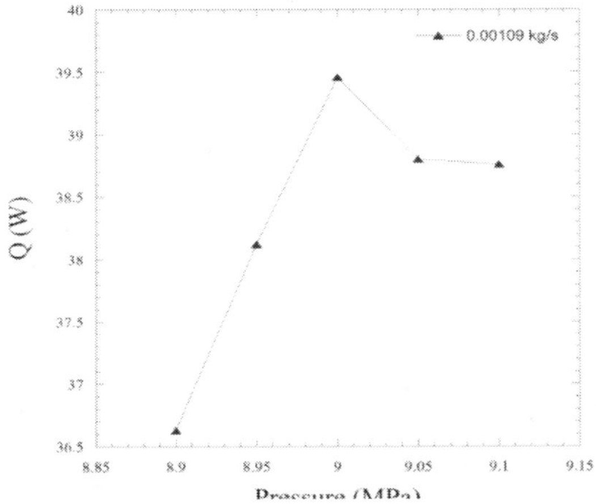

Figure 7. The optimal experimental heat extraction(mass flow rate is 0.00109 kg/s)

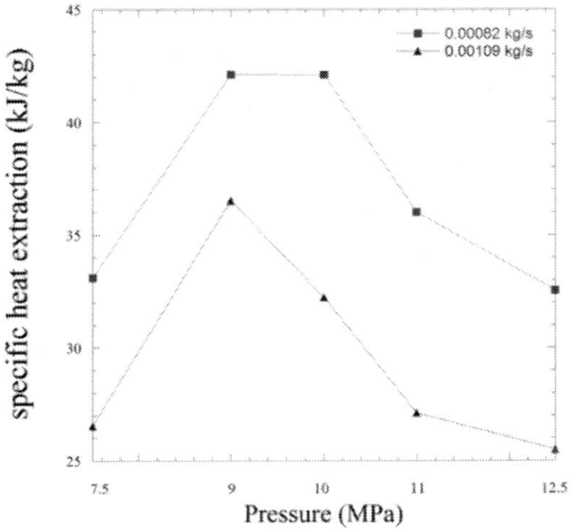

Figure 8. The specific heat extraction with various pressure and mass flow rate

4 Conclusions

In this study, the heat transfer performances for supercritical CO_2 flow in porous media are experimentally investigated and verificates the numerical simulations for the finding best combination of operated parameters. The results show that thesimulation model has been experimentally verifiedof the feasibility.The maximum heat extraction is happened at 9MPa.The maximum specific heat extraction happened at 9 MPaand flow rate of 0.00082 kg/s. We can observe that the heat transfer characteristics of supercritical CO_2affect the heat extractionstrongly.This study provides data regarding EGS operating conditions.We expect these results can enhance the heat extraction and reduce the cost of realistic test of enhanced geothermal system.

Acknowledgements

The authors are thankful for the financial support of this research from the Ministry of Science and Technology of R.O.C. through the grant no. MOST104-3113-M-024 - 001.The authors also express their gratitude to the Energy R&D Foundation of the Energy Bureau of the Ministry of Economic Affairs, Taiwan, for providing financial support.

References

1. S.R.Sigurdardottir, A.Valfells, H.Palsson, H.Stefansson, Geothermics**55**,171 (2015)
2. D.W.Brown, Geothermal Reservoir Engineering, Stanford University, California**25**,233(2000)
3. L. Pan, B. Freifeld, C. Doughty, S. Zakem, M. Sheu, B. Cutright, T. Terrall, Geothermics**53**, 100 (2015)
4. D.A. Atrens, H. Gurgenci, V. Rudolph, Geothermics**39**,161 (2010)
5. T. Xu, G. Feng, Y. Shi, Journal of Geochemical Exploration**144**, 179(2014)
6. J.C. Hsieh, David. T.W. Lin, C.H. Wei, H.J. Huang, EnergyProcedia**61**, 262 (2014)
7. L. Liu,Y. Suto, G.Bignall, N.Yamasaki,T. Hashida,Energ Convers Manage**44**, 1399 (2003)
8. T.A. Buscheck, J.M. Bielicki, M. Chen, Y. Sun, Y. Hao, T. A. Edmunds, M.O. Saar, J.B. Randolph, Energy Procedia **63**, 7619 (2014)
9. J.B. Randolph, M.O. Saar, J. Bielicki, Energy Procedia**37**, 6625(2013).
10. P.X. Jiang, B. Liu, C.R. Zhao, F. Luo, International Journal of Heat and Mass Transfer **56**, 741(2013)
11. G. Liu, Y. Huang, J. Wang, F. Lv, International Journal of Heat and Mass Transfer**91**, 640 (2015)
12. National institute of standards and technology, Reference Fluid Thermodynamic and Transport Properties,Available:http://webbook.nist.gov/chemistry/fluid/

Natural gas adsorption on biomass derived activated carbons: A mini review

Usman D. Hamza[1], Noor S. Nasri[2], Jibril Mohammed[3] and Zulkifil Abd. Majid[2]

[1,3]*Department of Chemical Engineering, Abubakar Tafawa Balewa University Bauchi, P.M.B. 0248, Bauchi Nigeria*
[2]*UTM-MPRC Institute for Oil and Gas, Resources Sustainability, Universiti Teknologi Malaysia, 81310 UTM Johor Bahru, Johor, Malaysia*

Abstract. Activated carbon materials are good candidates for natural gas storage due excellent textural properties that are easy to enhance and modify. Natural gas is much cleaner fuel than coal and other petroleum derivatives. Storage of natural gas on porous sorbents at lower pressure is safer and cheaper compared to compressed and liquefied natural gas. This article reviews some works conducted on natural gas storage on biomass based activated carbon materials. Methane storage capacities and deliveries of the various sorbents were given. The effect of factors such as surface area, pore characteristic, heat of adsorption, packing density on the natural gas storage capacity on the activated carbons are discussed. Challenges, improvements and future directions of natural gas storage on porous carbonaceous materials are highlighted.

1. Introduction

Natural gas is cheap, clean [1], safe and abundant energy source [2]. Due to these advantages, its storage, transportation and utilization have attracted global attention [3]. Natural gas (NG) is composed mainly of 85–95 mol% methane (CH_4), the remainder being carbon dioxide (CO_2), nitrogen (N_2) and small amounts of higher heavy gaseous hydrocarbons [1]. Natural gas from crude wells is known as associated gas, while those from gas wells or condensate is referred to as non-associated gas. NG from pores of coal is known as coalbed gas [2]. Until recently, Natural gas is stored as compressed natural gas (CNG) or liquefied natural gas (LNG). In (CNG), natural gas is stored as a compressed supercritical fluid at room temperature and at maximum pressure of about 200–250 bar [1]. Therefore, CNG requires expensive vessels and multi-stage compression [4]. Not only that, in terms of equal fuel-tank volume, the energy density of natural gas at pressure of 20 MPa, is only 29% that of gasoline [3]. NG is transported overseas as liquefied natural gas (LNG) in ships, LNG requires expensive cryogenic process [4]. It is usually stored as a boiling liquid at about (161^{0}C) in a cryogenic tank at a pressure of about 14.7 psia. LNG has volumetric energy density for about 72% of that of gasoline [5].

Due to high costs, safety and operational restrictions of CNG and LNG researchers are finding alternatives [1]. ANG (Adsorbed natural gas) is viewed as alternative technology to CNG and LNG. In ANG, the gas is stored on porous material packed into a vessel at much lower pressure [3]. In ANG the storage pressure is usually in the range of 35–40 bar. Therefore, there is improved safety and low compression costs in ANG [1].

Among solid adsorbents, activated carbons have been used and evaluated for methane adsorption. They are good candidates for natural gas storage due to large surface area, large pore volume and low density. Activated carbons with diversified microporous structures constitute an important category of potential adsorbents for natural gas storage [6]. Properties of activated carbon required for natural gas adsorption include: high adsorption capacity, high packing density, high adsorption desorption rate, ratio of amount desorbed at 0.1 Mpa and the amount adsorbed at 4 Mpa should be close to one [7], less susceptible to temperature changes during adsorption and desorption, extremely hydrophobic and it should be inexpensive for the final user. [10]

For ANG to be commercially viable, the ANG technology requires a methane storage target of 180 v/v at an ambient temperature, with the energy density of ANG comparable to that of current CNG technology [9]. Recently, the United States Department of Energy (DOE) reviewed the volumetric objective for adsorbed natural gas from the old value of 180 cm3 (STP: 273.15 K, 1 atm)/ cm^3 to a much ambitious value of 263 cm^3 (STP: 273.15 K, 1 atm)/cm^3 [10]. This paper reviewed the existing literature on natural gas/methane adsorption on biomass based porous carbons. Both single and gas mixture adsorption is discussed, challenges of ANG technology and improvements in the system are given.

2. Important ANG Adsorbent Parameters

For activated carbon to be considered as a good ANG material, certain parameters/properties such as adsorption capacity, deliverable gas, heat of adsorption, surface area, porosity and packing density needs to be assessed.

2.1. Deliverable Gas and Adsorption Capacity

Deliverable gas is the volume of gas delivered per volume of the storage container. It denotes the amount of gas released when pressure is reduced from high pressure mostly around 34 bars to atmospheric pressure. The storage capacity is usually greater than the delivered gas by about 15-30% [11]. The US Department of Energy (DOE) set target of absolute methane-adsorption capacity of 180 v/v (standard temperature and pressure equivalent volume of methane per volume of the adsorbent material) at 35 bars, under ambient temperature. CH_4 storage target was reviewed in 2012 to 230 v/v [12]. Methane adsorption on olive stones grain activated carbon chemically activated by $ZnCl_2$ led followed by physical activation with CO_2 gives 110V/V at 298K under 3.4MPa. Most of the studies conducted reported volumetric CH_4 capacities for activated carbons in the range of 100–170 v/v; therefore there is still need for improvement [12].

2.2. Heat of Adsorption

Adsorption of natural gas during charging into an adsorbent filled container is exothermic, while desorption is endothermic [13]. Increase in temperature brings about decrease in adsorption capacity of the sorbent. Therefore, fewer amounts CH_4 is stored during adsorption due to increase in temperature and more CH_4 is retained during desorption due to decrease in temperature [12]. Temperature changes in the ANG system can be managed by incorporating heat exchanger in the storage vessel, changes in the tank geometry, modification of the sorbent material and changes in nature and direction of flow. Higher thermal conductivities and heat capacity is required for the sorbent. Material with high heat capacity is less susceptible to temperature changes during charging and discharging of natural gas [12].

A study was carried out to determine the thermal effect of the adsorption heat on the discharge performance of an adsorbed natural gas (ANG) storage system [14]. It was found out that central region of the adsorbent bed suffers from the severest temperature fluctuation in a short period of discharge state; introduction of the hot water greatly reduce temperature fluctuation of the adsorbent bed, shorten the discharge process by 60% compared to that without supplemental heat [14].

2.3. Surface area and pore characteristics

Adsorbent for natural gas application is desired to have large surface area, predominantly microporous with low mesoporosity. Large surface area allows more contact between gas adsorbate and the solid adsorbent which translate to more adsorption. Low mesoporosity is required as feeder pores to provide easy access for the gas molecules to/from the adsorbent micropores [5]. Micropores are pores that are less than 2 nm, while mesopores are pores between 2 to 50 nm [15]. Surface area of activated carbon used for natural gas adsorption was in the range of few hundreds to 3000 m^2/g [5].

Biloe' [16] used Dubinin–Astakhov equation to determine the influence of the microporous characteristics of activated carbon on the performances of both charge and discharge of an ANG system. It was found out that for the charge stage, the activated carbon must be conductive with an average micropore width of 1.5 nm, and for the discharge, it should be permeable and sufficiently conductive with an average micropore width of 2.5 nm.

2.4. Packing Density

Packing density is defined as the mass of settled material per unit volume of storage space. A carbon sorbent material with high surface area and low parking density would still have low adsorption capacity [13]. Therefore, in addition to high surface area, high packing density is required for sorbent materials to be used as an ANG material. More porosity development in a carbonaceous sorbent lowers its packing density [17]. An adsorbents may have high uptake on a mass basis, but with the low packing density much of the potential advantage is lost and the volumetric energy densities will still be low [13]

For economic viability of ANG, it's good to have high energy density besides the high micropore volume on volumetric basis. In order to achieve that, the sorbent must have high packing density. Packing density of slit pores AC is higher than due to the additional space given for freedom of molecules between to parallel sides of slit pores [18]. [7] reported packing density of about 0.7 g/ml for activated carbon with BET surface area close to 2400 m^2/g.

2.5. Other properties

ACs for use as ANG materials are required to be hydrophobic, which prevents competition between water and adsorbate molecule. The adsorbent should be hard, not easily broken or powdered [5].

3. Adsorption of Methane on AC

Methane uptake is usually reported as adsorbed volume at standard pressure and temperature (cm3 (STP) per volume of sample (cm3 (STP) cm^{-3}), gravimetrically (cm3 (STP) g^{-1}), specific amount reported in mmol g^{-1}, or as sample- specific weight percent (wt.%) [19].

There are several studies conducted on use of activated carbon for methane adsorption. Activated carbon samples were prepared from Brazilian coconut shells modified by H_3PO_4 for methane adsorption [20]. The experiments were conducted at pressures between at 303 K. highest methane storage capacity was found to be 95 v/v at 303 K and 35 bar on the activated carbon samples. [21] prepared activated carbon (CAQF-30) from coconut shells by chemical activation with zinc chloride followed by physical activation. The performance of the prepared AC was compared with commercial AC (SRD-21) for adsorption of methane. The prepared activated carbon showed approximately 0.14 g/g CH_4 adsorption capacity, while the commercial sample showed 0.17 g/g CH_4 adsorption capacity.

Activated carbon from rice husk (AC-RH) was synthesized and its methane adsorption capacity was determined [22]. The experiment was conducted at constant temperature and pressure up to 2.5 Mpa. The AC-RH demonstrated 2350 m^2/g BET surface area and 0.95 pore volumes. The sorbent had 9.3 mmol/g adsorption capacities at 303.15 K under 2.5 Mpa. Activated carbons (ACs) were prepared by carbonization, impregnation with KOH and activation of sugarcane molasses [23]. The AC that had high surface area of 2202 m^2/g was evaluated for methane adsorption. Highest methane adsorption capacity of 197.23 mg/g (12.33 mmol g^{-1}) at 50 bars and 20 ^0C was recorded.

In another study, theoretical and experimental study was carried out to determine the amount of adsorption and desorption of methane on MERC commercial granular activated carbon [24]. The experiments were carried using volumetric method for pressure up to 500 psia at constant temperature of 25°C. The adsorption and desorption capacity of the MERC activated carbon reached 0.127 g/g and 0.091 g/g. It was found out that BET surface area, micropore volume, packing density, and pore size distribution have significant effect on the amount of methane adsorbed [24]. Delevar et al. (2010) investigated the capacity of granular activated carbon (GAC) for the storage of methane. The methane adsorption was determined at different pressures (0-50 bar) and temperatures (285.15 - 328.15 K). The CH_4 adsorption capacity on the activated carbon samples was around 8 mmol/g at 298K, 40 bar. I t was also revealed increasing the pressure and decreasing the temperature increases the amount of adsorption.

4. Adsorption of CH_4 and CO_2 Mixture

Some sources of natural gas are associated with CO_2, which is considered as an impurity. Presence of CO_2 reduces the energy content and heating value of natural gas, in the presence of water it corrupt transportation and storage system [25,26]. There is also need to remove CO_2 from the mixture in order to meet pipe line specification or to produce liquefied natural gas [26]. Adsorption on porous carbons is one of the techniques used to separate the CH_4/CO_2 gas mixture.

[27] used virgin palm shell activated carbon and PEI (polyethyleneimine) carbon molecular basket for methane and CO_2 adsorption. Virgin palm shell AC shows 3.03 and 10.79 (cm^3/gSTP) CH_4 and CO_2 adsorption capacity with 3.56 as CO_2/CH_4 ideal selectivity. PEI (polyethyleneimine) carbon molecular basket demonstrated 15.68 and 51.11(cm^3/gSTP) CH_4 and CO_2 adsorption capacity with 3.26 as CO_2/CH_4 ideal selectivity.

High-pressure adsorption of CO_2 and CH_4 in activated carbon was evaluated on Maxorb AC [26]. The researchers found out that CO_2 is more adsorbed than CH_4 in all the conditions tested. Maximum amount adsorbed (qmax) from Sips isotherm for CO_2 and CH_4 reached up to 31.97 and 15.02 mol/kg. There is deviation of up to 20% between results obtained from static and dynamic adsorption. In a similar study, Coconut activated carbons were prepared and used for CO_2 and CH_4 adsorption [28]. CO_2 adsorption capacity of 2.55 mmol/g at 200 kPa, CH_4 adsorption capacity of 1.93 mmol/g were recorded at 200 kPa.

In a similar study, methane adsorption was investigated on activated carbon obtained from coffee husks [29] at 298.15 K and 303.15 K and pressures up to 30.00 atm. maximum volumetric methane uptake by the synthesised carbon monoliths was observed to be 130 V/V at 298.15 K and 30.00 atm. It was observed that there was increase in amount adsorbed due to increasing surface area, total pore volume and micropore volume.

[30] studied adsorption equilibrium of carbon dioxide, methane and nitrogen on microwave activated carbon (MAC). The capacities of pure gas CO_2, CH_4 and N_2 were reported to be 2.13, 0.98 and 0.33 mmol g^{-1} at 298 K and the partial pressure of 100 kPa. [31] studied adsorption of methane and carbon dioxide on two microwave-activated carbon samples at different temperatures from 298 and 323 K. The results showed that the activated carbon modified by potassium carbonate sample has the higher equilibrium selectivity for CO_2 over CH_4. The adsorption capacities of CO_2 for unmodified and K_2CO_3 modified AC was 1.31 mmol/g and 1.63 mmol/g at 298 K and the pressure of 50 KPa, respectively. The ideal selectivity of CO_2/CH_4 is 2.66 and 7.42 for MAC and MAC/K_2CO_3 respectively.

5. Challenges and Improvements in ANG Technology

The first issue is that, according to the thermodynamic laws, during the process of adsorption, the temperature of the ANG system is increased, there by limiting the amount of natural gas uptakes. During desorption the temperature of the system diminishes which causes retention of some of the gas to be delivered [32]. Therefore, management of thermal effects as a result of adsorption heat on both charge and discharge processes of the ANG are still a bottleneck for developing the vehicular ANG storage technology [14]. MoreGasTech, a French company develop an efficient thermo-managing control system for cooling during adsorption or heating the adsorbent during discharge [32]. Another sorbent challenge is presence of impurities in the methane gas [19]. They compete favourably with CH_4 for pores and active sites. But issues related with impurity such as H_2O could be avoided by having a hydrophobic surface.

The volumetric storage of ANG still needs to be improved in order to favourably compete with other NG storage method. Increasing the packing density will help in this regard. Compacting the sorbents increases the density which in turn increases the adsorption capacity per unit volume. Another is problem is the difficulty in direct packing of the adsorbent carbon into the storage vessel. Briquetting, or immobilizing, the carbon was considered as an alternative. But the binder blocks some of the pores there by reducing storage due to inaccessibility to the micropores [32]. Configuration, design and nature of the carbonaceous sorbent also enhance the storage capacity. New adsorbents such as monolithic carbons and compressed active carbon fibers used in non-cylindrical vessels have made possible to store the same capacity of gas as CNG tank [33].

6. Conclusions

Biomass materials which are precursors for producing activated carbon material are cheap and readily available. Utilization of biomass as an ANG material brings about cost effectiveness and improved profitability in ANG related industries, with concomitant waste minimization. Choice of proper synthesis and modification conditions leads to sorbent with the desired properties for natural gas adsorption. Emphasis should be on having a material that is microporous with little mesopore; a material with high packing density, high adsorption and desorption capacity, low isosteric heat of adsorptiont, good mechanical strength, tolerant to impurities and cheaper. Biomass based AC have great potential as adsorbents for natural gas storage. But still there need to improve the volumetric capacity to the driving range to levels closer to that of gasoline vehicles.

Acknowledgement

The authors acknowledge and appreciate the financial support provided by the Ministry of Higher Education Malaysia through University Teknologi Malaysia (UTM), Johor, Malaysia, to fully undertake this research under the Research University Grant (RUG) Q.J130000.2509.10H89 and Q.J130000.2509.06H79.

References

1. Esteves A.A.C.I., Lopes M.S.S., Nunes, P.M.C., Mota J.P.B. (2008). Adsorption of natural gas and biogas components on activated carbon. Separation and Purification Technology, 62, 281–296.
2. Adewole J.K., Ahmad A.L., Ismail S., Leo C.P. (2013). Current challenges in membrane separation of CO2 from natural gas: A review. International Journal of Greenhouse Gas Control, 17: 46–65
3. Liu B., Wang W., Wang N., Au C.T. (2014). Preparation of activated carbon with high surface area for high-capacity methane storage. Journal of Energy Chemistry, 23, 662–668.
4. Wang Y., Hashim M., Ercan C., Khawajah A., Othman R. (2011). High Pressure Methane Adsorption on Granular Activated Carbons. *21st Annual Saudi-Japan Symposium Catalysts in Petroleum Refining & Petrochemicals Dhahran, Saudi Arabia,* November, 2011.
5. Menon V.C., and Komarneni S. (1998). Porous Adsorbents for Vehicular Natural Gas Storage: A Review. Journal of Porous Materials 5, 43–58.
6. Delevar M. Ghoreyshi A.A., Jananshahi M., Khalili, S, and Nabian N. (2012). The effect of Chemical Treatment on Adsorption of Natural Gas by Multi-Walled Carbon Nanotubes: Sorption Equilibria and Thermodynamic Studies. *Chemical Industry & Chemical Engineering Quarterly,* 18 (2), 193–207.
7. Lozano V.M.P. {n.d). Methane Storage in Activated Carbon Fibres. Universidat de Alicante.
8. Delavar M, Ghoreyshi A.A., Jahanshahi M., Irannejad M. (2010). Experimental Evaluation of Methane Adsorption on Granular Activated Carbon (GAC) and Determination of Model Isotherm. World Academy of Science, Engineering and Technology, 62, 47-50.
9. Ma, S. (2009). Gas adsorption applications of porous metal–organic frameworks. 2235 Pure Appl. Chem., Vol. 81, No. 12, pp. 2235–2251,
10. Casco M.E., Martínez-Escandell M., Gadea-Ramos E., Kaneko K., Silvestre-Albero J., and Rodríguez-Reinoso F. (2015). High-Pressure Methane Storage in Porous Materials: Are Carbon Materials in the Pole Position. Chem. Mater. 27, 959–964.
11. Judd R.W., Gladding D.T.M., Hodriean R.C., Bates D.R., Ingram J.P. M. Allen. (n.d). The Use of Adsorbed Natural Gas Technology for Large Scale Storage. BG Technology, Gas and Research Technology Centre, Loughbrough, U.K.
12. Mason, JA, Veenstra M. and Jeffrey R. (2014). Long Evaluating metal–organic frameworks for natural gas storage. Chem. Sci., 5, 32-51.
13. Mat T.H, Zakaria Z., Paouhe T.G. Development of Adsorbent Based Natural Gas Storage for Vehicle Application . Department of Chemical Engineering Faculty of Chemical And Natural Resources Engineering Universiti Teknologi Malaysia, VOT 72229. 1-1-100
14. Yang X.D., Zheng Q.R., Gu A.Z., Lu X.S. (2005). Experimental studies of the performance of adsorbed natural gas storage system during discharge. Applied Thermal Engineering 25: 591–601
15. IUPAC Manual of Symbols and Terminology, (1972). Appendix 2, 1, Colloid Surface Chem. Pure and Appl. Chem. 31, 587.
16. Biloe′ S., Goetz V., Guillot A. (2002). Optimal design of an activated carbon for an adsorbed natural gas storage system. Carbon, 40, 1295–1308.
17. Rufford T.E., Zhu J., Hulicova-Jurcakova D. (2014). Green Carbon Materials: Advances and Applications. CRC Press, 6 Mar 2014 - Technology & Engineering – pg .73.
18. Nalwa H S. (2001). Handbook of surfaces and interfaces of Materials. Volume 1, Surface and interface phenomena. Academic Press, San Francisco, USA. Pg. 337.
19. Bimbo N., Physick A.J., Noguera-Díaz A., Pugsley A., Holyfield L.T., Ting V.P., Mays T.J. (2015). High volumetric and energy densities of methane stored in nanoporous materials at ambient

temperatures and moderate pressures. Chemical Engineering Journal 272, 38–47.
20. Rios R.B., Silva F.W.M., Torres A.E.B., Azevedo D.C.S., Cavalcante Jr C.L. (2009). Adsorption of methane in activated carbons obtained from coconut shells using H_3PO_4. chemical activation. Adsorption, 15: 271–277
21. Bastos-Neto M., Torres A.E.B., Azevedo D.C.S., and Cavalcante Jr C.L. (2005). Methane Adsorption Storage Using Microporous Carbons Obtained from Coconut Shell. Adsorption. 11: 911–915.
22. Balathanigaimani M.S., Kang H., Shim W., Kim C., Lee J. and Moo H. (2006). Preparation of powdered activated carbon from rice husk and its methane adsorption properties. Korean J. Chem. Eng, 23(4), 663-668.
23. Sren´scek-Nazzal, J.S., Kamin´ska, W., Michalkiewicz, B., Koren, Z.C. (2013). Production, Characterization and Methane Storage Potential of KOH-Activated Carbon from Sugarcane Molasses. *Industrial Crops and Products,* 47, 153– 159.
24. Salehi E., Ghotbi V.T.C., Lay E.N., Shojaei A. (2007). Theoretical and Experimental Study on the Adsorption and Desorption of Methane by Granular Activated Carbon at 25°C. Journal of Natural Gas Chemistry, 16, 415–422
25. Xu X.L., Zhao X.X., Sun L.B., Liu X.Q. (2009). Adsorption separation of carbon dioxide, methane and nitrogen on monoethanol amine modified b-zeolite, J. Nat. Gas Chem. 18, 167–172
26. Grande C.A., Blom R., Moller A. and Mollmer J. (2013). High-pressure separation of CH_4/CO_2 Using Activated Carbon. *Chemical Engineering Science*, 89, 10–20.
27. Aroua, M.K. WanDaud, W.A., Yin, C.A. and Adinata, D. (2008). Adsorption Capacities of Carbon Dioxide, Oxygen, Nitrogen and Methane on Carbon Molecular Basket Derived from Polyethyleneimine Impregnation on Microporous Palm Shell Activated Carbon. *Separation and Purification Technology*, 62 (3), 609–613.
28. Yang H, Gong M, ChenY. (2011). Preparation of activated carbons and their adsorption properties for greenhouse gases: CH_4 and CO_2. Journal of Natural Gas Chemistry 20, 460–464.
29. Giraldo, L., Moreno-Piraján, J.C. (2011). Novel Activated Carbon Monoliths for Methane Adsorption Obtained from Coffee Husks. *Materials Sciences and Applications*, 2, 331-339.
30. Yi H., Li F., Ning P., Tang X., Peng J., Li Y. and Deng H. (2013). Adsorption Separation of CO_2, CH_4, and N_2 on Microwave Activated Carbon. *Chemical Engineering Journal*, 215–216, 635–642.
31. Ning P., Li F., Yi H., Tang X., Peng J., Li Y., He D., Deng H. (2012). Adsorption equilibrium of methane and carbon dioxide on microwave-activated carbon. Separation and Purification Technology, 98, 321–326.
32. Ginzburg Y. (2006). ANG Storage As A Technological Solution For The "Chicken-And-Egg" Problem Of NGV Refuelling Infrastructure Development. 1-17.
33. Vasiliev, L.L., Kanonchik, L.E. Mishkinis D.A. and Rabetsky M.I. (2000). Adsorbed natural gas storage and transportation vessels. Int. J. Thermal Sci., 39, 1047-1055.

AUTHOR INDEX

EDP Sciences – Web of Conferences
17, Avenue du Hoggar
Parc d'Activité de Courtabouf
BP 112
F-91944 Les Ulis Cedex A
France

ISBN 978-1-5108-2532-1

9 781510 825321

CURRAN ASSOCIATES INC.
proceedings.com

III International Symposium on Fundamental Aspects of Rare-Earth Elements Exploration, Mining and Separation and Modern Materials Engineering (REE-2016)

MATEC Web of Conferences Volume 96 (2017)

Almaty, Kazakstan
15 - 18 August 2016

edp sciences